THREE-DIMENSIONAL MACHINE VISION

**THE KLUWER INTERNATIONAL SERIES
IN ENGINEERING AND COMPUTER SCIENCE**

ROBOTICS: VISION, MANIPULATION AND SENSORS

Consulting Editor

Takeo Kanade

Other books in the series:

Robotic Grasping and Fine Manipulation. M. Cutkosky.
 ISBN 0–89838–200–9.

Shadows and Silhouettes in Computer Vision. S. Shafer.
 ISBN 0–89838–167–3.

Perceptual Organization and Visual Recognition. D. Lowe.
 ISBN 0–89838–172–X.

THREE-DIMENSIONAL MACHINE VISION

edited by

Takeo Kanade
Carnegie Mellon University

KLUWER ACADEMIC PUBLISHERS
Boston/Dordrecht/Lancaster

Distributors for North America:
Kluwer Academic Publishers
101 Philip Drive
Assinippi Park
Norwell, MA 02061, USA

Distributors for the UK and Ireland:
Kluwer Academic Publishers
MTP Press Limited
Falcon House, Queen Square
Lancaster, LA1 1RN, UNITED KINGDOM

Distributors for all other countries:
Kluwer Academic Publishers Group
Distribution Centre
Post Office Box 322
3300 AH Dordrecht, THE NETHERLANDS

Cover photograph by Bradley A. Hersch, Monroeville, Pennsylvania.

Library of Congress Cataloging-in-Publication Data

Three-dimensional machine vision.

  (The Kluwer international series in engineering and
computer science. Robotics)
  1. Robot vision.   I. Kanade, Takeo.
TJ211.3.T47 1987   629.8′92   86–27599
ISBN 0–89838–188–6

Printed in the United States of America

Table of Contents

Preface

A robot must perceive the three-dimensional world if it is to be effective there. Yet recovering 3-D information from projected images is difficult, and still remains the subject of basic research. Alternatively, one can use sensors that can provide three-dimensional range information directly. The technique of projecting light-stripes started to be used in industrial object recognition systems as early as the 1970s, and time-of-flight laser-scanning range finders became available for outdoor mobile robot navigation in the mid-eighties. Once range data are obtained, a vision system must still describe the scene in terms of 3-D primitives such as edges, surfaces, and volumes, and recognize objects of interest. Today, the art of sensing, extracting features, and recognizing objects by means of three-dimensional range data is one of the most exciting research areas in computer vision.

Three-Dimensional Machine Vision is a collection of papers dealing with three-dimensional range data. The authors are pioneering researchers: some are founders and others are bringing new excitements in the field. I have tried to select milestone papers, and my goal has been to make this book a reference work for researchers in three-dimensional vision.

The book is organized into four parts: 3-D Sensors, 3-D Feature Extractions, Object Recognition Algorithms, and Systems and Applications. Part I includes four papers which describe the development of unique, capable 3-D range sensors, as well as discussions of optical, geometrical, electronic, and computational issues. Mundy and Porter describe a sensor system based on structured illumination for inspecting metallic castings. In order to achieve high-speed data acquisition, it uses multiple light stripes with wavelength multiplexing. Case, Jalkio, and Kim also present a multi-stripe system and discuss various design issues in range sensing by triangulation. The numerical stereo camera developed by Altschuler, Bae, Altschuler, Dijak, Tamburino, and Woolford projects space-coded grid patterns which are generated by an electro-optical programmable spatial light modulator. Kanade and Fuhrman present a proximity sensor using multiple LEDs which are conically arranged. It can measure both distance and orientation of an object's surface.

Having acquired range data, the next step is to analyze it and extract three-dimensional features. In Part II, Ponce and Brady present the Surface Primal Sketch. They detect, localize, and symbolically describe the significant surface changes: steps, roofs, joints, and shoulders. Vemuri, Mitiche, and Aggarwal represent 3-D object surfaces by patches which are homogeneous in intrinsic surface properties. Sugihara describes how the knowledge of vertex types in polyhedra can be used to guide the extraction of edges and vertices from light-stripe range data.

An important goal of 3-D vision systems is to recognize and position objects in the scene. Part III deals with algorithms for this purpose. After describing object surfaces by curves and patches, Faugeras and Hebert use the paradigm of recognizing while positioning for object matching by exploiting rigidity constraints. Oshima and Shirai have long worked on a system for recognizing stacked objects with planar and curved surfaces using range data. Their paper presents procedures for feature extraction and matching together with recent experimental results. The matching strategy used by Bolles and Horaud to locate parts in a jumble starts with a distinctive edge feature of an object and grows the match by adding compatible features. Grimson and Lozano-Pérez discuss how sparse measurement of positions and surface normals may be used to identify and localize overlapping objects. Their search procedure efficiently discards inconsistent hypothetical matches between sensed data and object model using local constraints.

Finally, Part IV contains three papers discussing applications and system issues of three-dimensional vision systems which use range sensing. Beeler presents a precision 3-D measurement system for replacing damaged or missing tiles on space shuttle vehicles. The acquired shape description of a tile cavity is used to control a milling machine to produce a new tile. Nakagawa and Ninomiya describe two three-dimensional vision systems that use active lighting: one is for inspecting solder joints and the other for automatic assembly of electronic devices. The paper by Jarvis, arguing for a semantic-free approach to three-dimensional robotic vision, discusses various system issues in a real-time sensory-based robotic system, such as a hand-eye system.

In editing the book I have received great help from Richard Mickelsen and Magdalena Müller. Richard read all of the papers and brought consistency into the book. Maggie spent untiring effort for typing and perfecting the format. I am heartily grateful: without them, this book would not have been completed.

Takeo Kanade

PART I: 3-D SENSORS

A Three-Dimensional Sensor
Based on Structured Light

J.L. Mundy
G.B. Porter III
General Electric Company
Corporate Research and Development
Schenectady, NY 12345

Abstract

This paper describes the design of a 3-D range sensor using structured light to provide multiple range points by triangulation. The algorithms and sensor hardware are described along with the supporting theory and design strategy.

1. Introduction

The formulation of three-dimensional descriptions of objects from two-dimensional perspective images is a major goal for machine vision research. In many cases, the images are formed using unconstrained lighting and camera position. The interpretation of such scenes relies on features such as shadows and intensity discontinuities as the primitives from which the object descriptions are formed. It is now recognized that this process cannot be carried out successfully without detailed *a priori* models of the objects and the properties of illumination and image formation. The nature of such models and the design of efficient matching algorithms are still the focus of intensive research efforts in image understanding. Existing techniques are not yet robust enough to form the basis for a practical industrial inspection or robot guidance system.

An alternative approach is to carefully control the illumination and position of the objects to be described. In a calibrated configuration, it is possible to derive an independent description of object surfaces without additional *a priori* knowledge. This description is in the form of three-dimensional coordinate triples taken at intervals over the object surface. In many industrial applications this description is sufficient to determine product quality or object position for automated assembly. The various approaches for controlled illumination have often been presented under the name structured light or structured illumination.

This paper describes a sensor system that uses structured illumination to inspect metallic castings. The discussion focuses on optical design considerations and high speed processing techniques for converting intensity images into a range image. In addition, a number of examples of sensor performance are discussed.

2. Structured Light Principles

2.1. Simple Triangulation

The underlying principle of structured light methods is optical triangulation. This principle is illustrated with a single ray of light impinging a surface, as shown in Figure 1. The image of the ray is formed along an optical axis which is at an angle with the ray. This angle, θ, is known as the parallax angle. It follows that the image of the ray shifts in the image plane according to the position of the surface along the ray. This parallax shift can be used to calculate the surface location provided that the geometric configuration of the camera and ray are known. In practice, such a system can be calibrated by measuring a known object surface, such as a plane.

The relationship between the ray image and the surface position is given by

$$\Delta = \frac{\delta d_1}{d_2 sin\theta - \delta cos\theta} .$$

(1)

If the image field of view is limited so that the parallax angle can be considered constant, then the relation reduces to

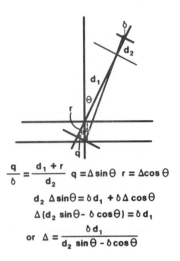

$$\frac{q}{\delta} = \frac{d_1 + r}{d_2} \quad q = \Delta \sin\Theta \quad r = \Delta \cos\Theta$$

$$d_2 \, \Delta \sin\Theta = \delta d_1 + \delta \Delta \cos\Theta$$

$$\Delta(d_2 \sin\Theta - \delta \cos\Theta) = \delta d_1$$

$$\text{or} \quad \Delta = \frac{\delta d_1}{d_2 \sin\Theta - \delta \cos\Theta}$$

Figure 1: The geometry of parallax triangulation. θ is the parallax angle, Δ is the height of the surface above the reference position, and δ is the shift in the image of the beam.

$$\delta = \Delta \left[\frac{d_2}{d_1} \right] sin\theta \; . \tag{2}$$

In this case, there is a linear relation between the ray image position and object surface position. This relation is conceptually useful, but not accurate enough for industrial measurement applications.

The object surface can be sampled by sweeping the ray over a range of angles within the field of view of the camera. If the triangulation relation of the ray image is calibrated for each ray position, it is possible to derive an array of surface positions within the field of view. This array is known as the range image. In this case, the range is defined as the distance from the unknown surface to a given reference surface such as a plane. The reflectance of the surface can be obtained by measuring the intensity of the ray image for each ray position.

2.2. More Complex Patterns

This simple approach to three dimensional measurement has been used in many industrial applications such as controlling the height of liquids and simple gauging instruments. However, the speed of such systems is limited by the rate at which the location of the ray image can be determined. A common sensor for this purpose is a solid state array camera which can operate at about 30 fields of view, or frames, per second; this corresponds to about 30 points per second of range information. This performance can be improved by using a one-dimensional image sensor or line array. These devices can be read out much faster, producing a range sample rate of several thousand points per second.

Another approach which can improve the range sample rate is to project a stripe of light rather than a single spot, as shown in Figure 2. Each row of

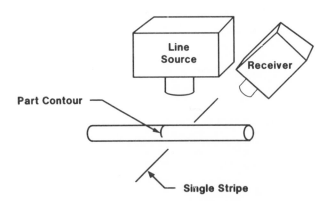

Figure 2: The projection of a single stripe on an object surface. The apparent location of the stripe in the image depends on the surface height due to parallax.

the image sensor is used to triangulate the position of the stripe along that row. This multiplies the throughput of range sensing by a factor corresponding to the number of rows in the image sensor. If the sensor array can resolve 128 stripe positions, the range sample rate is about 5000 samples per second.

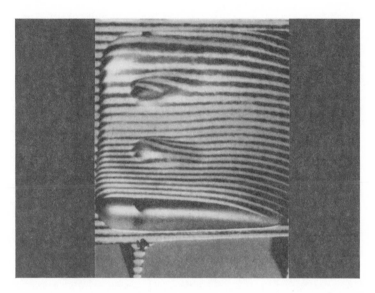

Figure 3: The projection of multiple stripes. The intricate surface profile of the casting is clearly revealed by the stripe contours.

Additional parallelism can be obtained by projecting more than one stripe, as shown in Figure 3. In this case, the details of the three-dimensional surface are clearly exposed. Again, it is possible to calibrate the relationship between each stripe position and the three-dimensional surface range. A major drawback to this method is that the stripes can be confused with each other. If the surface position is far enough from the reference position to shift the stripe image more than the stripe spacing, the range becomes ambiguous. The use of multiple stripes is attractive because, by virtue of its parallelism, it can produce a throughput of several hundred thousand range samples per second.

The ambiguity of the multiple stripes can be eliminated by projecting a number of stripe patterns in sequence. The set of stripes form a Grey code

sequence so that each surface point has a unique code value. Figure 4
illustrates such a set of patterns. An image point is recorded as illuminated

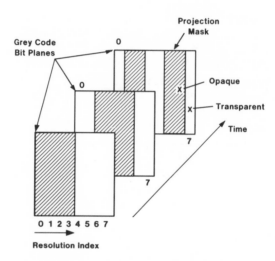

Figure 4:   A set of Grey code patterns for time multiplex projection.
The patterns define a unique binary code for each resolution element.

or unilluminated for each projected light pattern. The shift of each unique
code from the reference position is determined and a range image is
calculated. The number of stripe patterns required is proportional to the
logarithm of the number of image points along each row. For example, with
256 image points in each row, 8 stripe code patterns are required. At 30
sensor image frames per second, this approach provides about 200,000 range
points per second.

3. Limitations of the Structured Illumination Method

3.1. The Problem of Reflectance Variation

In the discussion above, it is assumed that the illumination ray has infinitesimal diameter, and the position of the ray image can be determined precisely. In practice, these ideal conditions cannot be met, and variations in surface reflectance produce errors in the range measurements. In this section the relation between reflectance variation and range error in the case of a single ray is analyzed.

The intensity cross section of the image of a laser beam is approximately a Gaussian function. That is,

$$I(x) = I_0 \, e^{-\left[\frac{(x-x_0)}{\sigma}\right]^2} \tag{3}$$

where x_0 is the beam center and I_0 is the intensity there. This function is shown in Figure 5(a). Note that the spot intensity function is circularly symmetric. For simplicity, this analysis will be carried out for the spot cross section, with the obvious extension for the symmetry. The analysis is, however, directly applicable to the practical case where a one-dimensional image sensor is used.

There are a number of methods for determining the location of the beam center in the image plane. The center location provides the parallax shift data necessary for the calculation of range values. The most common method is to compute the first moment of image intensity. If this is normalized by the zeroth moment, the result gives an estimate for the beam center, x_0.

For an ideal Gaussian, it follows that the ratio of the first intensity moment to the zeroeth moment is identically the beam center.

$$x_0 = \frac{\displaystyle\int_{-\infty}^{\infty} x e^{-\left[\frac{(x-x_0)}{\sigma}\right]^2}}{\displaystyle\int_{-\infty}^{\infty} e^{-\left[\frac{(x-x_0)}{\sigma}\right]^2}} \tag{4}$$

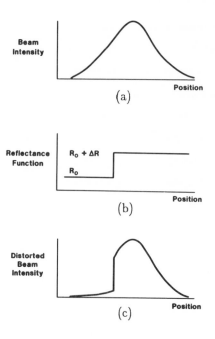

Figure 5: The effect of reflectance on the location of the center of a Gaussian beam. (a) The ideal Gaussian distribution; (b) A step in surface reflectance; (c) The resulting image intensity function. Note that it is no longer symmetrical about the beam center.

Actually, this result holds for any beam intensity cross section that is symmetric about x_0.

Now suppose that the incident beam is reflected from a surface that has variable reflectance, $R(x)$. The resulting beam image intensity is given now by

$$I(x) = P(x)\, R(x)\,, \tag{5}$$

where $P(x)$ is the incident illumination power density at the point x on the surface. The form of this modified intensity is plotted in Figure 5(c). The ratio of Equation (4) no longer corresponds to the center of the beam. This asymmetry introduces an error in the parallax shift measurement and, consequently, the resulting range is in error.

To get an idea of the magnitude of this error, a simple reflectance model is assumed. Suppose that the surface reflectance can be considered uniform everywhere except at the location x_r, where a step in reflectance from R_0 to $R_0 + \Delta R$ occurs as shown in Figure 5(b). To understand the effect of this variation in reflectance, let us examine a simple case in which the parallax angle is zero and the sensor is observing the surface at unit magnification. In this case, the coordinate systems of the sensor image plane and the object surface can be considered the same. The new location of the beam center is given by

$$
x_0' \approx \frac{\Delta R \int_{x_r}^{\infty} x e^{-\left[\frac{(x-x_0)}{\sigma}\right]^2} }{R_0 \int_{-\infty}^{\infty} e^{-\left[\frac{(x-x_0)}{\sigma}\right]^2} } + x_0 .
\tag{6}
$$

It has been assumed that ΔR is small compared to R_0. Carrying out the integrations, it follows that

$$
\frac{x_0' - x_0}{\sigma} \approx \frac{\Delta R}{2 R_0 \sqrt{\pi}} e^{-\left[\frac{x_r - x_0}{\sigma}\right]^2} .
\tag{7}
$$

Notice that the error in beam position is related to the location of the reflectance step within the beam cross section. If the step is far away from the beam center, then the error in beam location is small. The maximum error occurs when the reflectance step is right at nominal beam center. From this analysis we conclude that any nonuniformity in the surface reflectance will lead to a beam location error.

As another example, consider the case where the surface reflectance varies linearly over the region of interest. The resulting beam center error is given by

$$\frac{x_0{}' - x_0}{\sigma} \approx \frac{\sigma}{2R_0} \frac{dR}{dx}.$$ (8)

Here the error does not depend on the beam position, but only on the rate of change of surface reflectance and the beam spread, σ. In both of the examples, the width of the beam is the major factor that determines the magnitude of range error. Hence, the range errors can be reduced by using the smallest practical beam diameter.

To estimate the effect of these considerations on range accuracy, we will apply them to a typical case. Assume that the incident ray is an unfocused laser beam with a diameter of 1 mm and that the reflectance step is 10% of the incident energy. With the parallax angle at 30 degrees, the maximum range error is .05 mm.

The error can be much worse if the surface is metallic and has glint facets. High variations of reflected intensity due to glint are equivalent to radical variations in surface reflectance. It is possible to experience an intensity dynamic range of as much as 10^4 when imaging metallic surfaces with highly reflective facets. In this case, the measured beam center can be pulled far from the actual center if the glint source is at the periphery of the beam. As before, this problem can be reduced if the beam diameter is kept small.

3.2. Alternatives for Reducing the Effect of Reflectance

3.2.1. Normalizing the Reflectance

As mentioned above, the reduction in beam diameter is one important factor to consider for reducing reflectance effects. Another reasonable approach is to measure the reflectance and normalize the image intensity signal before calculating the parallax shift. This can be accomplished, in principle, by sampling the beam intensity over its cross section and computing the corresponding reflectance. However, in order to compute the surface reflectance function using the beam, the beam center, which can be shifted from the expected position by parallax, must be known. In other words, a single beam does not provide enough information to determine both range and reflectance. The additional information necessary can be measured by introducing an independent beam and appropriate image analysis. If both beams are illuminating the same surface point, the reflectance effects can be separated from parallax shifts.

The next sections will discuss several methods of obtaining the equivalent of two independent surface measurements in order to reduce the effects of surface reflectance.

3.2.2. Time Multiplexed Beams

If two beams are time multiplexed, two sequential image frames are acquired. As a simple example of how normalization can be achieved, suppose that one beam is shifted slightly with respect to the other, as shown in Figure 6(a). If the two images are subtracted, the resulting difference is as shown in Figure 6(b). Note that the difference crosses zero at the point where the nominal beam intensities are equal.

The position of this zero crossing is not dependent on the surface reflectance, since at the crossing both beams are identically affected and the difference remains zero. There is error introduced in actually determining the zero crossing since the difference signal about the crossing must be examined in order to locate its position precisely. The beams do not experience identical reflectance variations on both sides of the zero crossing, so that methods for determining the location based on sampling or interpolation will be in error. This effect is discussed further in a later section.

An additional source of error is any relative motion of the sensor and the unknown surface. If a motion that alters the appearance of the surface to the two beams occurs during the time that the two measurements are made, then the correction for reflectance is invalid. Unfortunately, in the case of

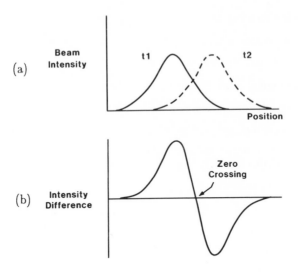

Figure 6: (a) The use of two time multiplexed beams; (b) **The** intensity difference between the two beams.

metal surfaces, only a small motion (or indeed, even a minor beam misalignment) can cause a significant error due to the faceted nature of metal surfaces.

3.2.3. Wavelength Multiplexing

An alternative to time multiplexing two beams is to use multiple wavelengths. The beams are projected simultaneously, and each beam has a discernible color spectrum. The sensor separates the image into appropriate color spectra using wavelength filters. The resulting images can be subtracted to form a difference image as described in the time multiplexing approach. The advantage of wavelength multiplexing is that both images can be collected at the same time using two aligned sensors with color filters. This avoids errors due to relative motion of the surface during the time the measurements are made.

There is some risk in the assumption in that the reflectance of the surface is the same at the two beam wavelengths. A difference can cause an erroneous shift in the location of the zero crossing and thus the range value. This effect does not present a problem for many industrial applications that involve metallic surfaces. Most metals have a reasonably flat spectral response over the range of wavelengths near the visible band. On the other hand, organic materials and specific color agents can cause a reflectance variation that produces a significant range error.

3.2.4. Other Multiplexing Possibilities

There are other physical phenomena which can be exploited to provide independent channels for reflectance normalization. For example, polarized light can be used to create two coincident beams. The image of the beams can be separated by polarizing filters. This method is not likely to be as robust as the wavelength technique since many materials cause a shift in the polarization. This shift causes crosstalk between the two channels and prevents proper normalization of the reflectance.

At present, it appears that time and wavelength multiplexing are the most promising methods. The next few sections discuss approaches to the normalization of reflectance variations using information from multiple beams.

3.2.5. Normalization Methods

Having established two separate channels of information using the multiple beam approach, the next issue is how to use the two intensity signals to normalize the effects of surface reflectance variations. There are many viable solutions, but the choice of an optimal one is still an open question. To determine the functional combination of the channels that leads to maximum accuracy in parallax shift detection, we must apply optimum signal theory analysis to a carefully honed model of the surface reflectance properties and the sensor system.

Two methods are presented that serve as a first approximation to satisfactory reflectance normalization. They use a simple model for the spatial variation of surface properties to remove the influence of reflectivity on the normalized signal.

3.2.6. Difference Ratio

The simplest normalization is given by the ratio of the difference of the intensity of the two channels to the sum of the intensities. If the two intensities are given by I_1 and I_2, the normalized signal is

$$N(x) = \frac{I_1 - I_2}{I_1 + I_2} . \tag{9}$$

The normalized result, $N(x)$, is independent of surface reflectivity under the following model. Suppose that the image intensity is given by

$$I_1(x) = R(x)\, P_1(x) , \tag{10}$$

$$I_2(x) = R(x)\, P_2(x) . \tag{11}$$

Here $P_i(x)$ represents the incident beam power spatial distribution for beam i. The contributions of sensor geometry and sensitivity, as well as the effective numerical aperture of the camera, are included in the reflectance, $R(x)$.

It is easy to show that the expression for $N(x)$ leads to complete suppression of the spatial variation of $R(x)$. Using the expressions for $I_1(x)$ and $I_2(x)$, $N(x)$ is given by

$$N(x) = \frac{P_1(x) - P_2(x)}{P_1(x) + P_2(x)} . \tag{12}$$

Notice that $R(x)$ appears in the numerator and denominator of $N(x)$ and is thus canceled. The normalized signal depends only on the incident beam power distribution and the sensor properties.

Consider the simple case of two Gaussian beams which are shifted slightly with respect to each other. The power distributions for the two beams are

$$P_1(x) = e^{-\left[\frac{x-x_0}{\sigma}\right]^2}, \tag{13}$$

and

$$P_2(x) = e^{-\left[\frac{x-x_0-d}{\sigma}\right]^2}, \tag{14}$$

where d is the shift in beam centers. Using a Taylor series expansion about x_0, $P_1(x)$ and $P_2(x)$ become

$$P_1(x) = 1 - \left[\frac{x-x_0}{\sigma}\right]^2 + \frac{1}{2}\left[\frac{x-x_0}{\sigma}\right]^4 - \cdots, \tag{15}$$

$$P_2(x) = 1 + \left[\frac{2d}{\sigma}\right]\left[\frac{x-x_0}{\sigma}\right] + \frac{1}{2}\left[\frac{2d}{\sigma}\right]^2 - 2\left[\frac{x-x_0}{\sigma}\right]^2 + \cdots. \tag{16}$$

If we represent

$$P_2(x) = P_1(x) + D, \text{ where } D \approx -\left[\frac{2d(x-x_0)}{\sigma^2}\right], \tag{17}$$

then $N(x)$ is given approximately by

$$N(x) = \frac{P_1-P_2}{P_1+P_2} = \frac{-D}{2\,P_1+D} \tag{18}$$

or,

$$N(x) \sim \left[\frac{d}{\sigma}\right]\frac{(x-x_0)}{\sigma} + \left[\frac{d}{\sigma}\right]^2\left[\frac{(x-x_0)}{\sigma}\right]^2 + \cdots \tag{19}$$

assuming that the shift between the two beams is small compared to the beam spread, σ. Note that $N(x)$ vanishes at x_0 and is approximately linear at the zero crossing. The slope at x_0 depends on the ratio $\frac{d}{\sigma}$, which implies that the accuracy of the zero location is improved either by a small beam radius σ, or by increasing the shift d. These conclusions only hold for small shifts since the Taylor series expansion is only accurate near x_0.

The assumption that the reflectance $R(x)$ is the same for both beams is quite reasonable for the time multiplex method. If wavelength multiplexing is used to produce separate beams, it is possible that a difference in reflectance for the two colors may result. This color variation introduces a reflectance dependency in the difference ratio normalization just discussed. The effect can be seen by assuming a different reflectance for each beam $R_1(x)$ and $R_2(x)$. Substituting into the normalization expression, we obtain

$$N(x) = \frac{R_1 P_1 - R_2 P_2}{R_1 P_1 + R_2 P_2}. \tag{20}$$

There is no obvious cancellation of the reflectance. In fact, the situation can become unsatisfactory if the reflectance in one wavelength band is much larger than the other. Suppose that $R_1 \gg R_2$. $N(x)$ becomes constant and no signal feature is computable to determine parallax shift. This problem can be reduced by using differential normalization.

3.2.7. Differential Normalization

Another form of normalization is given by the ratio of the spatial derivative of the intensity to the intensity itself. Consider this normalization function for a single illumination image intensity channel, $I(x)$.

$$N(x) = \frac{1}{I(x)} \frac{dI(x)}{dx}. \tag{21}$$

Given that $I(x) = R(x)P(x)$, it follows that

$$N(x) = \frac{1}{P(x)} \frac{dP(x)}{dx} + \frac{1}{R(x)} \frac{dR(x)}{dx}. \tag{22}$$

The first term is related only to the illumination pattern, while the second term gives the effect of reflectance.

The same normalization can be applied to the second illumination channel. By forming the difference of these normalized channel signals, the effect of reflectance variation may be further reduced. That is,

$$N(x) = N_1(x) - N_2(x) , \tag{23}$$

where $N_1(x)$ and $N_2(x)$ are the normalized signals for each beam. Expanding $N_1(x)$ and $N_2(x)$ we have

$$N(x) = \frac{1}{P_1(x)} \frac{dP_1(x)}{dx} - \frac{1}{P_2(x)} \frac{dP_2(x)}{dx} + \frac{1}{R_1(x)} \frac{dR_1(x)}{dx} - \frac{1}{R_2(x)} \frac{dR_2(x)}{dx} . \tag{24}$$

The effect of reflectance completely vanishes if $R_1(x) = R_2(x)$, as was the case for the difference ratio normalization form. The differential normalization goes further in that scaled reflectance forms are removed. For example, suppose that

$$R_1(x) = R_{10} \, f(x) , \tag{25}$$

$$R_2(x) = R_{20} \, f(x) . \tag{26}$$

We see that the reflectance terms in $N(x)$ cancel, leaving only a dependence on the illumination pattern.

There are reflectance variation effects that are not removed by differential normalization. For example, let

$$R_1(x) = R_{10} + r_1(x) , \tag{27}$$

$$R_2(x) = R_{20} + r_2(x) . \tag{28}$$

Then the effect of reflectance variation is not canceled by differential normalization unless

$$\frac{1}{R_{10}} \frac{dr_1(x)}{dx} = \frac{1}{R_{20}} \frac{dr_2(x)}{dx} . \tag{29}$$

It is unreasonable to expect that surface reflectance at different wavelengths will obey this relation for all materials. However, if the spatial variation of reflectance is small compared to the structured illumination pattern, the effects of reflectance can be ignored. The reflectance spatial slope, $\frac{dR}{dx}$, is rarely significant except from metallic surfaces with glint. However, metallic surfaces are reasonably neutral in color, so the reflectance terms cancel.

There is one second order effect associated with differential normalization that produces an error in range measurement. In the analysis above, it was assumed that the rate of change of the beam position from the reference position is small. This assumption is not valid at points where the range is changing rapidly, such as at surface height discontinuities. The effect can be demonstrated by considering the location, x, of the beam image to be a function of surface height. In other words, the modulation function, $f(x)$, can be expressed as $f(x_n+g(z))$, where $g(z)$ is an approximately linear function of z, and x_n is the nominal image coordinate (where nominal means the location of the beam image when the surface is at the reference position). The normalized form for $f(x)$ expands to two terms,

$$\frac{1}{f(x)}\frac{df(x)}{dx_n} = \frac{1}{f(x)}\frac{df(x)}{dx}\left[1 + \frac{dg(z)}{dz}\frac{dz}{dx_n}\right]. \tag{30}$$

If the surface slope $\frac{dz}{dx_n}$ is small, then the expression reduces to the original form. The second term can be large in the vicinity of step profile changes. In practice, the surface slope measurement is restricted by the limited resolution of the optical and sensor elements. This effect can be seen in Figure 20, where the profile step measurements exhibit a small overshoot at the step location.

3.2.8. Other Forms of Normalization

One can approach the problem of normalization from the standpoint of optimal signal theory. Most classical results have been developed for additive noise, while the effects of reflectance appear in product form. This can be handled by considering the logarithm of intensity rather than intensity itself. That is,

$$log\ I(x) = log\ R(x) + log\ P(x) \tag{31}$$

From this perspective, one can consider $log\ R(x)$ to be an additive noise term which interferes with an estimate of the signal, $log\ P(x)$.

This leads to consideration of classical noise elimination methods such as Weiner filtering. Such a filter is of the form

$$H(\omega) = \frac{P(\omega)}{P(\omega) + R(\omega)}. \tag{32}$$

The effectiveness of the filter depends on the degree to which the spatial frequency distribution of the reflectance variation is disjoint from that of the structured illumination.

At first, it would seem that this would be the likely case. For example, consider a periodic illumination pattern (stripes) which has a single distinct spatial frequency. The spatial frequency distribution corresponding to reflectance variations can be considered uniform, particularly for metallic surfaces with a rough finish. Thus, a notch filter can be constructed that only passes energy at the spatial frequency of the illumination pattern. This would eliminate most of the energy due to reflectance variations.

The fallacy in this line of reasoning is that the spatial frequency of the illumination pattern does not have a well defined distribution even if the projected pattern is a periodic set of stripes. The image of this pattern presents a stripe period which depends on the local surface slope. The spatial frequency can be higher or lower than the nominal stripe spacing depending on whether the surface is tilted toward or away from the sensor. For this reason, it is possible to encounter a band of spatial frequencies centered around the stripe frequency.

This observation would suggest that it is not feasible to approach the reflectivity normalization problem from an optimal filtering point of view without a model for the surfaces that will be examined. At the very least, it seems that a bandpass filter would be necessary and that the spatial response of the range sensor would be limited by the bandwidth of the filter.

An adaptive approach would be to form an estimate of the surface slope at each point of observation. It is not likely that this information will be available in advance. It may be possible to provide an iterative solution where a rough estimate is made of the surface range, and is used to compute surface slope. The slope is used to define a filter which provides a more refined estimate of surface range. This procedure can be repeated until the desired accuracy is achieved.

4. A Profile Sensor Based on Wavelength Multiplexing

4.1. Sensor Requirements

The basic principles for range sensing, described above, were exploited to implement a practical profile sensor to inspect metallic surfaces for small defects such as scratches and dents. In most cases, defect limits are related to the detailed profile of the defect. For example, maximum depth, radius of curvature, and lateral dimensions of a dent determine if it is unacceptable. All of these parameters can be computed from part surface profile measurements. The sensor must therefore perform a complete scan of the part surface. From this scan, an accurate measurement of the surface contour can be extracted.

As we have mentioned above, metallic surfaces can produce undesirable image conditions such as glint. These large excursions in surface reflectivity lead to a large dynamic range in image intensity. It has been shown [4] that the power reflected in a solid angle from metal varies exponentially as a function of incident and observation angles. This can lead to variations of several orders of magnitude in scattered light power over a small distance in the viewing plane. It follows that a sensor suitable for metal surface inspection must be tolerant to large and rapid variations in image intensity.

The scale of surface defects in the metal castings to be inspected is on the order of .001". That is, the depth variations due to defects must be determined with this accuracy. The lateral resolution required is ideally on the order of .002" to allow adequate presentation of small scratches and pits. This determines the pixel size and consequently the processing rate. A typical casting has ten square inches of surface area. If the pixel area is 4×10^{-6} square inches, 4×10^7 pixels are needed to cover the surface. In order to meet the inspection throughput requirements for a production environment, each casting must be processed in about 30 seconds. It follows that a rate of about 10^6 pixels/sec must be achieved. The sensor to be described here will not achieve all of these goals, but they serve as an important guide in the selection of design approaches and implementation strategy.

4.2. Theory of Sensor Operation

In order to achieve the aggressive throughput requirement, it is necessary to project a parallel pattern of parallax shift features. It is desirable to have the pattern as dense as possible so that many range readings can be obtained in one image capture. The density of the pattern is limited by the problem of shift ambiguity. If the surface range varies by an amount which causes a shift in the pattern by more than one stripe spacing, then the range becomes ambiguous.

The desire for high throughput rules out the use of time multiplexed patterns since the range pixel rate is ultimately limited by the sensor frame rate. These considerations dictate the choice of wavelength multiplexing since the channels can be detected in parallel with separate sensor arrays for each color. In addition, it is difficult to provide more than two color channels due to problems in wavelength filtering and sensor alignment. Thus the stripe spacing must be large enough to avoid range ambiguity over the range depth of field. In the application described, the range interval is .05". Larger excursions in depth are accommodated by a manipulation system which is guided by a nominal description of the part surface.

4.3. Sensor Normalization

The differential normalization method is used to minimize the effects of surface reflectance. In this design there are two wavelength channels, $P_1(x)$ and $P_2(x)$. The two patterns are projected in a complementary fashion which gives the best normalized signal amplitude. Assume that the patterns are complementary, that is

$$P_1(x) = s(x) , \tag{33}$$

$$P_2(x) = G\,(K - s(x)) \tag{34}$$

where G represents the different channel gain for $P_2(x)$, and K represents the overall illumination intensity. The normalized signal becomes,

$$N(x) = \left[\frac{1}{s(x)} \frac{ds(x)}{dx} + \frac{1}{R_1} \frac{dR_1}{dx} \right] - \left[\frac{-G}{G(K - S(x))} \frac{ds(x)}{dx} + \frac{1}{R_2} \frac{dR_2}{dx} \right] . \tag{35}$$

Assume for the moment that the reflectance terms cancel or are small compared to the illumination pattern term. Then, $N(x)$ is given approximately by

$$N(x) = \left[\frac{1}{s(x)} + \frac{1}{K - s(x)} \right] \frac{ds(x)}{dx} . \tag{36}$$

Also assume that the patterns are sinusoidal, which is the case in a practical optical projection system with limited spatial resolution. A suitable functional form for $s(x)$ is

$$P_1(x) = s(x) = 1 + m \cos x \tag{37}$$

where m is a modulation factor (it is assumed that the modulation factor is the same for both channels in this analysis).

$$P_2(x) = K - s(x) = (K - 1) - m \cos x . \tag{38}$$

Substituting this into Equation (36) for $N(x)$ and choosing $K=2+\delta$, we obtain

$$N(x) = - \left[\frac{(2+\delta)m \; sin \; x}{(1+m \; cos \; x) \; (1+\delta-m \; cos \; x)} \right] , \qquad (39)$$

which simplifies to

$$N(x) = - \left[\frac{(2+\delta)m \; sin \; x}{1-m^2 \; cos^2 \; x + \delta(1+m \; cos \; x)} \right] . \qquad (40)$$

For m reasonably less than 1 and δ small, this expression reduces to $-2 \, m \; sin \; x$. For $m=1$ and δ small,

$$N(x) = - \frac{2 \; sin \; x}{1 - cos^2 x} = -2 \; sec \; x . \qquad (41)$$

The first case represents a low level of modulation of the illumination. The resulting normalized signal is a sinusoid having easily detected features from which to determine the lateral position of the stripe image. The extrema and zero crossing locations give four distinct features for each projected stripe.

The second case represents maximum modulation of the illumination where the dark interval goes completely to zero intensity. The resulting normalized signal provides two zero crossings for each stripe with no other distinct features. Thus it is desirable to use a value of m small enough to produce distinct extrema at $x=(2n+1)\pi/2$. A suitable value for m is 1/2 where reliable detection of the features for each pattern stripe is achieved.

Note that both of these cases have the same phase, $sin \; x$, as long as δ is approximately zero. The phase of the peaks shifts slightly as δ becomes significant. For example, with $m=0.5$ and $\delta=0.1$, the phase of the peaks shifts about 1-1/3 percent. As we will see later, this is insignificant since only variations in phase with light level and time actually matter, and δ can be made relatively stable for both light level and time. Figure 7 shows the

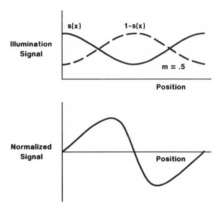

Figure 7: The illumination pattern intensity and the resulting normalized signal for a modulation factor of $m = 0.5$.

form of the illumination patterns and the normalized signal.

4.4. Sensor Optical Design

The ray diagram of the optical system is shown in Figure 8. The illumination projection system is conventional with the exception of the use of two wavelengths. The incandescent source is divided into two paths by the dichroic beam splitter, 5. The splitter has a wavelength cutoff at 800 nm. This divides the illumination into a visible component and a near infrared component. This choice leads to an equal division of effective power in the two wavelength bands. That is, when the wavelength distribution of the tungsten source and the detector response are taken into account, the detected signal from a neutral surface is equal for both channels. The two light beams are recombined by a patterned mirror, 13, providing a self-registered combination of the two wavelengths. The details of the mirror

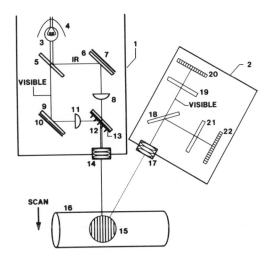

Figure 8: The optical system for the projection and detection of wavelength multiplexed patterns. The projector is 22 and the receiver 23.

configuration are shown in Figure 9. In the current implementation, the pattern is projected with a period that allows an unambiguous range depth of .05".

The receiver, 2, is composed of two linear photodiode arrays, 20 and 22. Each array has 512 detector elements spaced on .001" centers. The dichroic beam splitter, 18, splits the light into two images, one in the visible, the other in the near infrared. The receiver images at a magnification of .5X, resulting in an image of .002" for each detector element.

The overall sensor is shown in Figure 10. The internal views of the receiver and transmitter are shown in Figure 11. The optical design allows a nominal standoff distance of 3.5" from the sensor to the surface reference plane. The scan line covers a field of 1" with pixels on .002" centers. This scan is converted by special purpose hardware into an array of 128 distance

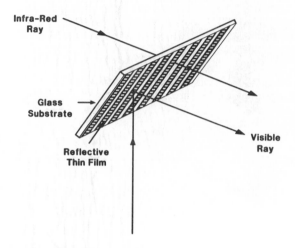

Figure 9: A detailed view of the patterned mirror principle. The two wavelength channels are automatically registered by the interleaved transmission and reflection of the deposited metal film pattern.

measurements on .008" centers. Since a one dimensional array is used, the sensor must be scanned mechanically along the axis orthogonal to the array. In this application, scanning is accomplished using a six-axis servo-controlled manipulator.

4.5. Normalization Signal Processing

The normalization ideas discussed earlier have been applied in the design of the sensor signal processing system. The functions to be described in this section have been implemented in special purpose hardware. The hardware implementation is necessary to achieve the design target of 60,000 range readings per second.

Figure 10: An overall view of the sensor. Note the parallax angle between the transmitter and receiver. There is an extendible mirror at the side for viewing at right angles to the normal optical axis of the sensor. This facilitates the observation of concave surfaces.

4.6. Sensor Calibration

The first step is to remove inherent distortions and artifacts in the intensity signals derived from an A/D conversion of the solid state diode array signals. The array signals are corrected for gain and offset differences between diodes as well as a periodic noise term related to the geometric layout of the detector array known as pattern noise. The pattern noise term is difficult to remove since its amplitude depends non-linearly on the illumination intensity. In the current implementation, the solid state arrays exhibit a noise pattern that repeats with a period of four diodes.

A block diagram of the sensor calibration function is also shown in Figure 12. The gain and offset corrections for each diode are saved in a local RAM. The pattern noise correction is also stored in a RAM table and the correction value is accessed using an address formed by combining the least

Figure 11:   Internal views. (a) The transmitter; (b) The receiver. The projection lens for focusing the incandescent source is an aspheric design. The objective lenses are standard components. The detector arrays are 512 element photodiode arrays. The wavelengths are split by a dichroic filter at 800 nm.

Out – Gain x (In + Pattern Noise Correction – Offset)

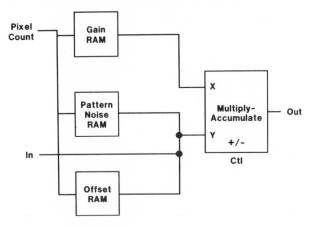

Figure 12: A block diagram of the sensor calibration and normalization process.

significant two bits of the diode location and the upper 7 bits of the signal amplitude. This allows correction to depend on the four diode period and the signal level, as required. The offset and pattern noise corrections are subtracted from each diode. The resulting signal is multiplied by the gain correction factor. The incoming signal is digitized to 10 bits and the corrected signal is carried to 12 bits which is the precision maintained throughout the remainder of the system.

4.7. Differential Normalization

As discussed earlier, reflectance normalization requires a computation of the form

$$\frac{1}{f(x)} \frac{df(x)}{dx} . \tag{42}$$

The first issue is the computation of the spatial derivative of the function $f(x)$. The approach here is to use one dimensional convolution with a mask selected to minimize the error in the estimate of the derivative. There has been extensive effort applied to the problem of optimal estimators for linear operations on signals [1], [2]. The first derivative of the Gaussian with suitable spatial scale is often suggested. In the implementation described here, the derivative mask coefficients are completely programmable so that any derivative function may be used. In the experiments to be described later, the first derivative was calculated using a least mean square estimator based on a second order polynomial model for the local signal behavior.

The hardware implementation allows five signal values to be used in the calculation of the derivative. The coefficients for the derivative estimate are

$$\frac{\hat{d}}{dx} = \frac{1}{10}[-2 \quad -1 \quad 0 \quad 1 \quad 2] \,. \tag{43}$$

These coefficients are suitably scaled to allow fixed point computation with 12 bit accuracy. The hardware carries out the 5 point convolution in one microsecond.

4.8. Signal Estimation

The signal itself is also processed with a linear estimator to obtain the denominator of the normalization form. The second order estimation for the value of the function, $f(x)$, is represented by the five point mask

$$\hat{f} = \frac{1}{35}[-3 \quad 12 \quad 17 \quad 12 \quad -3] \,. \tag{44}$$

4.9. Experimental Normalization Results

The normalization process requires four convolution steps to be performed in parallel. The division of $\frac{df(x)}{dx}$ by $f(x)$ is actually done by retrieving the reciprocal of $f(x)$ from a ROM table and then multiplying. The result from each wavelength channel is combined by subtraction to produce the desired normalized signal. The actual signals obtained by viewing a flat neutral surface are shown in Figure 13. Figure 13(a) shows the two wavelength signals superimposed. The visible channel is solid, the infrared channel is shown dotted. The variation in the image signal

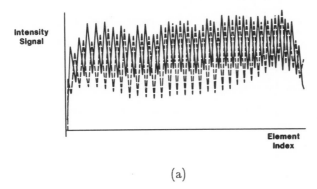

(a)

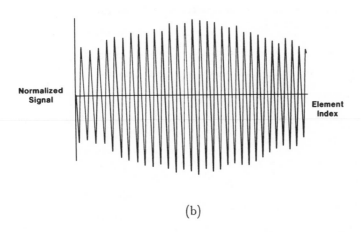

(b)

Figure 13: Experimental normalization results. (a) The two wavelength intensity channels. The infrared channel is shown dashed. (b) The normalized signal, $N(x)$. The variation in amplitude is due to lens distortions.

amplitude is due to spatial non-uniformity of the tungsten source and aberrations in the projection and receiver optics.

The resulting normalized signal is shown in Figure 13(b). The differential normalization is sensitive to the period of the illumination pattern since the period affects the magnitude of $\frac{df(x)}{dx}$. The lens used to project the illumination pattern introduces a small variation in the pattern period due to pincushion distortion. This appears in the normalized signal as a gradual variation in amplitude over the field of view with a peak in the center.

To illustrate the dynamic range of the normalization process, the same situation as in Figure 13 was repeated with the receiver optical aperture stopped down so that the signal level was reduced by a factor of 12:1. The illumination signals and the resulting normalized signal are given in Figure 14. Notice that there is only a minor variation in the normalized result for a decrease in image intensity of over an order of magnitude. (Note that the sensor intensity magnitude scales are not the same between Figures 13 and 14. The normalized signal scales are the same.)

4.10. Feature Detection and Selection

The next major step in the extraction of surface profile is the detection of the significant features in the normalized signal. The location of these features indicates the shift due to parallax, and thus is used to calculate surface depth.

In principle, each point in the normalized signal could be used to define a unique pattern location. If the normalization procedure were ideal, the value of $N(x)$ would map uniquely onto the corresponding location in the projected pattern. Thus the process of profile computation could be carried out for each pixel in the intensity sensor array.

In practice, this ideal condition is not realized due to many second order effects which lead to variations in the amplitude of the normalized signal

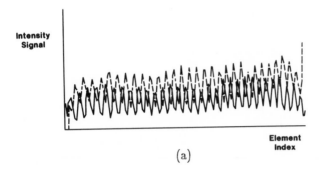

(a)

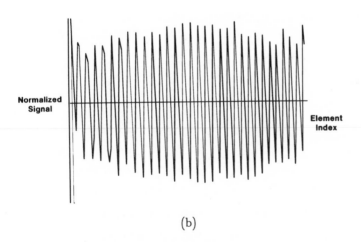

(b)

Figure 14: The same display as in Figure 13 except the receiver aperture is stopped down to reduce the signal level by 12:1.

that are not due to profile variations alone. The most stable features are the zero crossings and extrema of the approximately sinusoidal form of $N(x)$. The current design projects 32 stripes over the field of view so that each period of $N(x)$ contains 16 sensor array sample points. There are two extrema and two zero crossings in each period which are nominally separated by four sample intervals. This nominal spacing implies that 128 features are present in the field of view; thus 128 profile measurements can be obtained for each scan of the sensor array.

4.11. Peak Detection

The accurate location of peaks and valleys in the normalized signal is a problem which is encountered in many signal analysis applications. The approach here is to specify a number of properties that the signal extremum must satisfy and then implement a linear estimator for each property. This is done in such a way that the estimator vanishes if the property is satisfied. The final decision on the presence of the peak or valley is based on the sum of the absolute value of the estimators.

Three properties of a peak which should be satisfied at the maximum point of a signal, $f(x)$, are:

$$\frac{df}{dx}(x_m) = 0 \tag{p1}$$

$$\frac{df}{dx}(x_m+d) + \frac{df}{dx}(x_m-d) = 0 \tag{p2}$$

$$f(x_m+d) - f(x_m-d) = 0 \tag{p3}$$

These properties all reflect the basic idea that a signal is approximately symmetrical and constant in the vicinity of an extremum or inflection point. The conditions in (p2) and (p3) insure that the inflection point case is eliminated by requiring even symmetry in the distribution of the signal about the extremum location. The quantity, d, is a suitable distance on each side of the extremum. In the current design, $d = 2$ pixels.

4.12. Implementation of Symmetry Estimators

The estimators in (p1), (p2), and (p3) can all be calculated by convolutions, with the subtraction indicated in (p3) absorbed into the convolution coefficients. Using the same least mean square estimation methods as in the normalization, the masks are:

$$p1 = \begin{bmatrix} 0 & -3 & -2 & -1 & 0 & 1 & 2 & 3 & 0 \end{bmatrix}$$

$$p2 = \begin{bmatrix} -2 & -1 & 0 & 1 & 0 & -1 & 0 & 1 & 2 \end{bmatrix}$$

$$p3 = \begin{bmatrix} -17 & 5 & 17 & 19 & 0 & -19 & -17 & -5 & 17 \end{bmatrix}$$

These linear operators are also implemented with programmable coefficients so that alternative estimators can be tried. The convolution computation provides for up to nine coefficients. The computations are carried out in one microsecond using hardware multiplier circuits.

The scaled, absolute values of all three convolutions are added together to form the extremum signal. A peak or valley is indicated when this signal goes through a minimum. Ideally, the signal becomes identically zero at the extrema locations. The actual normalized signal is not perfectly symmetrical about the peak or valley so the properties in (p1) through (p3) do not all vanish exactly. However, the minimum location in the sum of the absolute values is a solid and reliable indication of the peak or valley location. In practice, the method works well, even for signals with barely discernable features.

4.13. Zero Crossing Detection

The detection of zero crossings is much more straightforward than for peaks and valleys. The approach here is simply to form an estimate of the normalized signal $N(x)$ and then take the absolute value of the estimate. It is seen that the resulting signal will go through a minimum at the location of the zero crossing. This method has the added advantage that the resulting signal has the same form as that obtained for the extremum detection. Thus, the treatment of the signals later in the feature identification process can be implemented with identical circuits. The same linear estimator for signal value is used as in the normalization case. That is,

$$\hat{f}(x) = \frac{1}{35} \left[-3 \quad 12 \quad 17 \quad 12 \quad -3 \right] \; . \tag{45}$$

4.14. Experimental Results

The results of the feature location operations are given in Figure 15; the conditions correspond to the setup for the normalized signal plot in Figure 13. Note that the detector signals for the extremum are interleaved with those of the zero crossing. The feature signals reach a minimum four times for each period of the normalized signal.

The effect of low image intensity is illustrated in Figure 16. Here the input light level has been reduced by stopping down the receiver objective. The normalized signal appears approximately the same as in Figure 15, but there are subtle differences which are clearly reflected in the feature signals. The positions of the minima are still quite stable and distinct. Note that the signal for the extrema feature is at a maximum when the zero crossing signal is at a minimum and vice versa. This property is exploited in the detection of the minimum in each signal.

4.15. Detection and Interpolation of Feature Position

The location of peaks, valleys and zero crossings in the normalized signal has been transformed into the problem of detecting and precisely locating minima in the feature signals. A crude location of the minima is provided by starting a sequential search along one signal when its value falls just below the other. This idea is clarified by Figure 17, which shows an expanded section of the feature signal plot.

To be specific, suppose that the minimum of the zero crossing signal is

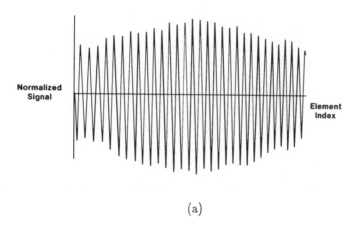

(a)

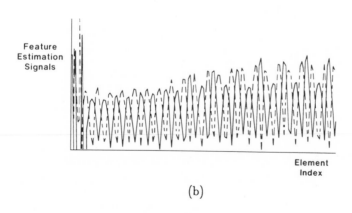

(b)

Figure 15: The feature location signals computed for the conditions of Figure 13. (a) The normalized signal of Figure 13(b); (b) The extremum and zero crossing signals. The extremum signal is solid and the zero crossing signal is shown dotted. The excursions at the left are due to an unilluminated portion of the detector array.

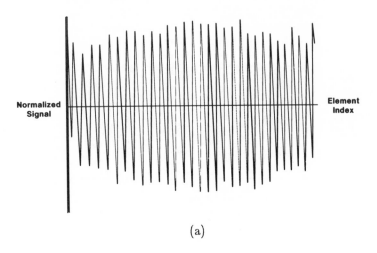

Normalized
Signal

Element
Index

(a)

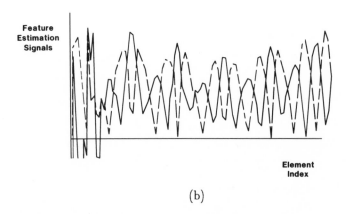

Feature
Estimation
Signals

Element
Index

(b)

Figure 16:   The same display as Figure 15 except the receiver aperture is stopped down as in Figure 14.

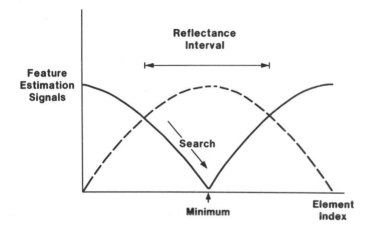

Figure 17: The procedure for locating the feature minimum. The search proceeds to the right from point A until the signal starts upward. This is taken as the starting point for the minimum interpolation.

desired. The zero crossing signal is scanned from left to right until it falls below the amplitude of the extremum signal. At this point a search is initiated for a minimum in the zero crossing signal. The minimum is assumed to be where the signal first reverses direction after falling below the extremum signal. The location of the minimum is thus known at the precision of the sampling interval of the feature signals, which in the current design is 512 samples per line.

The use of the complementary signal as a threshold to define the interval of search for the minimum in a signal is very robust since their amplitudes track together as the normalized signal degrades. There is, of course, the possibility that a signal could exhibit several minima over the interval where it is smaller than the complementary signal. This condition implies quite a high frequency phenomenon which should only occur in an extreme degradation of the normalized signal. Since there are only four sample intervals between each feature, a double minimum would require variation which is dangerously close to the Niquist sampling condition and should not be considered valid.

The location of the minimum can be refined considerably by interpolating around the minimum using a variant of Newton's method. The equation for interpolation is obtained by a Taylor series expansion of the feature signal in the vicinity of the minimum. Representing the feature signal by f(x), the approximation is

$$f(x) \approx f(x_{min}) + \frac{df}{dx}(x_{min})(x - x_{min}) + \frac{1}{2}\frac{d^2 f}{dx^2}(x_{min})(x - x_{min})^2 . \quad (46)$$

Now the minimum of $f(x)$ can be determined by setting

$$\frac{df}{dx} = 0 \qquad (47)$$

or, in terms of the approximation,

$$\frac{df}{dx} \approx \frac{df}{dx}(x_{min}) + \frac{d^2 f}{dx^2}(x_{min})(x - x_{min}) = 0 . \qquad (48)$$

In this case x_{min} is the approximate minimum location determined by the search described above. This linear equation can be easily solved for x, the interpolated value of the minimum,

$$x = x_{min} - \frac{\dfrac{df}{dx}(x_{min})}{\dfrac{d^2 f}{dx^2}(x_{min})} . \qquad (49)$$

With this interpolation it is possible to refine the location of the minimum to subpixel accuracy. Under ideal conditions with uniform reflectance surfaces and sufficient illumination, the standard deviation of the feature location is on the order of $\frac{1}{16}$ of a sample interval. Under poor illumination conditions and observing a metallic surface, the standard deviation can increase to $\frac{1}{2}$ a pixel. This is an extreme case with glint and low surface reflectivity so that the image intensity variations exceed the dynamic range of the sensor array.

4.16. Mapping of Feature Location Into Profile

The precise location of illumination pattern features is now complete. The next step is to map the shifted location of the features into the surface profile which is related to the shift of the features from its nominal position. Since the mapping function is non-linear and learned through calibration, it would be difficult to compute directly, and so it is implemented using table lookup methods.

The features are classified into four groups corresponding to the unique positions on the sinusoidal normalized signal. The classes in order of occurrence in the signal are:

- Z+ = Positive going zero crossing
- P+ = Positive peak
- Z− = Negative going zero crossing
- P− = Negative peak

The classes repeat in this sequence over the image. There are 32 groups of each of these four features or a total of 128 individual features. The translation of feature location into range is performed using the feature map which contains four regions corresponding to the four feature types. Each region is 512 elements long and each address corresponds to the location of a feature on the input image. Thus, if a P+ feature was found at pixel 47 of the input image, then the corresponding surface position is stored in address 47 of the feature map. Both X and Z coordinates are stored; the Z value is used to compute the actual range data and the X position is used as the X position of that range value. The Z value stored is the range displacement of the calibration surface which would result in the identified feature appearing in the measured position of the input image. This value is obtained by actually measuring the position of each of the features on a calibration surface. This method has the advantage of accounting for the first order anomalies of the optics, with one calibration point for each feature. The X position is a pointer to an output slot corresponding to the X position of the center of the feature in the input image. When a particular feature is found in the input image, its computed range or Z value is placed in the corresponding output or X position slot in the output range image.

The actual computation of the Z range is accomplished in three steps. First, an approximate (un-interpolated) Z value of the feature is retrieved from the address of the feature map corresponding to the input pixel location of the identified feature. In the example above, the value would be retrieved from address 47 of the feature map. The sign of the interpolator defined in Equation (49) is used to choose the adjacent pixel location in either the plus or minus direction. In our example, if the interpolator were negative, the value at pixel 46 would be retrieved. Finally, the interpolator magnitude is used to linearly interpolate between the two retrieved range values. The result is computed in fixed point arithmetic to twelve bits of resolution.

In addition to the determination of surface position (X and Z), the feature map also contains the address of the range of input pixels over which the reflectance of the corresponding range pixel should be computed. This information is passed to an averaging circuit which will be described next.

4.17. Acquisition of Surface Reflectance

The formation of surface profile readings is carried with a spatial resolution of four intensity samples (pixels) on the average. Provision is also made to acquire a measure of surface reflectance for each profile sample. In this design the reflectance is taken to be proportional to the sum of the intensities of each wavelength channel.

The intensity sum is averaged over the interval corresponding to each profile feature. This interval is illustrated in Figure 17, which shows a series of feature detection signals. The computation for R_i, which corresponds to profile sample z_i, is

$$R_i = \frac{2}{d_1+d_2} \sum_{j=i-\frac{d_1}{2}}^{j=i+\frac{d_2}{2}} (s_{1j} + s_{2j}) , \tag{50}$$

where s_{1j} and s_{2j} are the intensity signals in channels 1 and 2, respectively, d_1 is the number of pixels to the left of the current feature to the previous feature, and d_2 is the number of pixels to the right of the current feature to the next feature. The index j corresponds to the sensor sample index, while the index i is the profile feature location. There is not a fixed relationship between i and j because the profile feature location depends on the surface height, z.

The sum of the intensity channels does not exactly measure the surface reflectance since other factors can influence the intensity in each channel. For example, if the surface is not neutral, one of the wavelength channels will be smaller than the other and the sinusoidal pattern of illumination will not cancel. To see this effect, consider the following forms for $s_1(x)$ and $s_2(x)$:

$$s_1 = 1 + m \cos x , \tag{51}$$

$$s_2 = 1 + \delta K - (m + \delta m) \cos x . \tag{52}$$

The quantity δK represents the difference in optical transmission and δm represents the difference in modulation factor between channel 2 and channel 1. The sum of these two signals is

$$R(x) = 2 + (\delta K - \delta m \cos x) . \tag{53}$$

The reflectance measure should be constant (2), but instead is modulated with a sinusoidal component, $\delta m \cos x$ which is proportional to the imbalance in the channel modulation characteristics, and a constant offset, δK, which is proportional to the imbalance in the channel transmission. These two factors correspond to a first order approximation of the dynamic and static lumped response characteristics of the optics, the part surface, and the electronics.

In practice, for the application of the sensor to metallic surfaces, there is not a significant difference in the reflectance for the channels. The difference that does exist is reasonably uniform over the surface and can be removed by calibrating the gain constants for the channels to give equal intensity signals.

4.18. Experimental Results for Line Scans

Continuing the example of Figures 13 and 15, the results of profile and reflectance are now presented in Figure 18. The uppermost plot shows the original sensor intensity data for each channel. The profile readings should be nominally zero since these data were taken on the reference surface. Note that there is a small sinusoidal component in the reflectance signal. The standard deviation in profile is approximately .00015".

The corresponding result with the receiver objective stopped down is shown in Figure 19. Note that the reflectance signal is about one order of magnitude less than in Figure 18. The standard deviation in profile has increased to about .0005". The systematic slope to the profile data is caused by the change in optical characteristics of the receiver when the objective f number is increased. The rays are confined more to the center of the lens. As a consequence, the effective focal length of the lens is changed slightly, which produces a small change in image magnification. This change in feature scale is interpreted as a slope in surface height. Under normal operation, the lens settings are not changed and the optical characteristics of both the transmitter and receiver are incorporated into the profile mapping function.

Another example of sensor performance is shown in Figure 20. Here the sensor is measuring an adjustable step on a metal surface. The step is increased in steps of .002" for each scan. The overshoot in profile is due to the effect of surface slope on the differential normalization process. The nominal scatter in the profile readings is quite small in regions away from the step. In this example the deviations are on the order of .0005".

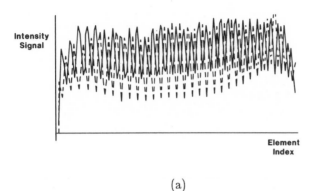

Intensity
Signal

Element
Index

(a)

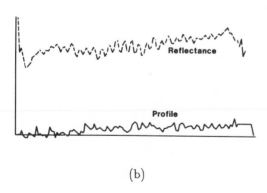

Reflectance

Profile

(b)

Figure 18: (a) The sensor intensity data of Figure 13; (b) The reflectance and profile results for the conditions in Figure 13. The dotted curve shows reflectance on a relative scale. The profile data is shown as a solid curve. Full scale is about .008".

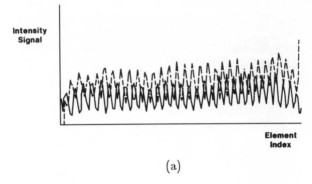

(a)

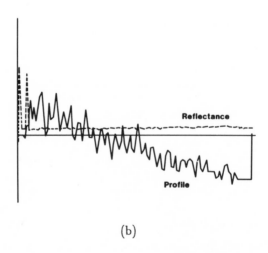

(b)

Figure 19: The same conditions as Figure 18 except the receiver aperture is stopped down to reduce the signal level. Note that the reflectance signal is about 1/12 of that in Figure 18.

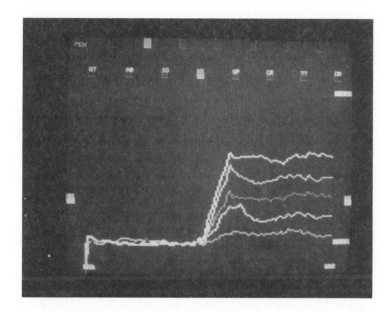

Figure 20:   The measurement of profile steps on a metal surface. Each   trace corresponds to a .002" increase in surface height.

4.19. Two Dimensional Profile Scanning

The sensor as described provides a single line of profile data. The data provide a profile sample every four sensor sample points, which corresponds to .008" using the average magnification of .5X. This magnification produces a sensor pixel image corresponding to .002" x .002" on the sample surface. The profile readings are computed using a normalized signal which is first averaged over two sensor lines for a physical domain of .008" x .004". This is again averaged for two lines which yields a square profile pixel of .008" x .008".

The input sensor rate is designed to be as high as 10^6 sensor pixels per second. After the indicated sampling and averaging, the resulting profile generation rate is 62,500 profile readings per second. This rate corresponds to an area coverage of four square inches per second, which is quite acceptable for inspection purposes.

In practice, this rate was not fully achieved. The sensor A/D conversion module limits the input sample rate to 800,000 samples per second. In addition, the low reflectance of many casting surfaces requires that the sensor integrate the incident intensity for longer than the time corresponding to the maximum line scan rate. The extra integration is necessary to achieve enough signal to noise ratio in the sensor data to allow computation of the normalized signal. For dark, oxidized metal surfaces, the maximum scan is limited by this effect to about one inch per second. Profiles can be measured at scan rates up to about three inches per second for moderately reflective surfaces.

Another limitation is imposed by the multi-axis manipulator, which provides the other direction of scan along the part surface. In the current system the six manipulator axes are divided between the sensor and the part, as shown in Figure 21. For best results, the sensor should have its optical

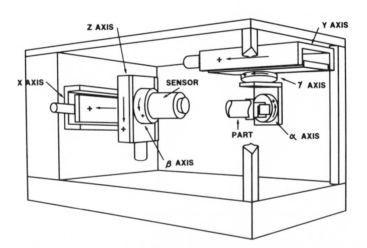

Figure 21: The configuration of manipulator axes for part surface scanning. The inertia is reduced by splitting the axes between the   part and sensor manipulation.

axis aligned with the normal to the object surface. If the the surface is highly curved, this condition requires large axial accelerations. The massive construction of the manipulator, required to maintain global positional accuracy, limits the maximum axial accelerations that can be achieved, so that for curved surfaces the scan rate is also limited to one inch per second.

4.20. Two Dimensional Surface Measurements

A typical surface scan is shown in Figure 22(a). This scan is taken of the surface of an airfoil casting as shown in Figure 22(b). The region corresponding to the profile surface is indicated by the arrow. A small step (.002") has been introduced in the data by moving the manipulator during the scan sweep. This step provides a calibration for the scale of the profile data. Another example is given in Figure 23, which shows the profile displayed as an intensity plot. The figure shows a small flaw which is about .025" across and .005" deep.

In Figure 24 a series of scans are taken on a recessed portion of a casting. The casting is shown in Figure 24(a) with the two regions to be scanned illustrated with arrows. A scan of the edge of the pocket is shown in Figure 24(b) as an intensity plot. The corresponding perspective view is in Figure 24(c). The lateral resolution of the sensor is illustrated by a scan of a portion of the raised serial number on the casting. The characters "80" are clearly legible in the profile intensity plot of Figure 24(d). The double images in some of the plots are a result of scanning back and forth over the area and recording profile images for both directions.

The effect of surface glint is illustrated by a scan made of a small metallic washer shown in Figure 25(a). The washer surface has many small glint sources and leads to false excursions in the profile data. The resulting profile is shown in an intensity plot in Figure 25(b) and the profile format in Figure 25(c). The profile noise here has a peak-to-peak variation of about .003".

The sensor has been applied to the inspection of casting surfaces for small pits and scratches. Extensive tests have indicated that defects as small as .020" across by .003" deep can be reliably detected for gradually curved

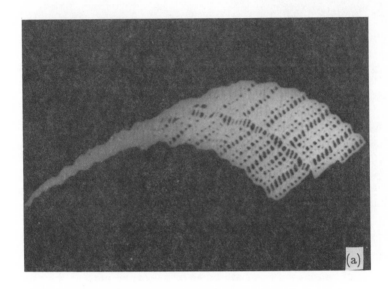

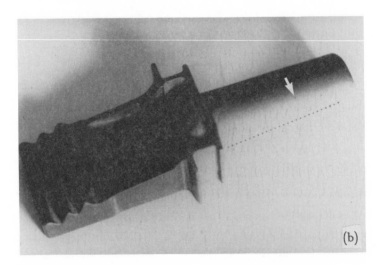

Figure 22: A two-dimensional scan of an airfoil surface. (a) The profile shown in perspective. The sensor was shifted .002" to show the scale. (b) The airfoil surface. The arrow indicates the area of the profile scan.

Figure 23: A small flaw displayed as an intensity plot. The flaw is about .025" across and .005" deep.

surfaces. The case of highly curved surfaces is still a problem because of the difficult mechanical problems in maintaining the sensor axis normal to the surface.

5. Future Directions

The effects of reflectance are diminished the most by projecting highly localized patterns. In the simple case of a single beam of illumination, this corresponds to reducing the beam spread as much as possible. It is therefore relevant to consider what optical phenomena ultimately limit the size of the beam.

5.1. Limits on Beam Spread

The effects of diffraction impose a final limit on the size to which a beam can be focussed. The main parameters are the wavelength of the light, λ, and the numerical aperture of the objective, NA. The definition of numerical aperture is given in Figure 26. Essentially, NA is a measure of the

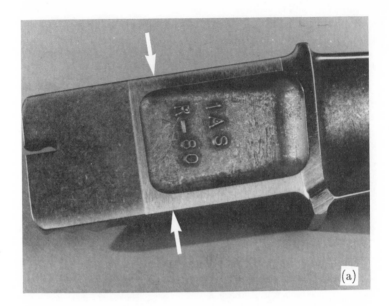

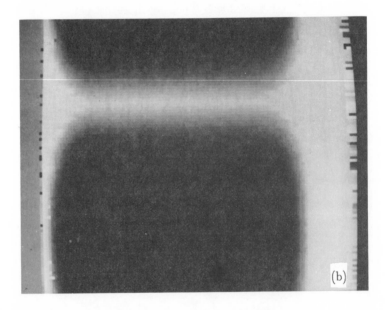

Figure 24: A scan of a recessed casting pocket. (a) A view of the casting with arrows indicating the location of the scans; (b) The profile of the edge of the pocket as an intensity plot. The surface was scanned in two directions which causes the repetition.

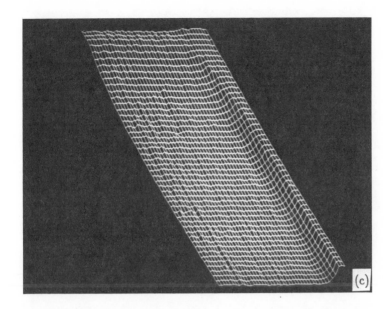

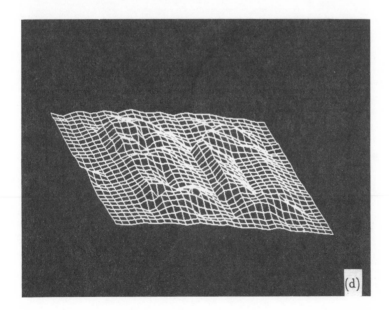

Figure 24: (c) The data in (b) shown as perspective; (d) Profile data in perspective showing the characters "80".

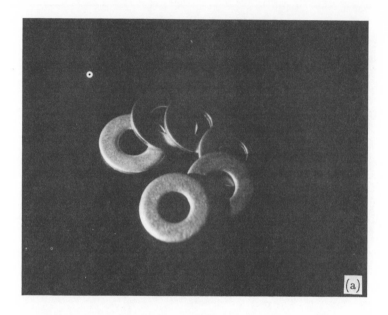

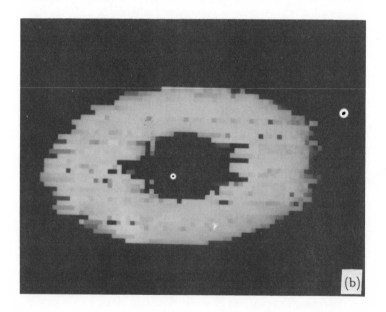

Figure 25: A scan of a metallic washer. (a) The washer; (b) The washer profile as intensity;

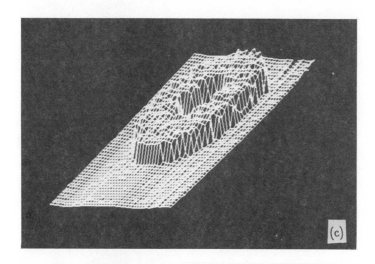

Figure 25: (c) The profile data in perspective.

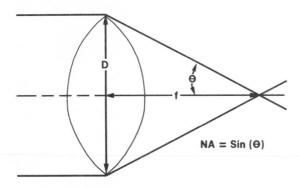

Figure 26: The definition of numerical aperture. f is the focal length of the lens, D is the lens diameter. θ is the angle subtended by the outermost ray intercepting the lens and the optical axis.

convergence angle of the rays towards the beam focus point. The beam spread, σ, at focus is given approximately by

$$\sigma = \frac{\lambda}{NA}.$$

(54)

The depth of field, δ, is also related to the numerical aperture of the objective lens. The focus spot diameter increases according a simple geometric interpretation of the divergence of the beam away from focus. That is,

$$\delta = \frac{\lambda}{NA^2}.$$

(55)

To get an idea of the magnitude of these quantities, consider a typical case where the objective lens has a .1 NA and the desired beam spread, σ, is .01 mm or .0004". The resulting depth of field is only .2 mm or about .010". This extremely limited range is difficult to maintain for manipulation over curved surfaces, so some form of profile feedback must be used to maintain the focus setting. On the other hand, with a beam spread of only .0004", the maximum profile error that can be caused by glint spikes is limited to about .001" for a parallax angle of 30°.

5.2. Dynamic Focus

To maintain the beam focus for such a limited depth of field, one approach is to provide two levels of depth resolution. An example of this is shown in Figure 27. Here two wavelengths are used to provide dual channels. One channel is focused with a small NA and correspondingly large depth of field. This provides a crude profile reading to precisely locate the focus for the second beam. The second beam is focused, with a large NA and, consequently, fine profile resolution.

It is difficult to see how an approach along these lines can be extended to a parallel pattern of stripes as in the sensor described earlier. Each point must be individually focused, which implies that sensor readings are taken for each sample beam location and focus position. In the scheme just described, at least two sensor readings must be taken for each profile

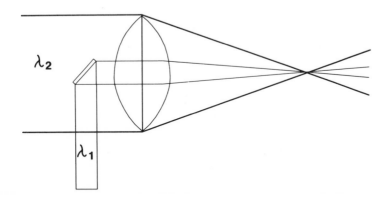

Figure 27: An arrangement for dual resolution beams. One beam is used to provide an approximate focus position for the other.

sample.

5.3. Structured Light and Stereo Matching

A more interesting prospect which avoids the need to maintain a finely focused system is the hybrid use of structured light and stereo feature matching. This idea is illustrated in Figure 28. Here, two receiver sensors are provided to observe the surface at the parallax angle with respect to the direction of the projected pattern.

If the reflectance of the surface is relatively uniform, the accuracy of profile measurements made by the differential normalization method will be good. In the presence of glint or other rapid variations in reflectance, the normalization procedure may degrade and lead to error in profile measurement.

However, this is exactly the condition in which stereo matching is effective. The rapid change in reflectance can be localized accurately in each sensor image using standard intensity edge detection methods [3]. The

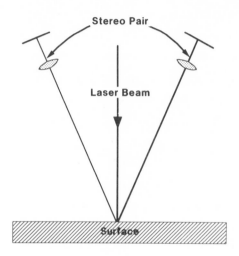

Figure 28: A hybrid arrangement using stereo matching and structured light.

location of these intensity features can be matched between the two views, which gives an accurate estimate of the surface position.

On the other hand, if the surface reflectance is uniform there are no intensity features to match in the stereo pair. The projected illumination pattern provides the needed triangulation features in this case. Thus the two methods nicely complement each other and are capable of high degrees of parallelism.

6. Conclusion

As we have described, the technology for direct profile sensing is complex and rapidly evolving. It is clear that the process of obtaining an accurate profile reading over rough metallic surfaces is computationally intensive. Fortunately, the present cost of computation is plunging rapidly through the application of VLSI technology. It is likely that in the near future many of the sensing and signal processing requirements described here will be available in highly integrated modules with correspondingly low cost.

In addition, other methodologies for profile sensing are becoming practical as a result of the improvement of semiconductor technology. For example, there is an increasing interest in time-of-flight methods. In these approaches, the round trip time or phase of a modulated laser beam is used to measure distance. It is not yet clear whether or not time can be measured with sufficient precision to compete with structured light triangulation methods for surface profile measurement.

References

[1]   Beaudet, P.
        Optimal linear operators.
        In *Proc. 4th International Joint Conference on Pattern
            Recognition.* 1979.

[2]   Canny, J. F.
        Finding edges and lines in images.
        Technical Report TR-720, MIT, Artificial Intelligence Lab, 1983.

[3]   Grimson, W. E.
        Computational experiments with a feature based stereo algorithm.
        Technical Report 762, MIT, Artificial Intelligence Lab, 1984.

[4]   Mundy, J. L.
        Visual inspection of metal surfaces.
        In *Proc. National Computer Conference*, pages 227. 1979.

3-D Vision System Analysis and Design

Steven K. Case, Jeffrey A. Jalkio, and Richard C. Kim

CyberOptics Corporation
2331 University Avenue S.E.
Minneapolis, MN 55414

and

University of Minnesota
Electrical Engineering Dept.
Minneapolis, MN 55455

Abstract

Structured light vision systems offer a practical solution to many 3-D inspection problems. This paper examines the tradeoffs involved in the design of 3-D inspection systems that use triangulation. General equations describing the behavior of such systems are derived. Problems common among structured light systems are explored and alternative solutions presented. The implementation of a multi-stripe system is discussed.

1. Overview

Vision systems are becoming an integral part of many automated manufacturing processes. Visual sensors for robot arm positioning, parts placement, and work cell security have greatly enhanced the utility of robots, while inspection systems employing similar sensors help ensure high quality products in many areas. However, standard vision systems provide only a two dimensional picture of the work area or finished product; unable to discern the distance to, or height of, the object viewed. Recently, various 3-D vision systems have been produced to alleviate this problem via several methods [10], [14]. The two major distance measurement techniques are triangulation and time of flight [12]. Triangulation systems can be further subdivided into active [11] (structured light) and passive [13] (stereo disparity) systems. Most commercially available 3-D vision systems employ structured light triangulation because this method currently has the greatest potential for fast, accurate, and inexpensive high resolution systems.

As shown in Figures 1 and 2, structured light triangulation systems determine distance by projecting light from a source onto an object and imaging the resulting illumination pattern onto a detector. Knowing the position of the image on the detector, as well as the lateral separation between the detector lens and the light source, and the projection angle of the source, allows the determination of the distance, z, to the object. Multiple measurements at different (x,y) coordinates on the object lead to a full 3-D image of the object's surface.

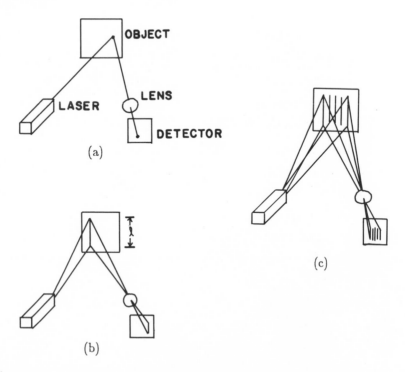

Figure 1:   Structured light triangulation. (a) Single point projection; (b) Single line projection; (c) Multiple line projection

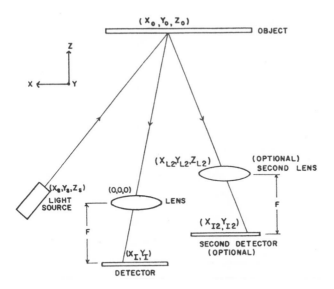

Figure 2: Geometric model for general triangulation system

In this paper we examine the design of structured light triangulation systems, starting with a discussion of illumination patterns, and an analysis of general triangulation geometry. We then consider the design of the illumination and detection subsystems (including light source and detector selection as well as the creation of illumination and receiving optics). Data processing techniques are addressed next, followed by a discussion of optical problems including a brief description of filtering techniques which minimize their effects. Finally, results from our prototype system are presented and our system's performance is discussed in the light of the original specifications.

2. Light Structures

The simplest structured light system projects a single point of light from the source onto an object for triangulation (Figure 1(a)). This method has the potential for delivering a great deal of light to that single point and can provide a high signal to noise ratio even when looking at dark or distant objects. Only a single point is measured per sampling interval so that $n \times m$ sample times are required to acquire data for an n by m lateral resolution element image. Single point triangulation systems usually use lateral effect photodiodes or linear arrays as detectors. If a full 3-D image is to be obtained, a bi- directional scanning apparatus is required [9].

A second class of 3-D imaging instruments operates by projecting a light stripe onto the object and using a two dimensional detector array (Figure 1(b)) [1]. Fewer frames are needed to scan over an object (n frames for an n by m element picture) and scanning need only be done in the direction perpendicular to the stripe. This results in a system with fewer moving parts. In fact no moving parts are required if the object being inspected moves past the vision system (as on a conveyor belt). The image of the illumination stripe is free, in principle, to move across the entire width of the detector array and therefore the system can have depth resolution at least equal to the number of pixels in a horizontal line of the detector array.

Whether all of this depth resolution will in fact be used depends on both the object and the optics used in front of the detector. For proper optical design, the length, l, of the stripe in Figure 1(b) should be imaged so as to match the vertical dimension of the detector array. With standard spherical imaging lenses and a 45 degree angular separation between source and receiver optics axes, a height change of approximately l must be encountered in order for the stripe image to move all the way across the width of the detector array. Many objects in the world are quasi-flat, (e.g., circuit boards) so that in going from minimum to maximum height we use only a fraction of the detector array unless expensive anamorphic optics are used.

We can make use of this disparity between potential depth variation (determined by the resolution of the detector) and the useful depth variation (determined by object size and opto-mechanical geometry) by projecting multiple stripes onto the object (Figure 1(c)). Since with n stripes projected onto the object, each stripe can traverse only $1/n$ of a detector scan line, the potential depth resolution has been decreased by a factor of n. However, as long as the useful depth variation for a quasi-flat object is smaller than the potential variation this is not a problem. A given lateral resolution can now be obtained with $1/n$ times the number of frames that a single stripe system would require.

3. Analysis of Triangulation Systems

Before designing a multi-stripe 3-D vision system we must understand the geometric constraints inherent in such a system.

3.1. Object Coordinate Calculation

In Figure 2 we show a general triangulation system, specifying the x,y, and z coordinates of the object, light source, and detector lens aperture centers. The coordinate system is centered on the primary detector lens. The (x,y) coordinates of an image point are given relative to the corresponding imaging lens center. A single illumination beam is represented at angles Θ_x and Θ_y to the z axis in the x and y directions respectively. It illuminates the object at the coordinates

$$x_o = x_s - (z_o - z_s)\tan\Theta_x , \tag{1}$$

$$y_o = y_s - (z_o - z_s)\tan\Theta_y . \tag{2}$$

An image of this point is formed on the primary detector at the coordinates

$$x_i = \frac{F}{z_o}x_o = \frac{F}{z_o}(x_s + z_s\tan\Theta_x) - F\tan\Theta_x, \tag{3}$$

$$y_i = \frac{F}{z_o}y_o = \frac{F}{z_o}(y_s + z_s\tan\Theta_y) - F\tan\Theta_y, \tag{4}$$

where F is the distance between the receiving lens and detector, which is nearly equal to the focal length of the receiving lens for long working distances (large z_o). If this image point is compared to that produced by an object at a calibration distance z_{ref} we find displacements of the image by

$$\Delta x_i = F(x_s + z_s \tan \Theta_x)(\frac{1}{z_o} - \frac{1}{z_{ref}}) \, , \tag{5}$$

$$\Delta y_i = F(y_s + z_s \tan \Theta_y)(\frac{1}{z_o} - \frac{1}{z_{ref}}) \, . \tag{6}$$

Using the x displacement to calculate the object distance we find

$$z_o = \frac{z_{ref}}{1 + \dfrac{\Delta x_i z_{ref}}{F(x_s + z_s \tan \Theta_x)}} \tag{7}$$

which implies that the relation between z_o and Δx_i is space variant (i.e., dependent on Θ_x) unless $z_s = 0$.

To simplify calculations and eliminate the need to buffer an entire frame, we would like image points to move only in the x direction (i.e., along a detector row or raster) for changes in object distance. For this we require $\Delta y_i = 0$ which implies $y_s = z_s = 0$, so that the effective source position must be on the x axis.

If we introduce a second camera, we can measure distance by comparing the relative separation of image points on two detectors instead of comparing the coordinates of a given image point to those corresponding to a previous reference point. As shown in Figure 2 the coordinates of the image on the second detector plane are

$$x_{i2} = \frac{F(x_o - x_{L2})}{(z_o - z_{L2})} \, , \tag{8}$$

$$y_{i2} = \frac{F(y_o - y_{L2})}{(z_o - z_{L2})} \, . \tag{9}$$

Using Equations (3) and (8) to eliminate x_o and solving for z_o we find

$$z_o = \frac{Fx_{L2} - z_{L2}x_{i2}}{x_i - x_{i2}} \, , \tag{10}$$

which tells us that unless $z_{L2}=0$ we will have space variant behavior. The vertical displacement is given by:

$$\Delta y = y_i - y_{i2} = \frac{Fy_{L2} - z_{L2}y_{i2}}{z_o},$$ (11)

indicating that for the desired condition of $\Delta y = 0$ we must have $y_{L2} = z_{L2} = 0$, so lens 2 lies on the x axis.

Finally, we must consider the determination of x_o and y_o. Clearly we could use Equations (3) and (4) to find them from x_i and y_i, but it might be simpler to report coordinates in the z_o, Θ_x, Θ_y system; since Θ_x is fixed for a given illumination stripe, it need not be calculated. Furthermore, we note from Equation (2) that if $z_s = y_s = 0$, Θ_y is uniquely determined by y_i and hence it need not be calculated either.

3.2. System Parameter Calculations

To demonstrate the procedure used to calculate the adjustable system parameters, we will assume that we wish to measure an object of overall dimensions x_{size}, y_{size} and varying in distance from z_{max} to z_{min}, as shown in Figure 3(a). We further assume that measurement resolutions of x_{res}, y_{res}, z_{res} are required. If we denote by $\Delta x_{i_{min}}$ the smallest measureable change in image stripe position on the detector (determined by both the pixel size and the stripe detection algorithm used), and by $\Delta x_{i_{max}}$ the change in image stripe position caused by the object distance changing from z_{max} to z_{min}, we find from Equation (5) that

$$\Delta x_{i_{min}} = \frac{Fx_s z_{res}}{z_{max}(z_{max} - z_{res})}$$ (12)

$$\Delta x_{i_{max}} = \frac{Fx_s(z_{max} - z_{min})}{z_{max}z_{min}}$$ (13)

Assuming that $\Delta x_{i_{min}}$ is known, $\Delta x_{i_{max}}$ can be found in terms of $\Delta x_{i_{min}}$ by eliminating Fx_s from Equations (12) and (13):

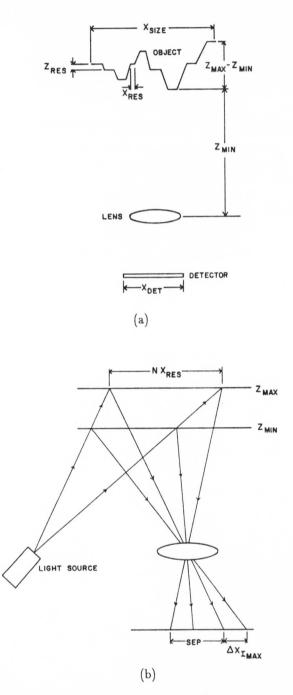

(a)

(b)

Figure 3:   Geometric model for the prevention of aliasing

$$\Delta x_{i_{max}} = \Delta x_{i_{min}} \frac{(z_{max} - z_{min})}{z_{res}} \frac{(z_{max} - z_{res})}{z_{min}}.$$ (14)

In Figure 3(b) we show two adjacent stripes projected from the source and imaged onto the detector. Shown are the image positions corresponding to maximum and minimum object distances. If we project M stripes onto the object and collect N frames of data wherein in each frame each stripe is shifted laterally on the object by an amount x_{res} from its position in the previous frame, we obtain a total lateral resolution of NM pixels, where to match the object resolution requirements

$$NM \geq \frac{x_{size}}{x_{res}}.$$ (15)

Adjacent, simultaneously projected stripes are separated by Nx_{res} and if we denote the separation between the images of these stripes on the detector by the distance sep, we have

$$sep = \frac{Nx_{res}F}{z_{max}}.$$ (16)

To image the entire object onto a detector of lateral extent x_{det}, y_{det} we impose the constraint

$$x_{det} > x_{size}F/z_{min}.$$ (17)

Eliminating F from Equations (16) and (17), we can find the number of frames required to obtain sufficient depth and lateral resolution:

$$N > \left(\frac{z_{max}}{z_{min}} \frac{sep}{x_{det}} \frac{x_{size}}{x_{res}} \right).$$ (18)

Once N is chosen, the number of stripes used in a single frame, M, can be found from Equation (15).

We define the ratio of the range of travel of a single stripe, $\Delta x_{i_{max}}$, to the separation between adjacent stripes, sep, and denote it by R. That is,

$$R = \frac{\Delta x_{i_{max}}}{sep} = \frac{x_s\left(z_{max} - z_{min}\right)}{N x_{res} z_{min}}.$$ (19)

Clearly we require $|R| < 1$ to prevent aliasing. The exact value of R chosen for a system must take into account the stripe finding algorithm used, since some algorithms require more dead space between stripes than others. Once a value has been chosen for R, we can solve Equation (19) for the separation between the light source and detector lens pupil:

$$x_s = \frac{R N x_{res} z_{min}}{\left(z_{max} - z_{min}\right)}.$$ (20)

Using this value in Equation (5) determines the focal length of the detector lens

$$F = \frac{z_{max} \Delta x_{i_{max}}}{R N x_{res}}.$$ (21)

If the dimensions of the object and detector are such that this focal length does not allow the entire y extent of the object to be imaged onto the detector, anamorphic optics can be used to provide a different focal length, and hence magnification, in the y direction, satisfying the relation

$$y_{det} = F_y y_{size} / z_{min}.$$ (22)

For systems employing two detectors, Equation (20) needs special interpretation. In particular, for the second detector x_s must be replaced by $x_s - x_{L2}$ and R_2 may differ from R_1. Combining the two versions of Equation (20), we can solve for the required lens separation x_{L2}:

$$x_{L2} = \frac{x_{res} N z_{min}}{z_{max} - z_{min}} (R_1 - R_2).$$ (23)

It should be noted that for the case of two detectors located symmetrically about the light source, we have $R_2 = -R_1$ and $x_{L2} = 2x_s$.

From the above derivations we can see that if the dimensions and resolution of the object and detector are known, all system parameters (i.e. number of stripes per frame, number of frames, lens focal length, and source and detector offsets) can be uniquely determined. As an example, our system was designed to view objects for which $x_{size} = y_{size} = 60$ mm, and $z_{max} - z_{min} = 12.5$ mm with $x_{res} = 0.5$ mm and $z_{res} = 0.25$ mm. We chose a working distance $z_{max} = 650$ mm. The active detector area has $x_{det} = y_{det} = 5.5$ mm, $\Delta x i_{min} = 0.001$ mm (using a super resolution algorithm) and we planned on R $= 1/6$ (see Section 6.2). Using these values and the above equations we find $\Delta x i_{max} = 0.051$ mm, $N > 6.8$ (we chose 8), $x_s = 34$ mm, and $F = 50$ mm.

3.3. Light Source Calculations

We have to this point presumed that stripes of light of suitable size could be accurately projected into the object space from the specified source position. Such does not happen by accident. In this section we will describe source design.

In Figure 4, a laser beam with Gaussian intensity profile impinges on a

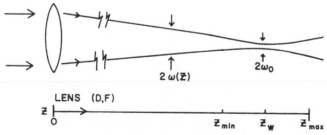

Figure 4: Focal properties for Gaussian beam

lens of diameter D and focal length F. The beam width as a function of position, z, is given by [15]

$$\omega(z) = \omega_o \left[1 + \left[\frac{\lambda(z - z_w)}{\pi \omega_o^2} \right]^2 \right]^{1/2}$$

(24)

where $w(z)$ is the distance from the center of the beam to the $1/e^2$ intensity point, and w_o is the minimum value of $w(z)$ (which occurs at z_w).

We require that the beam diameter

$$2w(z) \leq x_{res} \tag{25}$$

in the interval

$$z_{min} \leq z \leq z_{max} \tag{26}$$

in order to obtain the desired lateral resolution. Selecting

$$w(z_{max}) = x_{res}/2 \quad \text{and} \quad z_w = (z_{max} + z_{min})/2 \tag{27}$$

allows us to solve Equation (24) for w_o, the beam waist size, which in turn allows us to find $w(0)$, which is the minimum allowable lens radius. We find that

$$w_o = \frac{x_{res}}{2\sqrt{2}} \left[1 \pm \sqrt{1 - (\frac{4\lambda(z_{max} - z_{min})}{\pi x_{res}^2})^2} \right]^{1/2} . \tag{28}$$

Hence there are two beams that satisfy our requirements. The beam with the smaller value of w converges more rapidly than the other, hence it requires a larger lens diameter. For this reason we chose the larger value of w, in hope of minimizing the size of our optical system. For this beam, the lens diameter must have size

$$D = 2w(z{=}0) = 2w_o \left[1 + (\frac{\lambda z_w}{\pi w_o^2})^2 \right]^{1/2} . \tag{29}$$

As an example, with stand off distance $z_w = 650$ mm, depth of focus $z_{max} - z_{min} = 12.5$ mm, lateral resolution $x_{res} = 0.5$ mm, and $\lambda = 0.633 \times 10^{-3}$ mm, inserted into Equation (28) (with the positive radical) we find the beam waist diameter (full stripe width) to be

$$2w_o = 0.49990 \text{ mm} \approx 0.9998 x_{res} \tag{30}$$

so that the laser beam is only weakly convergent as it approaches the focus. Inserting Equation (30) into Equation (29) and applying our numerical values gives

$$D = 1.16 \text{ mm} \approx 2.32 x_{res} \quad .$$

4. Source Implementation

We saw in the previous section that it was advantageous for the fan beams producing our multiple illumination stripes to appear to diverge from a point which is only displaced in the x direction from the center of the detector lens pupil. Such an array of beams is easily produced via a multi-faceted hologram [3] which segments a single beam of light into M beams; these cross at a given distance to produce an effective source some distance from the hologram, and come into focus some distance later as shown in Figure 5. The cylinder lens diverges the beams to form stripes, although this function also could have been incorporated into the hologram.

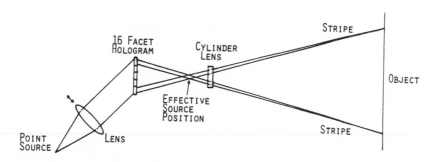

Figure 5:   Schematic of illumination subsystem

In order to obtain a horizontal resolution of 128 pixels, our system produces $M=16$ stripes which are laterally shifted in each of 8 consecutive frames. This motion could be accomplished by moving a lens or mirror within the source optical system or by using 8 adjacent light sources that can be switched on sequentially. LEDs and laser diodes are small and easily switched. LEDs have lower brightness than lasers but eliminate speckle problems (explained later). For our experiments we used a HeNe laser whose visible beam made optical alignment easier. Since it was not practical to use 8 HeNe lasers, our system moves a lens via a computer controlled translation stage to translate the stripes between frames. Our hologram facets were 1/16" = 1.6 mm in width and the illumination across each facet was of uniform irradiance so that depth of focus, etc. differed slightly from our above design analysis.

5. Detector Implementation

5.1. Detector Selection

Since our multiple stripe technique requires a two dimensional detector, we must choose between a solid state or vidicon type camera. Although solid state cameras are small, rugged, long life devices, our system uses a vidicon detector in order to obtain greater depth resolution. Not only does a vidicon camera offer higher resolution than currently available solid state cameras, but also the dead zones between the pixels of a solid state camera can cause distortion of narrow stripes, whereas data are obtained from a vidicon via smooth convolution with a scanning electron beam, which broadens the stripe but does not distort it.

Because of the scan instability inherent in vidicon devices, we chose to use a dual detector architecture to minimize the effects of jitter on our data. Comparing Equations (7) and (10) we see that in a single detector system, z data are obtained by comparing stripe positions in sequential frames, making the system quite sensitive to jitter (small changes in horizontal synchronization) and thermal drift, whereas with the dual detector architecture, z data are obtained by comparing two stripe images obtained at the same time. If the two detectors are produced via a double optical system in front of a single detector, the resulting system is immune to jitter effects. In fact, if the images from the dual optical system are interleaved so that the two image stripes corresponding to the same illumination stripe are adjacent, the resulting system is quite insensitive to fluctuations in the scan rate during a single frame. In the next section we will discuss several techniques for producing the virtual camera pair and compare their strengths and weaknesses.

5.2. Dual Detector Implementations

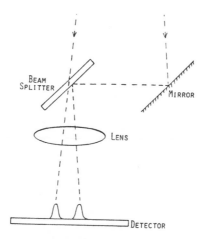

Figure 6: Mirror - Beamsplitter system produces asymmetric detector pair

One method for producing a pair of virtual detectors is to use a mirror and beamsplitter [4] as shown in Figure 6. With this arrangement the relative separation between the two stripe images increases with decreasing object distance in correspondence with Equation (5). This is inexpensive and simple to align, but it has three problems. First, the separation between the two detectors cannot be smaller than the aperture of the camera lens. Since there is an inverse relationship between the range of measurable distances, $z_{max} - z_{min}$, and the detector separation (Equation (20)), this gives us an undesirable trade-off between useful range and aperture size (light level). Second, the path length between the object and the camera is different for the two different paths. This means that the two images can never be simultaneously in focus. The third problem is also due to the path length difference, namely that the two virtual detectors do not lie in a plane perpendicular to the optical axis, leading to a space variant relationship between distance and image separation as shown in Equation (7). Since we want to reduce the computational complexity of the system this is a very undesirable side effect.

One way to reduce the detector separation, as required for long working ranges (while maintaining depth resolution and lens aperture size), is to create an equivalent arrangement of mirrors by sputtering a pattern of gold on both sides of a glass plate. For glass with an index of refraction of n, the pattern shown in Figure 7 behaves similarly to the mirror beamsplitter arrangement; half of the incident light which passes the first surface is transmitted directly to the camera while the other half is reflected twice to enter the camera at a position laterally displaced by

$$x_{L2} = 2w \sin \psi \,, \tag{31}$$

where l and w are related by

$$l = \frac{w}{3\sqrt{n^2 - 1}} \,. \tag{32}$$

This technique has different path lengths, but the difference can be so small as to not affect focus appreciably. However, the space variant effect arising in Equation (14) is unchanged, since it is the ratio between the detector separation in the x and y directions that determines the relative importance of this effect. This system is half as light efficient as the system shown in Figure 6.

Another way of producing two images using the same camera is to equip the camera with a dual aperture and operate away from perfect focus [2] as shown in Figure 8. The light from the two apertures will focus in a plane before that in which the detector is located, and hence form out of focus images in two different places on the detector plane. Since we do not want to operate far from correct focus, this method can be modified by placing a

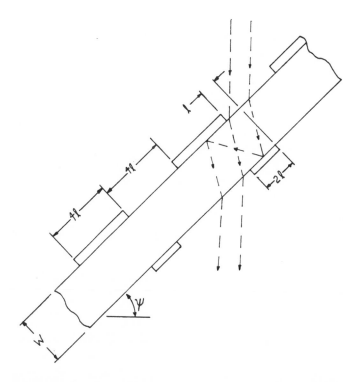

Figure 7: Multiple aperture asymmetric mirror system

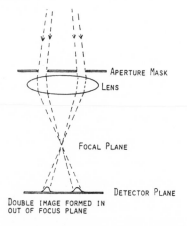

Figure 8: Dual aperture system producing defocused double images

chevron made of two glass plates in front of the apertures (Figure 9). In this manner a double, in-focus image is formed. Again the separation between

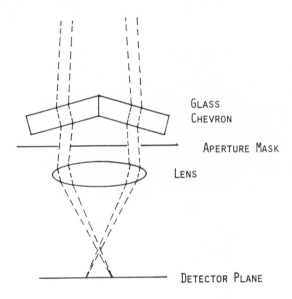

GLASS
CHEVRON

APERTURE MASK

LENS

DETECTOR PLANE

Figure 9: Dual aperture system employing a glass chevron to produce focused double images

the two cameras is forced to be greater than the lens aperture size. This can be overcome by using multiple chevron patterns adjoined to form the corrugated structure shown in Figure 10. However this severely limits the working range of the system since as the image moves far from proper focus the light from each segment of the glass 'wiggle' forms an 'image' at a different lateral position, leading to a great multiplicity of images.

All of the above techniques were investigated, and for our experiments we chose a symmetric modification of the first one mentioned above. Instead of a mirror and beamsplitter, four mirrors [7] are used as shown in Figure 11. This system provides symmetry for the two paths, simplifying focus and computation, but the separation between the two cameras is still constrained to be greater than the aperture size.

Figure 10: Multiple chevron

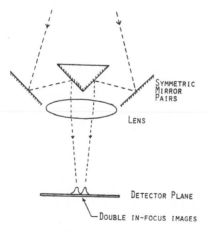

Figure 11: Symmetric mirror system

6. Data Interpretation

We must now face the question of how accurately the position of our stripes can be determined. The simplest approach to determining stripe position is to look for the highest intensity pixel in a region and assign its position to the center of the peak. For a noiseless system this gives an accuracy of $\pm 1/2$ pixel. This accuracy is rapidly degraded by noise; since the slope of the signal is zero at the peak, small amounts of noise can greatly affect the peak position.

If we assume that the peaks are symmetric or that the pair of peaks is symmetric about the center of the pair, we can determine the separation of the two peaks by noting where the intensity, $I(x)$, crosses a threshold, T. Assuming for a single peak that the threshold is crossed upward between pixel n and $n+1$ and downward between pixel m and $m+1$, the center of the peak, to a linear approximation, is at

$$C = \frac{1}{2}\left[n + m + \frac{T-I(n)}{I(n+1)-I(n)} + \frac{I(m)-T}{I(m)-I(m+1)} \right] \tag{33}$$

and the separation between two peaks is the difference between their centers. This technique is simple and effective for the peaks we have observed if the threshold is set at one half the maximum value of a Gaussian stripe image. It has lower noise sensitivity than peak detection because of the non-zero slope of the Gaussian at the threshold level.

A third technique uses the fact that the sampling theorem tells us that if a continuous bandlimited signal is sampled frequently enough, all of the details of the original signal can be reconstructed from the sampled data by convolution with a sinc function. Hence if the size of the peak on our detector is large enough to cover several pixels on the detector we should be able to determine the position of a noiseless signal exactly. This technique assumes that the sample values are not corrupted by noise, which is not a reasonable assumption for stripes reflected from real surfaces.

If the signal is not perfectly bandlimited, this reconstruction will not be ideal. Hence we expect that the error introduced by sampling will depend on the extent to which the high frequency components of the signal shift the peak position away from the position determined by the low frequency components. If this effect is slight (which we would expect for a Gaussian beam) we need not worry about it. Broadband noise will be a problem, however, since its effects cannot be eliminated by lowpass filtering. Some sources and potential solutions to this problem are discussed in the next sections.

A technique that reduces the effects of noise to obtain sub- pixel resolution is to perform a Gaussian least squares fit. Although this technique does provide a good estimate of the peak position, it is computationally intensive. Since one of our goals is to reduce the computational complexity of making height measurements, we did not choose to implement this algorithm as part of our system.

Yet another technique for achieving sub-pixel resolution is to view the intensity of the signal at a given pixel as the probability that the peak is centered at that pixel. Then averaging would give the mean position of the peak. The average is easily computed by the algorithm:

```
SUM = 0
SUM1 = 0
DO 10 x = N, 1, -1
SUM = SUM+I(x)
SUM1 = SUM1+SUM
10 CONTINUE
AVE = SUM1/SUM
```

producing a simple technique that works reasonably well. Our tests have shown that threshold crossing and averaging work equally well for the optical system described above.

7. Ubiquitous Optical Problems

Having discussed algorithms for improving the system's resolution, we must now turn to the optical problems that limit our ability to determine the object distance at a given point.

7.1. Surface Properties

The first two problems to be discussed are object surface roughness and varying object reflectivity. They have two deleterious effects on structured light triangulation systems. First they alter the shape of the stripe, so that the most intense point in the stripe image is not necessarily at the stripe center, and particularly dangerous, the images of the stripes in the two virtual detectors can have different intensity profiles. If the object has finite zones of locally planar surfaces as in Figure 12, then to overcome this

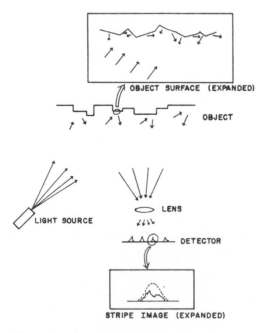

Figure 12: Effect of surface roughness

problem one is forced to make the illumination stripes on the object narrower than the scale of surface texture and reflectivity changes. (Then the problems arise only at regional boundaries). To accomplish this and still achieve sub-pixel resolution we must magnify the image of the stripes so that each stripe occupies multiple pixels on the detector, reducing the field of view. This may not be possible if the object is rough on a microscopic scale or if we cannot tolerate a reduced field of view. The second problem arising from roughness and reflectivity variations is the large dynamic range of the detector required by the large variation in peak intensities of the stripe images. We have found that our peak finding algorithms work acceptably over a one decade range of peak intensities. Since object reflectivities commonly vary over a four decade range, this leads to errors if there is a large variation in reflectivity in the field of view. One way to solve this problem is to take multiple frames of the same object with different illumination levels. If a log-intensity detector were available this would be ideal.

7.2. System Properties

The next problems are distortions introduced by the optical system itself. The imaging lens introduces aberrations and the detector can have a non-linear and/or space varying response to intensity. Both of these problems lead to space varying errors in distance measurements which do not vary with time, and they can be reduced by proper choice of lens and camera. If the additional processing time is tolerable, the most practical way of eliminating their effects entirely is by calibrating the system to take their effects into account. For example, detector non-uniformities can be compensated by uniformly illuminating the detector and using the resulting intensity readings as normalization factors for future reference measurements. Pincushion or barrel distortion can be compensated in a similar manner by measuring stripe image separation in a reference plane and using this measurement for calibration purposes.

7.3. Speckle

Speckle may be a problem in any system which tries to achieve sub-pixel resolution while using a coherent light source. Since the stripe's image must occupy several pixels to achieve sub-pixel accuracy, the detector must sample the image (and hence the speckle pattern). Several suggestions have been made to reduce this problem which include rotating the polarization of the source, so that the observed speckle pattern is the superposition of two uncorrelated patterns, and making slight perturbations in the position of the effective source to achieve similar effects. Both of these techniques require cycle times shorter than a single frame time, so that the speckle pattern captured in a single frame is really the superposition of the produced patterns. None of these effects have yet been tested sufficiently to report any significant results.

The goal is to make the speckle small with respect to the pixel size of the detector (25 speckles within 1 detector pixel is sufficient for speckle free imaging [6]). Speckle size is similar to the spot size (impulse response) produced by the detector optics [8]. For a focused image, the speckle size is diffraction limited so that it varies as $1/D$, where D is the lens diameter. However, for an out of focus image, the speckle pattern defocuses to produce patches whose size in the geometrical limit is proportional to D. There is therefore a lens diameter D_0 which minimizes the size of speckle over a given range of defocusing. For practical systems this optimal lens diameter results in such a high f that the light level is unsatisfactory. Hence our approach thus far has been to reduce the lens aperture as much as possible while maintaining adequate light levels.

8. Filtering Techniques

In the previous two sections we have mentioned several noise sources, these and others can be reduced somewhat through the use of filters. High frequency noise such as that due to object roughness, detector non-uniformities, and electronic noise in the detector system can be removed via an analog lowpass filter. The cutoff frequency for this filter can be readily determined by examining the image of a Gaussian stripe on the detector. For a Gaussian distribution with a variance of σ^2, lowpass filtering with a cutoff frequency of 2 σ will leave 95% of the spectrum of the stripe.

Low frequency noise due to variations in ambient lighting can be reduced by a narrow band filter in front of the camera lens. For our system this was entirely acceptable, for applications where it is not, we have found that local min - local max filtering nearly eliminates space variant biases due to ambient lighting.

Quantization noise can only be removed by using a high resolution A-D converter or avoiding the digital domain entirely. We have found that using an 8-bit A-D converter the noise introduced before quantization have always exceeded the quantization noise.

Once the distance data have been calculated, the noise in the resulting distance map can be reduced by using the assumption that distances vary smoothly and averaging distance data in small windows, or by assuming that distance features smaller than some minimal size are artifacts and eliminating them by median filtering [5]. However, it should be noted that severe noise in distance data usually corresponds to edges or holes where the stripes are not reflected back to the camera due to shadowing. Since edges are often important features of objects, it may be a good strategy to flag the absence of data caused by such features to provide data about possible object structure.

9. Experimental Results

A diagram of our prototype system is shown in Figure 13. We produce multiple illumination stripes via a multifaceted hologram, H, which spatially divides an expanded incident laser beam into 16 beams. The 16 beams collectively overlap at the position of lens L_3 (to form a small effective

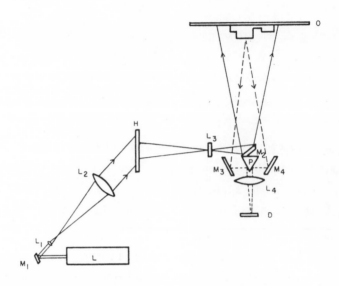

Figure 13:   Schematic of multi-stripe 3-D vision system

source) but individually come to focus at the object plane, O. A cylindrical lens, L_3, spreads each beam into a vertical stripe so that a set of 16 focused vertical stripes illuminate the object. Transversely translating lens L_2 in our prototype system, scans the stripes across the surface of the object. In a final system the HeNe laser would be replaced by a laser diode and multiple laser diodes would be pulsed in sequence to scan the stripes, thus eliminating all moving parts. A television camera (lens L_4 and detector D) directed at this pattern can generate the data required for a 128 by 128 element 3-D picture in 8 frames. Again only 8 diodes would be required to produce a picture with the resolution mentioned above. The apparent position of the light source is at the lens pupil so that the center position of any pair of stripes does not change as z is varied, which further simplifies data interpretation and also allows one to treat the system as two single triangulation systems in the event of hidden part problems.

Figure 14 shows a sample image in the detector plane of the stripes corresponding to the staircase object seen in Figure 15. Adjacent steps differ by 2.5 mm in distance from the detector, while the average range is 65 cm.

Figure 14: Image of stripes corresponding to step object in Figure 15

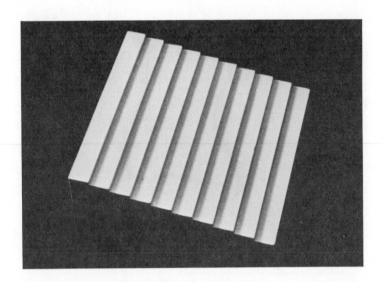

Figure 15: Step object corresponding to stripes in Figure 14

The green lines in Figure 14 are bin markers added by our computer to

Figure 16:   System output for step object in **Figure 15**

indicate the zone in which the stripes should be found. Figure 16 shows the
output of our system when viewing the staircase object with height encoded
in pseudocolor. A calibration scale has been added at the edge of the image
where blue corresponds to 0" object height (located at z_{max}) and red
corresponds to 1/2" height. Data acquisition for the image requires only 8
frame times. Doing all data processing in software, this picture was
generated in 2 minutes, and is displayed from a digital frame buffer at TV
frame rates. With a special purpose hardware processor the time required to
produce a single 3-D picture could be brought down to well under a second.
This step from software to hardware is facilitated by the low computational
complexity of our algorithms.

The remaining figures present the system's pseudo-color output when
viewing common objects in several application areas. Figure 17 shows a
printed circuit board. Note that the orientation depressions on the
integrated circuits are clearly visible. Figure 18 shows a pile of metal
washers. The topmost washer is pseudo-colored green; clearly distinguishing

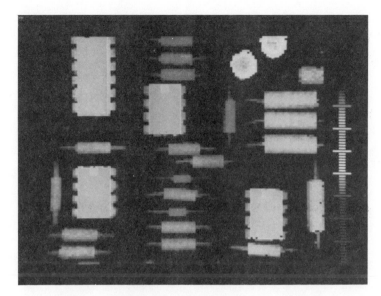

Figure 17: System output for a printed circuit board

Figure 18: System output for a stack of washers

it from lower washers. Such data would be quite useful in robotics applications where sorting or grasping must be performed. This photograph also demonstrates the problems caused by varying surface reflectivity and multiple reflections at the edges. Figure 19 is the median filtered version of Figure 18 showing why filtering is often desirable for human monitoring of the system's output. Median filtering preserves edge acquity but removes impulse noise. Figure 20 shows a side view of a nose. This problem was posed by persons in the medical field interested in the use of 3-D inspection systems for the data gathering phase of reconstructive facial surgery. Finally, Figure 21 shows a mapping of distance vs. displacement along a single vertical line of the nose showing how various displays can be generated from the digital range data within the system.

The above results show that 3-D vision can find useful applications in a number of areas. Multiple stripe techniques and dual detector architectures may prove to be particularly useful due to the properties described in this paper.

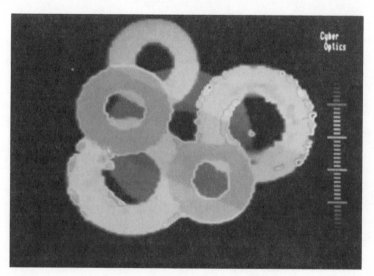

Figure 19:   Median filtered version of Figure 18

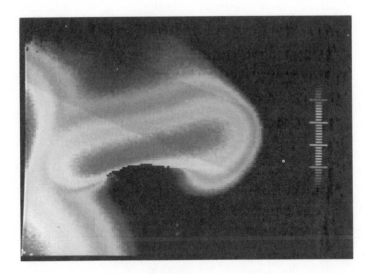

Figure 20:   Side view of a nose

Figure 21:   Distance data along a slice of the nose

Acknowledgments

This material is based upon work supported by a National Science Foundation Graduate Fellowship, a grant from the University of Minnesota Productivity Center, and by CyberOptics Corporation.

References

[1]   Agin, G. J. and Binford, T. O.
        Computer description of curved objects.
        In *Proceedings of the Third International Joint Conference on
            Artificial Intelligence*, pages 629-640. August, 1973.
        Also appeared in IEEE Transactions on Computers, 1976, Vol. C-25,
            pages 439-449.

[2]   Blais, Francois, Rioux, Marc, and Pousart, Denis.
        Very compact 3-D camera for robotic applications.
        Machine Vision, Status and Directions :WB4, 1985.

[3]   Case, S. K., Haugen, P. R., and Lokberg, O. J.
        Multi-facet holographic optical elements for wavefront
            transformations.
        Appl. Opt. 20:2670, 1981.

[4]   Connah, D. M. and Fishbourne, C. A.
        Using a laser for scene analysis.
        In *Proc. 2nd Int. Conf. Robot Vision and Sensory Controls*, pages
            233-240. 1982.

[5]   Gallagher, Neil.
        private communication, 1984.

[6]   George, N.
        private communication, 1984.

[7]   Geschke, C.
        Coordinated Science Lab Report.
        Technical Report R-837, Univ. of Illinois at Urbana-Champaign,
            December, 1978.

[8]   Goodman, J. W.
        Some fundamental properties of speckle.
        JOSA 66(11):1145, 1976.

[9]   Haugen, P. R., Keil, R. and Bocci, C.
        3-D active vision sensor.
        In *Intell. Robots and Computer Vision.* SPIE, 52, 1984.

[10]   Jarvis, R. A.
        A perspective on range finding techniques for computer vision.
        IEEE Trans. Pattern Anal. Machine Intell. PAMI-5:122-139,
            March, 1983.

[11]   Kanade, T. and Asada, H.
        Noncontact visual three-dimensional ranging devices.
        In B. R. Altschuler (editor), *3-D Machine Perception*, pages 48-53.
            SPIE, 1981.

[12]   Lewis, R. A. and Johnston, A. R.
        A scanning laser rangefinder for a robotic vehicle.
        Proc. 5th Int. Joint Conf. AI :762-768, 1977.

[13]   Nishihara, H. K.
        PRISM: A practical real time stereo matcher.
        In *Intell. Robots: Third Int. Conf. on Robot Vision and Sensory
            Controls*, pages 134-142. SPIE, 449, 1985.

[14]   Strand, T. C.
        Optical three-dimensional sensing for machine vision.
        Optical Engineering 24:33-40, 1985.

[15]   Yariv, A.
        Introduction to Optical Electronics.
        Holt, Rinehart & Winston, New York, 1976.

Robot Vision by Encoded Light Beams

Martin D. Altschuler
Kyongtae Bae

Department of Radiation Therapy, School of Medicine
and
Department of Bioengineering, School of Engineering
University of Pennsylvania
Philadelphia, Pennsylvania, 19104

Bruce R. Altschuler
Jerome T. Dijak
Louis A. Tamburino

Systems Avionics Division
Air Force Wright Aeronautical Laboratories
Wright-Patterson AFB, Ohio, 45433

Barbara Woolford

Man-Systems Division
Lyndon B. Johnson Space Center
National Aeronautics and Space Administration
Houston, Texas, 77058

Abstract

A robot eye is more than a camera and less than a complete vision understanding system. It is a device which can scan or sense objects in the three-dimensional environment and extract useful numerical information about those objects. The information obtained by the robot eye should be of a form that can be interrogated by a vision understanding system. This article describes the development of a dependable, robust, and versatile robot eye which can rapidly mensurate a surface in three dimensions. Using an active/passive camera pair, surface mensuration is achieved with fast electro-optic implementation of well-known stereophotogrammetric principles. We discuss the calibration of the robot eye and the application of the robot eye to exploring and mensurating 3-D objects.

1. Introduction

A computer vision system consists of both a robot eye (or preprocessor) and a vision-understanding (or analysis) component. The robot eye scans or images or otherwise records the environment (or 'scene'), and extracts information and patterns from the data. The vision-understanding component maintains a knowledge base of patterns (or mathematical functions) that may be present in the scene, interrogates the patterns obtained by the robot eye, and attempts to model the scene or classify the object using entities in its knowledge base. The robot eye can explore a scene according to pre-determined strategy or dynamically through feedback from the vision-understanding component.

A simple passive vision system consists of (1) a video camera (or image recorder) to view the scene, and software for image segmentation, and (2) an analyzer, essentially several 2-D patterns stored in memory, and software to recognize whether the observed image contains those patterns. The working environment of a simple vision system is usually strictly controlled with regard to lighting, the orientation and scaling of the camera image relative to the scene, the types of objects allowed in the scene, and the allowed orientations and spacings of those objects. Typical applications for such a vision system are to detect objects on an assembly line, to determine the width of a welding seam, and to inspect fabrics and LSI circuit masks [13].

More advanced passive vision systems attempt to extract 3-D information about a scene (in the form of models or object recognition) with the use of (a) camera images, (b) very sophisticated software for image processing (involving filtering, edge detection, segmentation, feature extraction, etc.), (c) the generation of data structures to organize patterns (such as polygons) extracted from the image, and (d) software to analyze or model the scene by comparing the information derived from the image with the information derived from known objects. When the scene contains many primitive objects that can occlude one another in an image, the problem becomes very difficult. Model building software may require a hierarchy of inference rules to account for many different image situations that arise with varying scene illumination, shadows, surface texture, and the occlusion of objects along the line of sight. Determining when a complete set of inference rules is obtained or obtainable is always a difficulty [37], [26], [47], [7], [28].

So far we have assumed that the passive robot eye extracts and organizes features of a 2-D image, and that image understanding software infers the 3-D scene by analyzing 2-D image features rather than 3-D information. It is possible, however, to derive 3-D information directly by using two passive stereo cameras separated by sufficient parallax. The robot eye is then a pair of passive stereo cameras that view the same scene. To triangulate a point in the 3-D scene, the corresponding image points must first be identified in the two camera images. The matching of corresponding stereo image points is a difficult correlation problem for a computer, since the light intensity patterns in the image of a scene depend on the camera angle. Humans can match corresponding image points for triangulation with their ability to perceive three-dimensions when the images of a scene are viewed stereoscopically. See [14], [44], [35], [16], [17], [21], [23], [36].

To overcome the complexities of passive vision systems and to obtain directly numerical 3-D information about the scene, active/passive camera systems, sometimes called 'structured light systems' have been developed. In these systems, the robot eye consists of a passive camera and an active element separated by sufficient parallax for triangulation of a scene. The active element emits light patterns which illuminate the scene. Bright points in the passive camera image can then be correlated automatically with rays or planes of light emitted by the active element, and triangulation of points in the scene can be effected. A robot eye that directly acquires and stores 3-D information about a scene obviates much of the complex software needed to extract 3-D information from 2-D images [38], [50], [2], [32], [34],

[39], [43].

Several types of structured light systems have been constructed. The simplest conceptually has a single laser beam sweep across the scene [20]. At any instant, the passive image sees a single illuminated spot of the scene, and triangulation of the spot location is immediate. To obtain a fast sampling rate of the scene, the passive image detector can be a special electronic chip which returns only the location (x^*, y^*) of the centroid of illumination of the image. Then, as the laser beam sweeps the scene, four numbers for each sample point i are obtained, namely the measured direction of the laser beam (u_i, w_i) and the corresponding passive image location (x_i^*, y_i^*). If the passive camera is calibrated to the active element, then each set of four numbers will triangulate a single point of the scene and provide the three coordinates (x_i, y_i, z_i) for the point's location. The simple laser scanner is as good as the chip and the accuracy with which it determines the centroid of illumination. In a noisy environment, where several laser beams or their reflections may be present, or in cases of a laser beam grazing a surface, several surface points may be illuminated simultaneously. Since the chip returns only the illumination centroid of the passive image rather than multiple image points, erroneous mensuration of positions in the scene may sometimes occur.

In a second type of structured light system, the active element projects a plane of light which intersects the scene as a bright stripe [50], [38], [2], [32], [34]. Each bright point appearing in the passive image determines a line from the passive camera to the bright stripe in the scene. Thus three numbers are recorded for any illuminated sample point i in the scene, namely, its passive image location (x_i^*, y_i^*), and the light plane u_i projected by the active element. If the passive camera is calibrated with the active element, each set of three numbers will triangulate a single point of the bright stripe in the scene. After sufficient points of the illuminated stripe on the scene are triangulated, a new stripe can be generated by a parallel light plane from the active element. In this way the stripe can be swept across the scene, and many sample points of the scene triangulated. Spatial resolution is limited by the thickness of the light stripe. Inaccuracies may result if the parallax is small or if random errors occur in measurement of u_i, x_i^*, or y_i^*.

The difference between the simple laser beam scanner and the stripe projection scanner lies in the precision of the triangulation. The laser beam scanner provides four equations (one each for u_i, w_i^*, x_i^*, y_i^*) to determine the three spatial coordinates of the illuminated point. The equations are overdetermined, hence inconsistent, since two lines in 3-D space need not intersect (and generally won't if there are random errors in image position or active element pointing). The overdetermined equations, however, permit a least-squares triangulation solution, which for the same parallax is less sensitive to random errors of measurement than is the consistent solution (3 equations for 3 unknowns) obtained with the light stripe generator. To achieve the same precision with the light stripe scanner, we can sweep the scene twice, once with vertical and once with horizontal stripes, thereby obtaining four numbers (u,w,x^*,y^*) for each scene point and overdetermining the triangulation problem as before.

Rather than sweep a light stripe across the scene so that N light stripes are necessary to mensurate the scene, we can illuminate (and observe) the entire scene with a sequence of only $\lceil \log_2 (N+1) \rceil$ binary-coded striped-light patterns. For image point (x_i^*,y_i^*), viewing sample point i of the scene, we can decode the brightness variation in the sequence of images and determine the coordinate u_i of the light plane (stripe) which illuminates point i. Thus eight binary coded light patterns and passive images provide the same information for triangulation (namely u_i,x_i^*,y_i^* for each sample point i) as do 128 individually-projected stripes and passive images. Although just $\lceil \log_2 N \rceil$ light patterns are needed to binary code N stripes, one stripe, zero, is always unilluminated in the binary code.

In this paper we describe the development over the past few years of an advanced robot eye of the structured light type [3], [4], [33]. The eye contains an active element which can convert a beam from a single laser source into an interferometric array (currently 128 × 128) of laser 'beamlets' diverging from a common focus, then project and spatially modulate the distinct laser beamlets. For robot vision purposes, the beamlets are essentially rays. Thus the active element obeys exactly the same mathematics as a passive camera with an array (128 × 128) of pixels. We will call the active element an 'active camera', and the active/passive camera pair a 'numerical stereo camera' or 'NSC'. The computer controller can be programmed to turn on and off individual rays from the active camera, so that the NSC can simulate a single laser beam scanning the scene, a light stripe projector, or a projector of binary coded light patterns. Although we briefly describe the NSC hardware, we are concerned primarily with the algorithmic and software effort to make the NSC into as reliable and intelligent a robot eye or vision system preprocessor as possible. The goal is to make a robot eye that can adapt to different ambient light environments, can be calibrated simply and quickly, and is intelligent enough to decide on new viewing perspectives when concavities or holes are poorly observed from initially chosen perspectives. A robust robot eye which mensurates a 3-D scene can be an off-the-shelf item applicable to many different industrial and biomedical problems without being dependent on the vision understanding component of the system [19], [9], [5], [53]. Such an independent entity which directly obtains 3-D information could lead to vision understanding modules which are less complex and more standardized, require less development time for a particular application, require less time for computation in operation, and probably require less memory for pattern recognition of 3-D objects.

Some recent developments in direct acquisition of 3-D scene information also show promise:

1. Structured light systems which project multi-wavelength (i.e., multi-color) light patterns for stripe coding [52].

2. Radar-type rangefinders which emit picosecond pulses of laser light and record their echoes from the scene; this approach does not use triangulation [24], [43].

3. Passive fixed cameras which observe the same scene as it is illuminated from different directions; here some knowledge about the texture of the reflecting surface is useful, and the goal is to obtain the spatial orientation of surface elements (that is, partial 3-D information) [40], [18], [51].

We will not discuss these techniques further.

Over the past several years, the USAF has developed an advanced prototype NSC. This work began at the USAF School of Aerospace Medicine in the late 1970's and was transferred to the Wright Aeronautical Laboratories (AFWAL) in 1982. An advanced prototype system was constructed at the AFWAL Avionics Laboratory where it is still being developed and evaluated. Dr. J. Taboada (USAF School of Aerospace Medicine), a co-inventor of the NSC, has been and remains a key contributor to the development of the NSC optical subsystems. The integrated NSC hardware/software system developed at AFWAL is somewhat different from that described in this paper, and will be described in detail elsewhere.

2. System Description of the NSC

The NSC version of the robot eye consists of three components: an active camera, a passive camera, and software.

The active camera transmits an array of discrete laser beams outward from a common focus. Each beam can be identified and addressed by its row and column in the array of beams. The beams intersect an 'image plane' that controls which of them may pass through to illuminate the environment. We can use the same terminology for both the active and passive cameras by considering each discrete beam of the laser array to be a 'pixel' of the active camera image plane.

The active camera hardware consists of (a) a single laser, (b) a lens to expand the single beam into a larger solid angle, (c) an optical assembly to generate (by reflection or refraction methods) an interference pattern of $M \times N$ discrete and orthogonally arranged laser beamlets, (d) optics to position the pattern so that the $M \times N$ laser beamlets are collimated and aligned to correspond precisely to optical window elements of a planar light modulator, (e) an electro-optic programmable spatial light modulator device (or PSLM) to modulate the projection of each of the discrete laser beamlets into the environment [10], [8], [22], [30].

All the laser beamlets have the same plane polarization orientation. A PSLM can control the transmission of beamlets because the plane of polarization of electro-optic material in the PSLM can be rotated as a function of an applied electric field, and because the applied electric field in the PSLM can be made to vary with position across the wavefront of beamlets. The presently implemented PSLM is a composite of two PSLMs, one to modulate the N columns and the other to modulate the M rows of the beamlet array, with $M = N = 128$. A command by computer can cause a (preprogrammed) set of columns in a PSLM to rotate optically in polarization, thereby permitting either transmission or absorption of the corresponding incident beamlets (depending on the relative orientation of the polarized beamlets to that of the PSLM electro-optic material). A polarization rotator at the laser source is adjusted to maximize the transmission/absorption ratios of the PSLM in the optical system.

The generation of the initial interference pattern which creates the orthogonal array of beamlets is the product of diffraction effects. Individual beamlets are part of the interference pattern, so that range of projection is limited by the diffraction limits of the entire projecting lens aperture, and not by the aperture size of each beamlet.

The PSLM has window apertures greater than the beamlets, and light is so aligned and collimated that each beamlet passes through (or is blocked by) a corresponding window with minimum far field diffraction by the PSLM. Thus beamlets at infinite focus can project significantly into the far field with little degradation.

Lenses can converge the laser beamlet array for the mensuration of microscopic surfaces or expand the array for the mapping of large surface areas. Each beamlet retains its relative spatial position and size in the array, and the entire pattern expands or contracts proportionately. Limitations to the system are set by the wavelength of the light and by the S/N ratio; the latter involves laser beam power, quality of the active camera hardware and optical alignment, monochromaticity, sensitivity and electronic characteristics of the passive camera, ambient light, and noise in the environment.

The passive camera is an optical image recorder which views the scene from a given perspective as different patterns of light are emitted by the active camera. The passive camera obtains a numerical image of the scene (that is, an 8-bit number for gray level for each pixel) for each pattern of scene illumination. Some types of passive camera can obtain an image directly in digital form; others must pass an analog image through an A/D converter. Buffer storage to hold a sequence of images, or else a device to transfer rapidly image data to and from a computer, is needed. The optical recorder must be synchronized with (the transmission of patterns by) the electro-optic light modulator of the active camera. To obtain a good signal to noise ratio, optical filters are used to make the passive camera specific to the wavelength of the laser light transmitted. The choice of wavelength is largely dependent on the application and the available lasers and sensors.

The region of the scene illuminated by a single beamlet should be observed by several passive camera pixels to ensure resolution. Ultimate mapping resolution is more dependent on a higher resolution video camera than on the laser array size.

Although available active cameras can create huge laser beamlet arrays, PSLM technology is presently limited to modulating 128×128 arrays. With video cameras and recorders of 2048×2048 on the verge of practicality, mapping scenes illuminated with 512 × 512 beamlet arrays should be reasonable as soon as appropriate PSLMs become available.

When the first NSC was implemented several years ago, much laser light was lost in generating interference fringes by multiple reflection, thus lowering the signal to noise ratio. As a result, it was often difficult to determine whether or not a passive image pixel was viewing a laser illuminated part of the scene. Also, the active and passive cameras had relatively long focal lengths (that is, weak convergence toward the focal point of the camera), so that determining those calibration coefficients containing perspective information was an ill-conditioned (error sensitive) problem. An intense software effort was initiated to improve the reliability of the NSC and to simplify wraparound mapping of a 3-D object. Recently patented optical hardware to generate interference fringes by refraction, and other significant optical improvements in PSLM alignment, have led to very high efficiencies of laser light during array generation in the system. The array now has extremely discrete and sharp laser beamlets, and a much improved S/N ratio. With improved hardware and software, the NSC is becoming a reliable and versatile device. In this paper we discuss the improved algorithms developed for the NSC; the new hardware improvements will be described elsewhere.

Since we discuss only the robot vision applications rather than the engineering details of the system, we hereafter use the word 'beam' or 'ray' instead of 'beamlet'. Keep in mind, however, that the 'beams' we refer to are peaks of an interference pattern on a single laser beam, and cannot mutually overlap or interfere with increasing range of projection.

3. Mathematical Concepts

We consider a beam of the laser beam array to be simply an image point (or pixel) of the active camera. We can then use the same terminology for both the active and passive cameras, thus the terminology of stereophotogrammetry. A correlation of a specific point in the passive image with a specific beam in the laser array is considered to be a correlation of image points (or pixels) in the active and passive cameras.

3.1. Triangulation

The point (x,y,z) on the surface of interest and its passive camera image (x^*,y^*) are related by the perspective transformation [14], [44], [35]:

$$(T_{11}-T_{14}\,x^*)x + (T_{21}-T_{24}\,x^*)y + (T_{31}-T_{34}\,x^*)z + (T_{41}-x^*) = 0 \;, \qquad (1)$$

$$(T_{12}-T_{14}\,y^*)x + (T_{22}-T_{24}\,y^*)y + (T_{32}-T_{34}\,y^*)z + (T_{42}-y^*) = 0 \;. \qquad (2)$$

If the scene-to-image transformation matrix T is known, we have for each illuminated point visible to the passive camera the known quantities T_{ij}, x^*, y^* and two equations for the three unknowns (x,y,z). We need one more equation for triangulation. The active camera point (or laser beam) (u,w) which illuminates the point (x,y,z) of the surface of interest is also given by a perspective transformation

$$(L_{11}-L_{14}u)x + (L_{21}-L_{24}u)y + (L_{31}-L_{34}u)z + (L_{41}-u) = 0 \;, \qquad (3)$$

$$(L_{12}-L_{14}w)x + (L_{22}-L_{24}w)y + (L_{32}-L_{34}w)z + (L_{42}-w) = 0 \;, \qquad (4)$$

where L is the scene-to-active-image transformation matrix. If we sequence the electronic shutter to project certain specific light patterns (that is, a spacecode) with the active camera columns, we can associate each bright image point (x^*,y^*) in the passive camera with a value of u corresponding to the column of a beam in the projected array of laser beams. Since only three equations are necessary for triangulation, we can solve for each of the laser-illuminated surface points with only column coding, making use of Equations (1)-(3). However, if we spacecode the optical shutter for rows and columns, we can improve the precision of the triangulation. The full set of four (usually inconsistent) Equations (1)-(4) can be solved for a 'best' value of (x,y,z) by least squares. In computation, the singular-value-decomposition method (SVD) provides for least-squares answers with more significant figures for the same computer word size [15], and is preferred when the condition number of the matrix is large, as is the case when the triangulation parallax is small. A further justification for spacecoding both rows and columns lies in the detection of surface regions that are illuminated by the active camera but not seen by the passive camera.

Thus the problem of triangulation is: Given a (passive) image point (x^*, y^*), a corresponding active image point (u, w), and the constants $\{T\}$ and $\{L\}$, find the three-dimensional location (x, y, z) of the point being illuminated.

Although the triangulation problem is conceptually straightforward, the practical problem of programming the numerical stereo camera is not trivial. In a subsequent section we discuss the software needed to make the numerical stereo camera a practical device.

3.2. Spacecodes

The correlation of corresponding image points in cameras at two different perspectives can be made automatic with the use of active and passive cameras. The array of laser beams emanating from the laser-interference-electro-optical projector is equivalent to an array of pixels of an active camera. When the laser beam array is turned on and off in certain light patterns, and the passive camera, which is synchronized to the active camera, receives those light patterns in sequential images, the corresponding pixels of the active and passive cameras (that is, a stereo pair) can be determined for each observed surface point. For example, a simple binary spacecode can be projected to encode the address of each column of a 128 column laser array as follows:

Pattern	Columns transmitting (columns are numbered 0-127)
1	64-127
2	32-63,96-127
3	16-31,48-63,80-95,112-127

4	8-15,24-31,40-47,...,104-111,120-127
5	4-7,12-15,20-23,...,116-119,124-127
6	2-3,6-7,10-11,...,122-123,126-127
7	1,3,5,...,125,127 (i.e., every other column)
8	0,1,2,...,126,127 (i.e., every column)

Pattern 8 is not used for the binary code but to identify the regions of the scene illuminated by laser column 0. (See also Figure 1.)

From the table above, we see that if a pixel of the passive image is bright in the first image, dim in the second, bright in the third and fourth, dim in the fifth and sixth, and bright in the seventh, in accord with the pattern 1011001, then the pixel is viewing a surface position illuminated by a laser beam from column $(2^6+2^4+2^3+1=)$ 89 of the laser beam array.

Although binary coding of a scene by light patterns is conceptually simple, image noise can result from the hardware, from the nature of the surface of the objects in the scene, and from the chosen perspectives of the cameras. To compensate for noise, additional light patterns may be projected and imaged, and additional processing steps may be added.

3.3. Focal Point of a Camera

The focal point of a camera with calibration coefficients T_{ij} is

$$(x_f, y_f, z_f) = -(A/D, B/D, C/D) \tag{5}$$

where the values of A, B, C, D are given by the determinants

$$A = \begin{bmatrix} T_{41} & T_{21} & T_{31} \\ T_{42} & T_{22} & T_{32} \\ T_{44} & T_{24} & T_{34} \end{bmatrix} \quad C = \begin{bmatrix} T_{11} & T_{21} & T_{41} \\ T_{12} & T_{22} & T_{42} \\ T_{14} & T_{24} & T_{44} \end{bmatrix}$$

$$B = \begin{bmatrix} T_{11} & T_{41} & T_{31} \\ T_{12} & T_{42} & T_{32} \\ T_{14} & T_{44} & T_{34} \end{bmatrix} \quad D = \begin{bmatrix} T_{11} & T_{21} & T_{31} \\ T_{12} & T_{22} & T_{32} \\ T_{14} & T_{24} & T_{34} \end{bmatrix} \tag{6}$$

and $T_{44} = 1$.

Equations (5) and (6) derive from the perspective projection

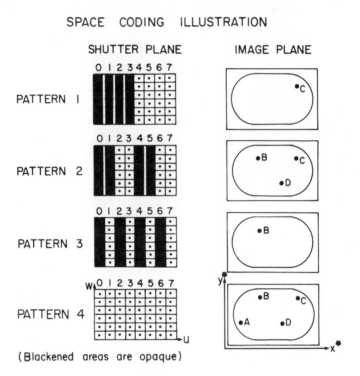

Figure 1:   The array of laser beams is space coded by the sequence of
binary patterns (shown at the left) obtained by controlling the opacity of
the columns of the shutter plane. The intersections of the laser beams
with the scene provide bright spots on the scene which can be imaged by
a passive camera from a suitable perspective.   The passive camera is
synchronized with the active camera so that for each projected light
pattern a passive image is obtained (shown at the right). In the figure,
the passive image sees four bright spots of the scene.   The column
location u of a beam in the active image plane can be determined by the
space coding received by the passive image plane. Point B is seen in
passive images 2 and 3 but not in image 1. The binary code of the
column of the corresponding active beam is therefore $u = 011 = 3$. The
fourth (open) pattern illuminates column zero.

$$T_{11}x + T_{21}y + T_{31}z + T_{41} = Hx^* = X \tag{7}$$

$$T_{12}x + T_{22}y + T_{32}z + T_{42} = Hy^* = Y \tag{8}$$

$$T_{14}x + T_{24}y + T_{34}z + T_{44} = H \tag{9}$$

If $H \neq 0$, these three equations reduce to Equations (1) and (2). Thus the case $H \neq 0$ corresponds to a line or ray from the camera focal point to the point (x^*, y^*) of the image plane and then outward into the 3-D space of the scene. The focal point is on every ray of the camera, and cannot depend on the values of x^* or y^*. Only for the case $H = 0$ do Equations (7)-(9) give a solution for a point rather than a line; thus the focal point is given by the case $H=0$. Equations (5) and (6) correspond to the solution of Equations (7) through (9) with $H = 0$.

3.4. Camera Calibration

We now discuss camera calibration, the determination of the matrix coefficients connecting the coordinates of the scene and the image plane. In this section (only), the same notation T_{ij} and (x^*, y^*) is used for both the passive and the active cameras. To obtain the calibration mathematics in the notation used previously for the active camera, one need merely replace T_{ij} with L_{ij} and (x^*, y^*) with (u, w). By image plane we mean either the image plane of the passive camera or the electro-optic shutter plane of the active camera.

The values of the camera calibration coefficients T_{ij} depend on the orientation, translation, focal distance, and scaling of the camera relative to the 3-D scene. There are two independent coordinate systems: (x^*, y^*) for an image point and (x, y, z) for a point in the scene. The calibration procedure determines the T_{ij} values that satisfy Equations (1) and (2) when we are given a set of known points $\{(x_i, y_i, z_i)\}$ in the scene and their known corresponding image points $\{(x_i^*, y_i^*)\}$, where $i=1, \ldots, n$ and $n \geq 6$.

We can set $T_{44}=1$ [44], [35] and obtain the $2n$ $(n \geq 6)$ (inconsistent) inhomogeneous equations for the 11 unknown values of T_{ij}:

$$
\begin{bmatrix}
x_1 & y_1 & z_1 & 1 & 0 & 0 & 0 & 0 & -x^*_1 x_1 & -x^*_1 y_1 & -x^*_1 z_1 \\
. & . & . & . & . & . & . & . & . & . & . \\
. & . & . & . & . & . & . & . & . & . & . \\
. & . & . & . & . & . & . & . & . & . & . \\
x_n & y_n & z_n & 1 & 0 & 0 & 0 & 0 & -x^*_n x_n & -x^*_n y_n & -x^*_n z_n \\
0 & 0 & 0 & 0 & x_1 & y_1 & z_1 & 1 & -y^*_1 x_1 & -y^*_1 y_1 & -y^*_1 z_1 \\
. & . & . & . & . & . & . & . & . & . & . \\
. & . & . & . & . & . & . & . & . & . & . \\
. & . & . & . & . & . & . & . & . & . & . \\
0 & 0 & 0 & 0 & x_n & y_n & z_n & 1 & -y^*_n x_n & -y^*_n y_n & -y^*_n z_n
\end{bmatrix}
\begin{bmatrix}
T_{11} \\
T_{21} \\
T_{31} \\
T_{41} \\
T_{12} \\
T_{22} \\
T_{32} \\
T_{42} \\
T_{14} \\
T_{24} \\
T_{34}
\end{bmatrix}
=
\begin{bmatrix}
x^*_1 \\
. \\
. \\
. \\
x^*_n \\
y^*_1 \\
. \\
. \\
. \\
y^*_n
\end{bmatrix}
\tag{10}
$$

or simply

$$A T = D \tag{11}$$

where A is the $2n \times 11$ matrix, T is the 11×1 column vector, and D is the $2n \times 1$ column vector.

The matrix Equation (11) can be solved for 'best' values of T_{ij}. The normal equation method would involve solving

$$A^t A T = A^t D \tag{12}$$

where A^t is the transpose of A, the matrix $A^t A$ is an 11×11 symmetric matrix, and $A^t D$ is an 11×1 column vector. However, for the same word size, the normal equation method of Equation (12) loses more significant figures than other methods such as singular value decomposition [15].

Significant figures are important in this calibration procedure because the matrix A may not be very well conditioned if the known points in the scene (x_i, y_i, z_i) or the camera points (x_i^*, y_i^*), are not sufficiently separated. The calibration coefficients T_{14}, T_{24}, T_{34} derive from the converging of rays to the focal point; without significant separation of sample points and careful solution of Equation (11) they cannot be extracted accurately.

The problem of calibrating a camera is thus: Given a set of known noncoplanar points in 3-D and their corresponding image points in the camera, find the constants $\{T\}$. Thus given $\{(x_i, y_i, z_i, x_i^*, y_i^*); i=1, \ldots, n; n \geq 6\}$, find $\{T_{ij}; i=1, \ldots, 4; j=1, 2, 4\}$.

4. Fundamental Software Problems

The fundamental software problems that must be solved to make the NSC robot eye a relatively independent preprocessor for a vision system are:

1. Signal Detection: To determine whether or not an observed bright point seen in a passive image corresponds to a laser-illuminated region of the scene, and if so, to determine the bit of the spacecode contributed by that light pattern at that point of the passive image.
2. Passive Camera Calibration: To calibrate the passive camera; that is, to determine the constants T_{ij} of Equations (1) and (2).
3. Active Camera Calibration: To calibrate the active camera; that is, to determine the constants L_{ij} of Equations (3) and (4).
4. Motion Recalibration: To recalibrate rapidly the passive and active cameras if the surface of interest or any of the cameras are displaced by a known amount.
5. Triangulation: To triangulate a point in the scene reliably with the minimal loss of significant figures.
6. Missing Region Detection: To detect regions of the scene that were not triangulated because of missing data (due to blocking of beams or occlusion of lines of sight by intervening hills and valleys).
7. Information Recovery: To pick new perspectives to attempt to mensurate regions of missing data in the scene.
8. Wraparound 3-D Mapping: To effect wraparound mapping of any 3-D object by merging in the same 3-D coordinate system the triangulation data of different stereo perspectives.

5. Signal Detection: the Problem of Determining the Spacecode

5.1. The Nature of the Problem

To stereo-correlate the active and passive camera images and triangulate points in the scene, the sequence of images acquired by the passive camera must be analyzed to determine the binary code received at every pixel of the passive image. To achieve this task, the passive camera must distinguish laser-illuminated areas of the surface from other areas. Unfortunately, there is no obvious single threshold value in the brightness distribution of the passive image to flag those image regions viewing laser-illuminated surface. The reason is that considerable variations of brightness may occur from one passive image to the next, because (1) regions of the surface of interest may reflect differently (anisotropically) for different angles of beam incidence, (2) different regions of the scene may have different textures and scatter properties, (3) the shutter configuration of each different projected light pattern may have different scatter and transmission properties, and (4) the environment may be noisy because of other lasers, grazing beams, or reflections. Detection (or thresholding) errors mean incorrect values for the spacecode, thus the creation of non- existent beams with concurrent loss of existing beams in the database created for 3-D stereo triangulation.

5.2. Algorithmic Considerations

A number of ways are available to reduce and correct errors. These methods are similar to the error detecting and correcting codes of information theory. All of them require additional projected light patterns and recorded passive images. Two methods have been tried for error detection and correction with the numerical stereo camera.

In the first method, the scene is illuminated by an all-laser-beams-on pattern and the passive image is scanned for brightness peaks and valleys. A single (global) threshold is determined to separate the bright regions (which are presumably seeing laser illuminated surface) from the rest of the passive image. These 'valid' regions are remembered and followed in the subsequent sequence of images containing the light code. For each passive image in the sequence, the same brightness threshold determines the spacecode bit for each valid pixel; if a pixel is brighter than threshold, the spacecode bit contributed by that image is one. The method gives reliable spacecodes when there is a large S/N ratio and the image contrast does not change with the projected light pattern (shutter configuration). Experience has shown, however, that S/N ratio may vary over an image because the angles of incidence of the passive and active rays vary over the surface, and that the contrast at the same point in the passive image plane may vary significantly from one coded image to the next because the shutter transmission may vary with the configuration. Under such circumstances, a single global brightness value to threshold all the passive images in the coded sequence is not reliable.

The second method, developed by L. Tamburino, determines the spacecode received at each pixel of the passive image, and, in addition, a measure of statistical reliability for the spacecode [45]. Since 1983, several versions of the Tamburino detection algorithm have been implemented on the prototype NSC at the AFWAL Avionics Laboratory, and all demonstrate marked improvements in computational and spacecode accuracy [46], [12].

6. Calibrating the Passive Camera

In Section 3.4 we discussed the mathematics of camera calibration. For the case of the passive camera, the simplest calibration procedure is to (1) insert a known rectangular prism into the scene, (2) let the vertices of the prism define the scene coordinate system, (3) pick the 6 or 7 visible vertices of the prism and record their image positions as seen on the graphics screen, and (4) solve Equation (11) by singular value decomposition (SVD) for the

T_{ij} coefficients.

A sequence of two software programs is required. The first allows the user to define the prism coordinates and pick the vertices visible to the passive camera. The second solves for the values of T_{ij} by SVD.

This section and the next describe possible approaches to passive and active camera calibration. A somewhat different approach has also been implemented [11].

6.1. Picking Calibration Points

The first program records the spatial coordinates of six or seven points in the calibration prism and their image coordinates in the passive camera.

The algorithm steps are as follows:

S1: Enter the name of the data file for the sequence of coded images. The name acts as a tag to be associated with all the files associated with this problem.

S2: Enter the dimensions of the sides of the calibration prism. The first dimension is assumed to be along the x-axis; the second to be along the y-axis; and the third to be along the z-axis.

S3: Enter the 3-D locations of the calibration points on the prism. The points chosen are usually six or seven of the vertices of the prism since their images are easiest to identify in the passive camera image. The eight vertices of the prism can be conveniently chosen to be $(0,0,0)$, $(a,0,0)$, $(a,b,0)$, $(0,b,0)$, $(0,0,c)$, $(a,0,c)$, (a,b,c), $(0,b,c)$, where a, b, c are the known dimensions of the prism for the x, y, z axes respectively. With an opaque calibration prism, at most seven of these eight vertices can be expected to be visible from any given perspective.

S4: Enter the (2-D) locations of the image points corresponding to the calibration points of the prism. In the program a corresponding image point is entered immediately after the chosen calibration point.

S5: Create a data file which contains the values

$\{(x_i, y_i, z_i, x_i^*, y_i^*);\ i=1,6\}.$

6.2. Determining the Passive Camera Calibration Coefficients

The second program finds the values of the calibration coefficients T_{ij} of the passive camera.

The algorithm steps are as follows:

S1: Read the data file generated by program 1.

S2: Find the coefficients T_{ij} . The input data are used to create a $2n \times 11$ matrix and a $2n$ vector, where n is 6 or 7. The 11-vector containing the values of T_{ij} is then solved by SVD. If the parallax is so small that the converging of the rays to the camera focal point cannot be detected, the condition number of the solution will exceed a threshold determined by the number of significant figures for single precision real numbers. In that case, the program will reduce the $2n \times 11$ matrix to a $2n \times 8$ matrix and the 11-vector to an 8-vector, the rotation and translation components of T_{ij} will be computed, and the converging-ray components set to zero.

S3: Find the location of the focal point of the passive camera if the parallax permits (i.e., if the SVD provides sufficient significant figures for the available parallax). The focal point of the passive camera is used primarily as a check on the reasonableness of the calculation, and later in the program for active camera calibration to determine the visible faces of the calibration prism.

S4: Find the maximum and minimum values of $\{x_i^*\}$ and $\{y_i^*\}$. In the calibration of the active camera (next section), these values allow us to limit the search range in the passive image and to ignore background scatter (noise) from points outside the calibration prism.

S5: Write an output file which contains the values of T_{ij}, the focal point location, and the limits of the calibration prism in the passive camera.

7. The Problem of Calibrating the Active Camera

7.1. Algorithmic Considerations

The problem of calibrating the active camera requires that the constants L_{ij} of Equations (3) and (4) be determined. Just as for the passive camera, the values L_{ij} depend on the orientation, translation, focal distance, and scaling of the active camera relative to the 3-D scene.   There are two independent coordinate systems: (u,w) for an active image point (i.e., for the address of a laser beam in the array of laser beams) and (x,y,z) for a point in the scene. The calibration procedure determines the L_{ij} values that satisfy Equations (3) and (4) for a set of known points (x_i, y_i, z_i) in the scene and their known corresponding active image points (u_i, w_i) where $i=1, \ldots, n$ and $n \geq 6$. We have tried four ways of calibrating the active camera.

Method 1: Turn on one beam in the laser array at a time and locate in 3-D by direct measurement where the beam falls on a surface. With six or more noncoplanar measured points (x_i, y_i, z_i) and their corresponding beams (u_i, w_i) , we can solve for the L_{ij} using the same technique that we used for the passive camera calibration. This method has the advantage that the accuracy of the active camera is independent of the accuracy of the passive camera. The disadvantage is that the location of the intersection of the laser beam (u_i, w_i) with the 3-D surface at (x_i, y_i, z_i) must be physically measured in 3-D space.

Method 2: Place a rectangular prism of known dimensions in the scene.   The six surfaces of the prism might, for example, satisfy the equations $x=0$, $x=a$, $y=0$, $y=b$, $z=0$, $z=c$ where the values for a, b, c are the lengths of the prism edges. Observe (sequentially) six or more laser-illuminated non- coplanar spots on the calibration prism (due to six or more separate laser beams of the array) and record each laser beam (u_i, w_i) and the corresponding position (x_i^*, y_i^*) in the passive image. Identify and record the passive image positions of the (observable) vertices of the prism (or retrieve them from the files of the passive camera calibration). Then, use the image positions of the vertices to determine the face of the prism that each spot illuminates, and to locate in 3-D (by interpolation in the image) each of the illuminated spots on the prism. We then have $\{(x_i,y_i,z_i,u_i,w_i); i{=}1, \ldots ,n; n \geq 6\}$ and can determine the L_{ij} calibration coefficients of the active camera. The advantage of the method is that no measurement need be made in the 3-D scene. The disadvantage is that correctness and accuracy depend on the accuracy of the pixel discretization of the passive camera, the absence of distortions in the passive image, and the foresight to ensure that the selected active camera beams do not all intersect a single plane surface in the 3-D scene.

Method 3: Turn on one beam (u_i,w_i) of the laser beam array and locate the illuminated prism point (x_i,y_i,z_i) by using (1) the known T_{ij} values previously calculated for the passive camera, (2) the known image point $(x_i{}^*,y_i{}^*)$, and (3) the equations for the faces of the rectangular prism. To do this, trace the ray from the passive camera and find the intersections of this ray with the planes of each of the prism faces. Choose the correct intersection point automatically by considering the visibility of the prism faces relative to the passive camera. Repeat this procedure for $n \geq 6$ non-coplanar illuminated points on the prism (from n different laser beams), and determine a set of active camera and spatial points $\{(x_i,y_i,z_i,u_i,w_i); i{=}1, \ldots ,n; n \geq 6\}$ to provide the calibration. The advantages and disadvantages of this method are similar to those of method 2. Now the accuracy also depends on the accuracy of the values of the T_{ij}. However, since this method determines the 3-D positions of the calibration points i from the ray and plane equations in space and not from interpolation in 2-D images, it is preferred if there are distortions in the passive image; this follows because even a distorted image plane corresponds point by point to rays radiating outward from the focal point, whereas interpolation in a distorted image must be done in non-Euclidean coordinates.

Method 4: A variation of method 3, but now instead of imaging one active camera point (i.e., laser beam) at a time, we sequence through a spacecode pattern of illumination, correlate each (u_i, w_i) with the appropriate (x_i^*, y_i^*) which appears in the passive camera (provided (u_i, w_i) hits the surface at a point visible to the passive camera). Here as many as 128×128 points may be used to calculate the L_{ij} by least squares. Since a plane has the equation

$$A x + B y + C z = D \qquad\qquad (13)$$

and the values of A, B, C, D are known for each face of the rectangular prism, we can solve Equations (1), (2), and (13) for (x, y, z) when we are given (x^*, y^*) and the T_{ij} .

The problem of calibrating the active camera is thus: Find a set of known points in 3-D that are illuminated by known laser beams (active image points) of the active camera. This may be obtained from knowledge of the prism faces, the passive image points, the passive camera calibration, and the spacecodes of the passive image points. Then calculate the L_{ij} .

7.2. Calculating the Coefficients of the Active Camera

We have implemented a program using method 4 above to find the values of the calibration coefficients L_{ij} of the active camera.

The algorithm steps are as follows:

S1: Retrieve the data files generated by previous programs for the prism parameters and for the passive camera calibration coefficients T_{ij}.

S2: From the passive camera images of a sequence of coded light patterns, derive the spacecode received at each pixel of the passive camera, and create both a column and row coded image. The column coded image is a 512×512 table of bytes, one byte for each pixel of the passive camera. The numerical value of each byte is either the (reliable) spacecode of an active camera column, or zero if the space code at the pixel of the passive image is not reliable. (In a normal image, the numerical value of each byte corresponds to the intensity or gray level at a pixel; thus the spacecoded image should be distinguished from a normal picture image!) Similarly, the row coded image is a 512×512 table of bytes, each byte containing the spacecode for the active-camera row seen by a pixel of the passive image.

S3: Ask for an optional speed-up factor. A speed-up factor of 2 will process every other pixel of a coded image, allowing shorter times of computation at the cost of poorer statistics.

S4: For the column and row coded images separately, check each pixel. For each pixel with a non-zero (reliable) spacecode, trace back the ray from the pixel to the calibration prism. The ray of (x^*, y^*), determined by Equations (1) and (2), intersects the known planes (see Equation (13)) of the calibration prism. If an intersection point is found on a prism face visible to the passive camera, then the matrix and vector used to derive the active-camera calibration coefficients are updated. In effect, the matrix and vector contain both the 3-D locations of points on the calibration prism and the corresponding beams from the active camera (derived from the column or row spacecode values detected at the passive camera pixel). The number of intersection points found for each face of the calibration prism is recorded. At least two prism faces must be sampled for the active camera to be calibrated.

S5: Use the matrix and vector generated by the data to determine the values of the calibration coefficients L_{ij} of the active camera. Singular value decomposition is used to solve the large matrix equation, with one matrix row for each reliable spacecode byte.

S6: Calculate and report the focal point of the active camera. The location of the focal point checks the reasonableness of the calculation and helps pick the face of the prism intersecting a line of sight from the passive camera.

S7: Store the values of L_{ij} .

8. Rapid Recalibration for Changing Perspectives

8.1. Algorithmic Considerations

To map the surface of a truly 3-D object, we must view the object from several different perspectives. With the calibration technique so far described, the wrap-around mapping procedure would be as follows: (1) Set up the active and passive cameras; (2) Insert a calibration prism into the scene and calibrate the cameras; (3) Remove the calibration prism from the scene and insert the object of interest; (4) Map the object from this perspective; (5) Move the object or cameras to a new perspective; (6) Repeat steps 2 through 5 until the entire 3-D surface of the object is mapped. This calibration procedure is slow and error prone because the object of interest must be removed prior to each new perspective to allow camera calibration and then replaced in exactly the same position it was originally.

If a precision positioning device can be attached to the camera mounts or object platform, the calibration procedure can be somewhat simplified to: (1) Set up the active and passive camera; (2) Insert a calibration prism into the scene and calibrate the cameras; (3) Move the prism or camera to a new perspective, note the position readings, and calibrate the cameras; (4) Repeat step 3 for a sufficient number of perspectives; (5) Remove the calibration prism from the scene and insert the object of interest; (6) Map the object from each of the previously calibrated perspectives. The advantage of this method is that the calibration is done initially for all the perspectives so that the object of interest can be mapped in 3-D without disturbing its position on the platform. The disadvantage is that we must decide on the viewing perspectives in advance.

If a precision positioning device keeps track of the translations and rotations of the object and cameras, we need only calibrate the cameras and object initially. Thereafter we can recalibrate the cameras with a mathematical transformation without actually inserting a physical prism into the scene. The advantage is that we can choose new perspectives as the topology of the surface dictates; we need not restrict ourselves to previously-chosen perspectives of uncertain efficacy. Thus the simplest calibration procedure with a precision positioning device is: (1) Set up the active and passive camera; (2) Insert a calibration prism into the scene and calibrate the cameras; (3) Remove the calibration prism from the scene and insert the object of interest; (4) Map the object from this perspective; (5) Move the object or a camera to a new position; (6) Recalibrate the cameras mathematically from the motion information, in accord with the discussion below; (7) Repeat steps 4 to 6 for a sufficient number of perspectives. (See also [42] and [41] for more on calibration techniques.)

Let T be the perspective operation $(x,y,z,1) \rightarrow (X,Y,Z,H)$ or

$$[x \; y \; z \; 1]T = [x^* \; y^* \; z^* \; 1]H$$

and let P be the projection operation onto the $z=0$ plane

$$[x \; y \; z \; 1]P = [x \; y \; 0 \; 1] \; .$$

Then the operation $T P$ (i.e., first T then P) is

$$[x \; y \; z \; 1]T P = [x^* \; y^* \; 0 \; 1]H \; . \tag{14}$$

Equation (14) is shorthand for Equations (7) to (9), or (1) and (2). Let R be the (known) displacement operator which rotates or translates points (x,y,z) to (x',y',z') so that

$$[x \; y \; z \; 1]R = [x' \; y' \; z' \; 1] \; .$$

Then the 3-D points would project onto new (primed) camera points by

$$[x \; y \; z \; 1]R \, T P = [x^{*\prime} \; y^{*\prime} \; 0 \; 1]H' \tag{15}$$

Because we want to map a 3-D object from multiple perspectives, we want coordinates (x,y,z) fixed to the object regardless of any motions of the cameras or the object. We thus interpret Equation (15) by saying that there exists new calibration coefficients T' where

$$T' = R\,T/k \tag{16}$$

$$H'' = H'/k \tag{17}$$

and k is a normalization constant, so that Equation (15) becomes

$$[x\ y\ z\ 1]T'P = [x^{*\prime}\ y^{*\prime}\ 0\ 1]H'' \tag{18}$$

We can now triangulate using the primed (new perspective) image points and the known calibration coefficients $T'P$ and continue to locate the surface points of the object in the same object coordinate system as before. Thus

$$T'P = \frac{R\,T\,P}{k}$$

$$= \frac{1}{k}
\begin{bmatrix}
R_{11} & R_{12} & R_{13} & 0 \\
R_{21} & R_{22} & R_{23} & 0 \\
R_{31} & R_{32} & R_{33} & 0 \\
D_1 & D_2 & D_3 & 1
\end{bmatrix}
\begin{bmatrix}
T_{11} & T_{12} & 0 & T_{14} \\
T_{21} & T_{22} & 0 & T_{24} \\
T_{31} & T_{32} & 0 & T_{34} \\
T_{41} & T_{42} & 0 & T_{44}
\end{bmatrix} \tag{19}$$

where the R_{ij} are rotation terms, the D_i are displacement terms, and the T_{ij} are the original calibration coefficients. Since we assigned $T_{44} = 1$ (normalization) to reduce the number of coefficients T_{ij} from 12 to 11, we similarly require $T_{44}{}' = 1$ for the new calibration coefficients T'; thus

$$k = D_1\,T_{14} + D_2\,T_{24} + D_3\,T_{34} + 1\ . \tag{20}$$

Note that the normalization factor k is only non-unity when translation occurs (that is, D_1, D_2, D_3 are not all zero).

The same type of transformation applies to the active camera, namely

$$L' = R\,L/k\ . \tag{21}$$

On the other hand, if the object remains stationary but the camera is displaced, we use the inverse of R for the camera that moves and obtain the same form of Equation (18). But now only the camera that moves requires new calibration coefficients. If both cameras move, then two different (inverse) operators of the form of R are used. A sequence of motions for the camera or object leads to a sequence of new calibration coefficients given by

Equation (16) and (21) but with

$$R = R_1 R_2 R_3 \cdots \tag{22}$$

8.2. Rapid Recalibration After Relative Motion

A program has been developed to find the values of the new calibration coefficients T_{ij} and L_{ij} after relative motion between the object and the passive and/or active cameras. If the relative motion is known precisely, there is no need to insert the calibration prism back into the scene or to disturb the cameras or object of interest. The recalibration can be done mathematically rather than physically.

The algorithm steps are as follows:

S1: Read the data files for the present calibration points and camera calibration coefficients T_{ij} and L_{ij}. These files contain information about the calibration prism, the original calibration coefficients for the passive and active cameras, and the calibrations done at each step in a chain of previous motions chosen by the user or system. That is, every time the object or a camera is moved and a recalibration is done, the new calibration information is stored. If a string of motions is carried out, calibration is continually updated. Error buildup is less if the motions are known more precisely.

S2: For interactive mode, display a menu of options for the user:
1. rotate passive camera
2. rotate active camera
3. rotate object
4. translate passive camera
5. translate active camera
6. translate object

After a motion option is completed, the program returns to display this menu, and the user can continue to move the object and/or a camera until he chooses an exit option.

S3: To rotate the object, specify the axis and angle of rotation. The axis specification requires any point on the axis and a vector parallel to the axis.

S4: Calculate new calibration coefficients for the passive and/or active cameras directly from the old coefficients and the given rotations and translations (see Equations (19) and (20)). The case of translation has the complication of requiring a renormalization in the calculation for the new calibration coefficients (cf. Section 8.1). The accuracy of the new calibration coefficients does not depend on the parallax; although there is little point in doing so, one can move the object or camera miles apart and still recalibrate accurately.

S5: Determine the new focal points of the active and passive cameras directly from the calibration coefficients.

S6: Although the calibration prism has been physically removed from the scene, we can imagine the vertices to be still attached to the object and can follow them mathematically. If the object is rotated or translated, determine the new positions for the vertices. From the 6 or 7 vertices of the calibration prism, calculate the corresponding image points in the passive and active cameras. If only the passive (or active) camera is rotated, calculate new image points only for the passive (or active) camera. From these image points, calculate the new limits of the image of the (imaginary) calibration prism on the passive camera. In this way we can keep track of the field of view.

S7: Store the new calibration coefficients, and the locations of the calibration prism vertices and their images, in a user-designated file.

9. Triangulation

When the passive and active cameras have been calibrated by the methods described in Sections 6 to 8, the NSC can be used to triangulate multiple points in the scene. With a 128×128 array of laser beams, up to 16,384 points of the scene or of a surface can be mensurated in 3-D space. The basic mathematics of the triangulation problem was discussed in Section 3.1.

The major problem that arises in actual practice is that the parallax between the passive and active cameras is often small, so that the solution of Equations (1) through (4) becomes ill-conditioned. When this happens, the number of significant figures in the solution decreases and one must be careful not to end up with (meaningless) answers that have no significant figures. One approach is to use double precision real numbers in the calculation. The other is to use SVD [15] to solve the four overdetermined equations. Conceptually, the loss of significant figures corresponds to the error-sensitivity of the triangulated position in the direction perpendicular to the baseline between the active and passive cameras.

Since the T_{ij} and L_{ij} coefficients of the cameras are known from the calibration procedures, the position (x^*,y^*) of a passive image point (or pixel) is known, and the corresponding laser beam (u,w) is known from the spacecode received at the pixel, Equations (1) through (4) can be written as the matrix equation

$$T x = b . \tag{23}$$

Here T is a 4×3 matrix containing the expressions in parentheses in the first three terms of each of the Equations (1) through (4), b is the 4×1 vector containing the negative of the expressions in parentheses in the last term of each of the equations, and x is the 3×1 column vector containing the unknown coordinates (x,y,z) of the scene point to be triangulated.

If the parallax is small, then Equation (23) represents a set of four ill-conditioned inhomogeneous linear equations. These four equations are usually inconsistent because of measurement errors, and correspond to planes in space which intersect in a tetrahedron rather than at a single point. If we were to solve the equations by creating the 3×3 square matrix $T^t T$ and solving for x, we would lose many more significant figures than if we simply solved Equation (23) directly by SVD. The least squares solution obtained by SVD corresponds to a point in the tetrahedron satisfying a condition of minimum error.

Constraint equations can be added to Equation (23) to weight the solution and move the solution point around within the tetrahedron. If the tetrahedron is elongated away from the cameras because of ill-conditioning, then the additional constraints may significantly affect the location of the solution. One should introduce constraints into the least squares solution only if one has some prior knowledge as to how the solution should behave. With large parallax and well-conditioned equations, the tetrahedron is generally small, and any point in the tetrahedron should provide an adequate solution for triangulation.

In the algorithm and software that we have developed, the SVD method is used to flag cases with a very high condition number (that is, too few significant figures for the accuracy desired).

10. Missing Region Detection

Stereo cameras, separated by parallax, view a scene from different perspectives. As the parallax increases, the problem of triangulating points in the scene becomes more well-posed (less error sensitive) but the number of points in the scene that are imaged simultaneously by both cameras generally decreases. In stereophotogrammetry, an observer views a stereo pair of photographs and perceives which points of the scene are visible to both cameras and thus can be triangulated. If some scene regions are not viewed by either camera because mountains or valleys obstruct the line-of-sight from a camera perspective, the observer may use or request images taken from yet other perspectives.

The same problem of missing regions can occur with the NSC, except that now missing regions can be detected by an algorithm rather than by human perception. For the NSC, there are two separate cases, depending on which camera (active or passive) is obstructed.

Case 1: Light projected by the active camera does not illuminate a part of the scene because of a surface convolution (obstructing mountain, cliff, or hole). The passive camera viewing the unilluminated part of the scene sees a region of unreliable spacecodes in the image plane because there is little change in light intensity received when either a light pattern or its complement is on. Thus, if the passive image has a region which (1) contains pixels with mostly unreliable spacecodes, and (2) is surrounded by regions which contain many pixels with reliable spacecodes; then we conclude that lines of sight from the active camera to the corresponding region of the scene are occluded because of intervening mountains or holes.

Case 2: Light reflected from an illuminated part of the scene does not reach the passive camera because of an intervening surface convolution. The passive camera may obtain reliable spacecodes in the image plane, but may see spacecode values which change discontinuously from one observed bright spot to the next across some unseen boundary in the image plane, indicating some spacecode values are missing. Thus if the passive image sees two adjacent regions, each containing pixels with mostly reliable spacecodes, and a sudden discontinuity in the value of the spacecode from one region to the next, then we conclude that some lines of sight from the scene to the passive camera are occluded because of intervening mountains or holes.

In different regions of the same passive image both cases 1 and 2 may arise (Figure 2).

We have two tasks in this section: (1) to detect which regions of the scene are not illuminated by the active camera, or viewed by the passive camera, and (2) to estimate a minimum depth of any hole or minimum height of any hill in the unmensurated regions of the scene.

We assume that the active camera can sample the scene sufficiently; that is, every significant wiggle or convolution in the scene can be sampled at least twice. Equivalently, the separation of laser bright spots on the scene nowhere exceeds half the smallest wavelength of any 3-D spatial harmonic contained in the scene. Without sufficient sampling by the active camera, the mensuration of the scene could be aliased. (See for example [27].)

In practice the grid of pixels in the passive image has much finer spatial

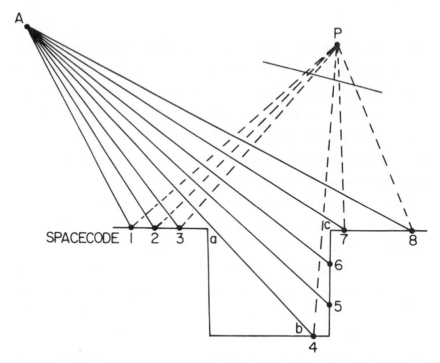

Figure 2: Region ab is unilluminated by the active camera so that the passive image gets an unreliable spacecode there (thus Case 1). The bright dots within region bc are not seen by the passive camera; the camera sees bright dots at b and c and a jump in the spacecode values from 4 to 7 (Case 2).

resolution than the grid of pixels in the active image. Thus, the passive image may have (1) pixels which view the same bright spot of the scene and thus detect the same spacecode, and (2) pixels which view regions between the bright spots of the scene and thus have unreliable spacecode. Pixels in the second group do not correspond to obstructed rays from the active camera, and must be distinguished from them.

It is therefore convenient to define a neighbor of a passive image pixel as follows. Consider a passive image pixel which has a reliable spacecode corresponding to active camera beam (p,q). Then a neighbor of that pixel is a passive image pixel which has a reliable spacecode corresponding to beam $(p',q')=(p+i,q+j)$ in the allowed domain of the active camera beam array such that p, q, p', q', i, j are integers and (i,j) is in one of the classes

$(i > 0,0)$, $(0,j > 0)$, $(i < 0,0)$, $(0,j < 0)$ called the directions of the neighbor. A neighbor of order n is a neighbor such that the magnitude of either i or j is n. Note that adjacent passive pixels which view the same bright spot are not neighbors by the definition above, but share the same neighbors. See Figure 3.

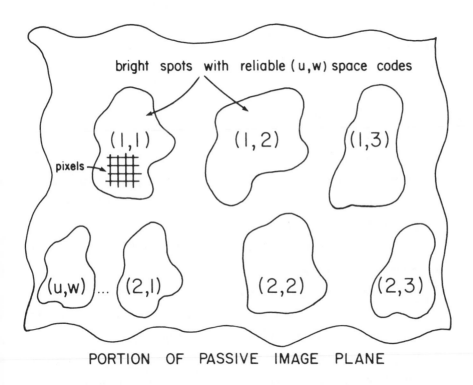

PORTION OF PASSIVE IMAGE PLANE

Figure 3: All pixels within the labelled regions (or spot images) have reliable spacecodes and are seeing the reflections of laser beams from the surface of interest. Every pixel in region (1,1) is a neighbor of order 1 to every pixel in regions (1,2) and (2,1), a neighbor of order 2 to every pixel in region (1,3), and not a neighbor to pixels of regions (2,2) and (2,3). Pixels in regions (1,2) and (2,1) are not neighbors.

Task 1a: To detect regions of the scene that are not illuminated by the beams of the active camera (case 1):

S1: In the passive image, find the mean and standard deviation of the separation distance (in pixel units) of pixels from their neighbors of order 1.

S2: Flag all passive pixels which are separated by two or more standard deviations from (at least) one neighbor of order 1. Remove duplication by keeping only the median pixel of any group of pixels which view the same illuminated spot on the scene (and have the same spacecode).

S3: Determine which of the flagged pixels are neighbors of order 1 and are separated by more than two standard deviations. These should form the boundary of the region of unreliable spacecodes in the passive image.

Task 1b: To detect regions of the scene that are illuminated by the active camera but are not seen by the passive camera (case 2):

S1: Find all beams of the laser array whose spacecodes are missing from the passive image.

S2: In the passive image, record the locations of all pixels which have (1) a reliable spacecode and (2) a neighbor of order $n \geq 2$ in a certain direction but no neighbor of order 1 in that direction.

S3: Find the contours in the active image plane that correspond to the loci of pixels found in step 2.

S4: Find which of the missing beams of step 1 lie within these loci in the active image. This subset of the missing beams reaches the scene, but the rays from the scene to the passive camera are obstructed.

Task 2: We have now delineated the regions of missing (blocked or obstructed) rays in the active and passive images. However, the boundaries of these missing ray regions can be used in triangulation of the scene. The boundary of an unmensurated region appears in the passive image plane for case 1 and in the active image plane for case 2. Since the mathematics is the same for both cases, we simply say the image boundary of the unmensurated region is in the image plane.

The steps to determine the minimum depth of an unmensurated region of the scene are as follows (see Figure 4):

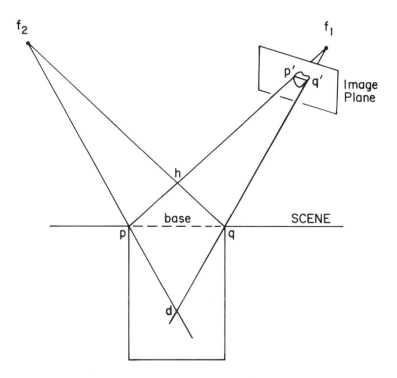

Figure 4: Cameras are at f_1 and f_2 and an unmensurated region is outlined in the image plane of camera 1. Image points p' and q' at the image boundary of the unmensurated region are correlated by the NSC with points of the other camera so that the boundary points p and q in the scene can be triangulated. The quadrilateral *phqd* in the plane $f_1 f_2 p$ gives the height of the possible hill or the minimum depth of the possible hole that could cause obstruction of rays.

S1: Pick an image point p' on the image boundary of the unmensurated region; this image point is associated with a scene point p which can be triangulated.

S2: Determine the plane f_1, f_2, p containing that scene point and the focal points of both the active and passive cameras.

S3: Determine the line of intersection $p'q'$ of that plane with the image plane. The image point chosen in step 1 is on this line.

S4: If the point q' is on the image boundary of the unmensurated region, we have a line segment in the image plane which cuts through the image of the unmensurated region. Find the scene points p,q corresponding to the endpoints of the line segment in the image plane; these scene points will be considered the base of a triangle.

S5: Draw rays from the focal points of the active and passive cameras to these measured endpoints in the scene. These four rays intersect in a quadrilateral. Calculate the altitudes of the two triangles containing the measured scene points as a base. These altitudes determine the minimum possible height of a hill or the minimum possible depth of a hole (in the plane of the rays) to explain the unmensurated region.

S6: Choose a neighboring image point of p' on the image boundary of the unmensurated region and repeat steps S1-S5. Repeat for consecutive points on the image boundary until every point has been considered. Choose the greatest minimum height and greatest minimum depth derived from all the line segments through the image of the unmensurated region in the image plane. These values correspond to the minimum possible height and minimum possible depth consistent with the entire unmensurated region, as determined from the perspectives of the active and passive cameras.

11. Information Recovery

Once we have detected a missing region of the scene and have found the minimum values of height and depth for that NSC perspective, we can choose new perspectives to allow triangulation of points in that region.

We begin on the assumption that the unmensurated region is a hole. Since we know nothing about the hole, we assume it extends as a right cylinder in a direction perpendicular to the least squares plane of the hole boundary. We find the principal axis of the hole boundary (projected) on the least squares plane, and measure depth along the central axis which bisects the principal axis and parallels the cylinder of the hole. Since we do not know the depth of the hole, we must consider the tradeoffs in positioning the cameras:

1. If we position the cameras so that both are within the cylinder delineated by the hole boundary, we can look directly into the

hole. Unfortunately, triangulation becomes more ill-conditioned as the parallax between the cameras decreases. Moreover, the error is greatest in the direction of depth into the hole. Thus for every possible depth of the hole bottom, the parallax angle must exceed some minimum if triangulation is to provide a given number of significant figures.

2. To map the floor of the hole with several sample points, the passive image must resolve the bright dots illuminated by adjacent beams of the active camera. For laboratory situations, there will be little difference in resolution from object surface to hole bottom.

3. For a hole of a given depth, maximum parallax is obtained as follows: Consider a triangle whose base is the principal axis of the hole and whose third point lies on the central axis at the depth of the hole (Figure 5). Extend the sides of the triangle from the third point and let the cameras be within the angle of these lines and in the plane of the triangle. Position the cameras as near as possible to the sides of the extended triangle. Note that the triangle becomes more acute as the hole depth increases.

The strategy we choose is as follows:

S1: For each unmensurated region, begin by assuming it is a hole.

S2: Decide on the minimum depth for which repositioning cameras is worthwhile. In particular, if the hole opening is so narrow that only a small depth into the hole can be measured without losing significant figures in the triangulation calculation, moving the cameras may not be efficacious.

S3: To position both cameras within but at the edge of the hole cylinder, calculate the parallax of measurement to the hole bottom for each possible depth. Determine the maximum depth for which some desired number of significant figures can be obtained with the available parallax.

S4: To position both cameras to maximize parallax (Figure 5), calculate the maximum parallax for different possible depths (from the extended triangles of tradeoff (3) above). Again calculate the maximum depth for the hole consistent with significant figures in triangulation.

S5: If neither S3 nor S4 allow probing of the hole to any significant depth, the hole is too small; do not reposition the cameras. If S4 allows probing to a significant depth but S3 does not, then reposition the cameras in accord with S4. If S3 allows probing to a significant depth, reposition the cameras in accord with S3.

S6: If it should turn out that the mensuration at the new camera positions is poor and indicative of a mountain rather than a hole, then the camera positions should be shifted (in the plane of the central and principal axes of the region boundary) so that both cameras are first on one side of the unmensurated region and then on the other.

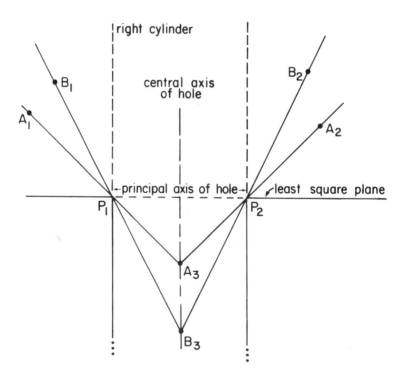

Figure 5: The cross-section of an unmensurated hole is used to illustrate how the maximum parallax decreases when we want to see deeper into the hole. Let P_1P_2 be the principal axis (thus the widest opening of the hole). To probe the hole to depth A_3 we can separate the cameras by a maximum parallax angle of $A_1A_3A_2$. To probe to depth B_3, the maximum parallax angle $B_1B_3B_2$ is less.

12. Fully Three-Dimensional Wrap-Around Surface Mapping

To reconstruct the surface of a three-dimensional object, we need images from multiple perspectives around the object.

12.1. Data Acquisition

The first step in reconstructing the complete surface of a 3-D object is to triangulate numerous surface sample points from each of several perspectives. The numerical stereo camera is capable of illuminating and mensurating up to 128×128 different regions (spots) of a surface at each perspective. For real-time 3-D data acquisition, several active and/or passive cameras may be used simultaneously at various perspectives around an object; the reflections from the different active cameras can be distinguished if each laser operates in a different spectral region. When real time is not required, a single setup can be used sequentially at different perspectives (for example, by placing the object on a turntable). Since object and camera repositioning need not require recalibration, cameras can be relocated to any perspective to achieve better measurements.

The problem then is to combine the mensurations made at each different perspective to achieve the overall surface map.

12.2. Representation of the Surface Data Points

Suppose, for an object or scene, that the surface points which have been triangulated from different perspectives or camera positions are given by

$$\{(x_{ip}, y_{ip}, z_{ip}) : i=1,2, \ldots ,I(p); p=1,2, \ldots ,P\}$$

where P is the number of camera perspectives used in the mensuration, and $I(p)$ is the number of surface points measured from perspective p. A vision preprocessor or robot eye should be able to specify the complete surface of the object or scene from these sample points. This involves two tasks: (A) the surface locations determined from the different perspectives must be consistent, and (B) the surface measurements must permit an accurate representation of the surface on the global scale. It is left for the vision-understanding component of the system to segment the 3-D information and identify the object or scene components.

As long as the camera positions and their changes from perspective to perspective are accurately known, and the parallax between the active and passive cameras of the NSC is sufficient to provide well-conditioned solutions, triangulation from the different perspectives will be consistent with one another. In fact, any inconsistency for a surface location measured with overlapping camera perspectives is itself evidence of either a systematic error in the determination of camera positions or insufficient parallax. Experimental work with the NSC indicates that task (A) has been achieved. We are therefore concerned primarily with task (B) of the robot eye.

To obtain a useful global representation of a surface, it is necessary to sample every 'significant structure' of the surface at least twice by bright dots (laser beam/surface intersections) from some active camera perspective. The passive camera is always assumed to sample a surface of interest more finely than the active camera. If some surface features are smaller than twice the spacing between sample points on the surface, then the reconstructed surface derived from the triangulated bright dots will be undersampled (aliased). An aliased representation of a 3-D surface not only loses the fine structures (that is, the high harmonics), but provides an incorrect global (low harmonic) map as well. Great care must be taken to ensure that surface mensuration is not done with aliased data.

A complication is that 'significant structure' is determined not only by the spatial harmonic spectrum of the object but also by the minimum information required by the vision-understanding component to recognize the object. If one needs to know only that an object of a certain size is present, then the 'significant structure' of the object can be obtained with only a few surface sample points. On the other hand, if one needs to distinguish one human face from another or to detect small shape abnormalities for anthropometry, then very high harmonic information, thus fine sampling, is needed to obtain the 'significant structure' of the surface. This complication requires feedback loops between the robot eye and vision-understanding modules of the system. In the case of the NSC, collecting high density sampling data is only logarithmically more expensive than collecting low density data. (In fact, the data collection time, even for high density data, is only about 1 second at TV camera scan rates. Since the solid state shutter switches in microseconds, data collection times of 1 ms are conceivable with faster and more sensitive cameras.) The processing of the data, however, is linear with the number of sample points when a sequential computer is used. Hence to apply the NSC in a 3-D vision system, the number of sample points to be processed for any application would be read from the vision- understanding component. Thus a single number for the sampling rate would feed back from the higher understanding level to the level of surface data acquisition. The higher level may also decide where the robot eye should look, that is, the boundaries of the scene, but once this is done, the 3-D data acquisition for the delineated scene is done at the preprocessor level. The modular integrity of the eye and the understanding components is thereby maintained.

Some surfaces of interest may have undersampled fine structure only over relatively small patches. For example, a small grating in a street may be undersampled, when the rest of the street is adequately sampled. A good global representation of a surface should not be unduly affected by small undersampled patches.

We are studying several approaches to the representation of a 3-D object from local position measurements. The first method uses voxels [6], the second uses the convex hull [29], [49], and the third uses surface patches [35], [31]. All of the methods are reasonably insensitive to small areas of undersampling. (See also [1], [48].)

12.2.1. Simple Voxel Approach

S1: For each surface region, choose the smallest dot spacing from among those active camera perspectives which sample the region.

S2: Choose the voxel size to be twice the largest of these smallest dot spacings of the different surface regions.

S3: Fill the three-dimensional space (which contains the object or scene of interest) with voxels of the chosen size.

S4: For each sample point (bright dot) on the surface, find the address of the voxel which contains it. The voxels which contain bright dots constitute the surface of the object or scene. The addresses of these voxels are the only numbers that need be stored to describe the surface.

S5: Because of the way we chose the voxel size there should be few gaps (empty voxels) in the object surface. Filled (surface) voxels are connected if they share a surface, edge, or point. Close any gaps in the mensurated surface which are at most a voxel thick.

12.2.2. Convex Hull Approach

S1: Find the convex hull of all the triangulated surface points. The convex hull is chosen as the initial approximation of the object's surface.

S2: For each point within the convex hull, effect a local connection to the hull to achieve a better approximation of the object's surface. Continue this procedure until all mensurated points are attached to the surface. Several methods have been proposed to do this based on various optimization goals such as minimum area [29], or minimizing a cost function of various local geometric properties [49].

12.2.3. Bicubic Spline Approach

S1: Construct a separate surface patch for each perspective of the active and passive cameras.

S2: If the edges of every patch overlap with edges of other patches so that an extended path from a patch always lies within some patch, we can construct a global representation of the surface of the 3-D object.

This method is used in aerial photogrammetry, and in mapping moons from interplanetary fly-bys [25].

13. Appendix: Singular Value Decomposition

To solve least squares problems with a minimum loss of significant figures, the method of singular value decomposition has been used throughout this work. Here we review briefly the salient ideas of the method.

The problem is to solve the linear system of equations

$$A\,x = b \tag{A1}$$

where A is an $m \times n$ real matrix with $m \geq n$, the vector x is an unknown n-vector, and b is a given m-vector. The classical method of solution is to form and solve the normal equations

$$A^t\,A\,x = A^t\,b \tag{A2}$$

where $A^t\,A$ is an $n \times n$ symmetric matrix.

To solve this equation, one can apply the well-known diagonalization algorithms such as the Gaussian elimination method and the QR decomposition. There is, however, a serious fundamental disadvantage in using the normal equations. The computation of $A^t\,A$ involves excessive and unnecessary loss of significant figures, thus numerical inaccuracy, because of its ill-conditioned nature.

The most reliable method for computing the solutions of least squares problems is based on a matrix factorization known as the singular value decomposition (SVD). Although there are other methods which require less computer time and storage, they are less effective in dealing with errors and linear dependence [15].

The SVD is based on the following theorem. Any $m \times n$ matrix A with $m \geq n$ can be factored into the form

$$A = U D V^t \tag{A3}$$

where

$$U^t U = V^t V = V V^t = I \ \text{ and } \ D = diag(d_1, \ldots, d_n). \tag{A4}$$

The diagonal elements of D are the non-negative square roots of the eigenvalues of $A^t A$; they are called 'singular values'.

The SVD is a little more complicated factorization than that given by Gaussian elimination or by the QR decomposition but it has a number of advantages. The idea of the SVD is that by proper choice of orthogonal matrices U and V, it is possible to make D diagonal with non-negative diagonal entries.   The orthogonal matrices have highly desirable computational properties; they do not magnify errors because their norm is unity.

In the least squares problem for the solution of $Ax=b$, we require the vector x for which $norm(b-Ax)$ is a minimum. Then

$$norm(b-A \ x) = norm(U^t(b-A \ x)) = norm(U^t \ b - D \ V^t \ x)$$

$$= norm(c - D \ y) \tag{A5}$$

where

$$c = U^t b \ \text{ and } \ y = V^t \ x \ . \tag{A6}$$

Since D is a diagonal matrix with elements d_i ,

$$norm(c - D \ y) = \sum_i (c_i - d_i y_i)^2 + \sum_i c_i^2 \ . \tag{A7}$$

Therefore the minimal solution is

$$y_i = c_i/d_i \ \text{ for } \ i = 1, \ldots, n; \ \ y_i = 0 \ \text{ for } \ i = n+1, \ldots, m \ . \tag{A8}$$

The most fundamental question in the least squares problem is the reliability of the computed solution x. In other words, how can we measure the sensitivity of x to changes in A and b? The answer can be obtained by the idea of matrix singularity. If matrix A is nearly singular, we can expect small changes in A and b to cause very large changes in x. Since the singular values d_i of A are sensitive to perturbations in the values of A, we can determine the measure of nearness to singularity by defining the 'condition number' of A to be

$$cond(A) = d_{max}/d_{min} \tag{A9}$$

where

$$d_{max} = max\{d_i\} \text{ and } d_{min} = min\{d_i\} . \tag{A10}$$

That is, the condition number of a matrix is defined to be the ratio of its largest and smallest singular values.

If $cond(A) = 1$, then the columns of A are perfectly independent. If $cond(A)$ is very large, then the columns of A are nearly dependent and the matrix A is close to singular. The condition number can also be interpreted as a relative error magnification factor in computation.

References

[1]   Aggarwal, J. K., Davis, L. S., Martin, W. N., and Roach, J. W.
        Survey: Representation methods for three-dimensional objects.
        In L. N. Kanal and A. Rosenfeld (editor), *Progress in Pattern
            Recognition*, pages 377-391. North-Holland Pub., Amsterdam,
            1981.

[2]   Agin, G. J. and Binford, T. O.
        Computer description of curved objects.
        In *Proceedings of the Third International Joint Conference on
            Artificial Intelligence*, pages 629-640. August, 1973.
        Also appeared in IEEE Transactions on Computers, 1976, Vol. C-25,
            pages 439-449.

[3]   Altschuler, M. D., Altschuler, B. R., and Taboada, J.
        Laser electro-optic system for rapid three-dimensional topographic
            mapping of surfaces.
        Optical Engineering 20:953-961, 1981.

[4]   Altschuler, M. D., Posdamer, J. L., Frieder, G., Altschuler, B. R., and
        Taboada, J.
        The numerical stereo camera.
        In B. R. Altschuler (editor), *3-D Machine Perception*, pages 15-24.
            SPIE, 1981.

[5]   Altschuler, M. D., Posdamer, J. L., Frieder, G., Manthey, M. J.,
        Altschuler, B. R., and Taboada, J.
        A medium-range vision aid for the blind.
        In *Proc. of the Int. Conf. on Cybernetics and Society, IEEE*, pages
            1000-1002. 1980.

[6]   Artzy, E., Frieder, G., and Herman, G. T.
        The theory, design, implementation, and evaluation of a three-
            dimensional surface detection algorithm.
        Computer Graphics and Image Processing 15:1-24, 1981.

[7]   Ballard, D. H. and Brown, C. M.
        Computer Vision.
        Prentice-Hall, NJ, 1982.

[8]   Bleha, W. P., Lipton, L. T., Wiener-Avnear, E., Grinberg, J., Reif,
        P. G., Casasent, D., Brown, H. B., and Markevitch, B. V.
        Application of the liquid crystal light valve to real-time optical data
            processing.
        Optical Engineering 17:371-384, 1978.

[9]   Cornelius, J. Gheluwe, B. V., Nyssen, M., and DenBerghe, F. V.
        A photographic method for the 3-D reconstruction of the human
            thorax.
        In *Applications of Human Biostereometrics*, pages 294-300. SPIE,
            1978.

[10]   Cutchen, J. T., Harris, Jr., J. O., and Laguna, G. R.
        PLZT electro-optic shutters: Applications.
        Applied Optics 14:1866-1873, 1975.

[11]   Dijak, J. T.
        *Precise three-dimensional calibration of numerical stereo camera
            systems for fixed and rotatable scenes.*
        Technical Report AFWAL-TR-84-1105, Air Force Wright
            Aeronautical Labs. (AAAT-3), September, 1984.

[12] Dijak, J. T.
System software for the numerical stereo camera.
Technical Report AFWAL-TR-85-1078, Air Force Wright
Aeronautical Labs. (AAAT-3), 1985.

[13] Dodd, G. G. and Rossol, L. (eds.).
Computer Vision and Sensor-Based Robots.
Plenum Press, New York, 1979.

[14] Duda, R. O. and Hart, P. E.
Pattern Classification and Scene Analysis.
Wiley, New York, 1973.

[15] Forsythe, G. E., Malcolm, M., and Moler, G. B.
Computer Methods for Mathematical Computations.
Prentice-Hall, 1977.

[16] Gennery, D. B.
A stereo vision system for an autonomous vehicle.
In *5th Int. Joint Conf. on AI*, pages 576-582. 1977.

[17] Helmering, R. J.
A general sequential algorithm for photogrammetric on-line
processing.
Photogrammetric Engineering and Remote Sensing 43:469-474,
1977.

[18] Horn, B.K.P.
Obtaining shape from shading information.
In P.H. Winston (editor), *The Psychology of Computer Vision*,
pages 115-156. Mc Graw-Hill, New York, 1975.

[19] Hugg, J. E.
A portable dual camera system for biostereometrics.
In *Biostereometrics '74*, pages 120-127. American Society of
Photogrammetry, Falls Church, VA, 1974.

[20] Kanade, T. and Asada, H.
Noncontact visual three-dimensional ranging devices.
In B. R. Altschuler (editor), *3-D Machine Perception*, pages 48-53.
SPIE, 1981.

[21]   Karara, H. M., Carbonnell, M., Faig, W., Ghosh, S. K., Herron, R. E.,
         Kratky, V., Mikhail, E. M., Moffitt, F. H., Takasaki, H., Veress, S. A.
         Non-topographic photogrammetry.
         Manual of Photogrammetry.
         Amer. Soc. of Photogrammetry, Falls Church, VA, 1980, pages
             785-882.

[22]   Land, C. E.
         Optical information storage and spatial light modulation in PLZT
             ceramics.
         Optical Engineering 17:317-326, 1978.

[23]   LaPrade, G. L., Briggs, S. J., Farrel, R. J., Leonardo, E. S.
         Stereoscopy.
         Manual of Photogrammetry.
         Amer. Soc. of Photogrammetry, Falls Church, VA, 1980, pages
             519-544.

[24]   Lewis, R. A. and Johnston, A. R.
         A scanning laser rangefinder for a robotic vehicle.
         Proc. 5th Int. Joint Conf. AI :762-768, 1977.

[25]   Light, D. L., Brown, D., Colvocoresses, A. P., Doyle, F. J., Davies, M.,
         Ellasal, A., Junkins, J. L., Manent, J. R., McKenney, A., Undrejka,
         R., and Wood, G.
         Satellite photogrammetry.
         Manual of Photogrammetry.
         Amer. Soc. of Photogrammetry, Falls Church, VA, 1980, pages
             883-977.

[26]   Mero, L. and Vamos, T.
         Medium level vision.
         Progress in Pattern Recognition.
         North-Holland Publ., Amsterdam, 1981, pages 93-122.

[27]   Montgomery, W. D.
         Sampling in imaging systems.
         Journal of the Optical Society 65:700-706, 1975.

[28]   Nevatia, R.
         Machine Perception.
         Prentice-Hall, NJ, 1982.

[29]   O'Rourke, J.
         Polyhedra of minimal area as 3-D object models.
         Proc. Int. Joint Conf. on AI :664-666, 1981.

[30]   Oliver, D. S.
       Real-time spatial modulators for optical/digital processing systems.
       Optical Engineering 17:288-294, 1978.

[31]   Pavlidis, T.
       Algorithms for Graphics and Image Processing.
       Computer Science Press, Rockville, MD, 1982.

[32]   Popplestone, R. J., Brown, C. M., Ambler, A. P., and Crawford, G. F.
       Forming models of plane-and-cylinder faceted bodies from light
           stripes.
       In *Proc. 4th Int. Joint Conf. AI*, pages 664-668. Sept., 1975.

[33]   Posdamer, J. L. and Altschuler, M. D.
       Surface measurement by space-encoded projected beam systems.
       Computer Graphics and Image Processing 18:1-17, 1982.

[34]   Rocker, F. and Kiessling, A.
       Methods for analyzing three dimensional scenes.
       Proc. 4th IJCAI :669-673, September, 1975.

[35]   Rogers, D. F. and Adams, J. A.
       Mathematical Elements for Computer Graphics.
       McGraw-Hill, 1976.

[36]   Ryan, T. W. and Hunt, B. R.
       Recognition of stereo-image cross-correlation errors.
       Progress in Pattern Recognition.
       North-Holland Publ., Amsterdam, 1981, pages 265-322.

[37]   Shapira, R. and Freeman, H.
       Reconstruction of curved-surface bodies from a set of imperfect
           projections.
       5th Int. Joint Conf. on AI :628-634, 1977.

[38]   Shirai, Y. and Suwa, M.
       Recognition of polyhedrons with a range finder.
       In *Proceedings of the Second International Joint Conference on
           Artificial Intelligence*, pages 80-87. 1971.

[39]   Shirai, Y.
       Three-dimensional computer vision.
       Computer Vision and Sensor-Based Robots.
       Plenum Press, NY, 1979, pages 187-205.

[40]   Shirai, Y. and Tsuji, S.
       Extraction of the line drawings of 3-dimensional objects by sequential
           illumination from several directions.
       2nd Int. Joint Conf. on AI :71-79, 1971.

[41]   Slama, C. C. (ed.).
       Manual of Photogrammetry
       4th edition, American Soc. of Photogrammetry, 1980.

[42]   Sobel, I.
       On calibrating computer controlled cameras for perceiving 3-D
           scenes.
       3rd Int. Joint Conf. on AI :648-657, 1973.

[43]   Strand, T. C.
       Optical three-dimensional sensing for machine vision.
       Optical Engineering 24:33-40, 1985.

[44]   Sutherland, I. E.
       Three-dimensional input by tablet.
       Proc. IEEE 62:453-461, 1974.

[45]   Tamburino, L. A.
       *Complementary pair detection algorithm and improved noise
           immunity for numerical stereo camera systems.*
       Technical Report AFWAL-TR-83-1117, Air Force Wright
           Aeronautical Labs., Wright-Patterson AFB (AAAT-3), August,
           1983.

[46]   Tamburino, L. A. and Dijak, J. T.
       Error correction in the numerical stereo camera system.
       Technical Report AFWAL-TR-85-1077, Air Force Wright
           Aeronautical Labs., Wright-Patterson AFB (AAAT-3), 1985.

[47]   Thomason, M. G. and Gonzalez, R. C.
       Database representations in hierarchical scene analysis.
       Progress in Pattern Recognition.
       North-Holland Publ., Amsterdam, 1981, pages 57-91.

[48]   Tomita, F. and Kanade, T.
       A 3-D vision system: Generating and matching shape description.
       First Conf. on AI Applications, IEEE :186-191, 1984.

[49]   Uselton, S. P.
       Surface reconstruction from limited information.
       PhD thesis, Univ. of Texas at Dallas, 1981.

[50]   Will, P. M. and Pennington, K. S.
       Grid coding: A preprocessing technique for robot and machine vision.
       2nd Int. Joint Conf. on AI 2:319-329, **1971**.

[51]   Woodham, R. J.
       A cooperative algorithm for determining surface orientation from a
           single view.
       5th Int. Joint Conf. on AI :635-641, **1977**.

[52]   Yang, H. S., Boyer, K. L., and Kak, A. C.
       Range data extraction and interpretation by structured light.
       First Conf. on AI Applications, IEEE :199-205, **1984**.

[53]   Young, J. M. and Altschuler, B. R.
       Topographic mapping of oral structures -- problems and applications
           in prosthodontics.
       In B. R. Altschuler (editor), *3-D Machine Perception*, pages **70-77**.
           SPIE, **1981**.

A Noncontact Optical Proximity Sensor for Measuring Surface Shape

Takeo Kanade and Michael Fuhrman

Computer Science Dept. and Robotics Institute
Carnegie-Mellon University
Pittsburgh, Pa. 15213

and

Equipment Development Division
Alcoa Laboratories
Alcoa Center, PA 15069

Abstract

We have developed a noncontact multi-light source optical proximity sensor that can measure the distance, orientation, and curvature of a surface. Beams of light are sequentially focused from light emitting diodes onto a target surface. An analog light sensor - a planar PIN diode - localizes the position of the resultant light spot in the field of view of the sensor. The 3-D locations of the light spots are then computed by triangulation. The distance, orientation, and curvature of the target surface is computed by fitting a surface to a set of data points on the surface.

The proximity sensor that has been built uses 18 light sources arranged in 5 conical rings. The sensor has a range of 10 cm where it can measure the distance of approximately 200 discrete points per second with a precision of 0.1mm, and then compute surface orientation with a precision of 1.0°. This sensor may be used by a robotic manipulator to home in on an object and to trace the object's surface.

1. Introduction

Noncontact proximity sensors that can measure the range and shape of a surface have useful robotic applications such as surface inspection, seam or edge tracking, obstacle avoidance, and homing a manipulator to an object. The simplest devices are optical switches that detect the presence of a nearby surface [3], [11]. In these devices, a beam of light originating from a light source on the sensor illuminates a surface. The closer the surface is to the sensor the greater the intensity of the light reaching the sensor. By assembling the light emitter and the detector on small moving platforms the range of this device can be continuously altered [14]. However, this technique for measuring distance is sensitive to the orientation and reflectivity of a target surface as well as the distance of a surface. One method that was developed to overcome this limitation uses a single detector and several light sources. Distance is measured by comparing the relative intensities of the light reflected from a surface [13].

When light detectors sensitive to light spot position are used in a proximity sensor, distance can be measured by triangulation. The measurement then does not depend on the amount of reflected light reaching the sensor. Linear or planar CCD arrays [5], and linear or planar PIN diodes [1], [2], [6], [9], [8], [12] have been incorporated into proximity sensors. For example, in one application, a mirror deflects a beam of light into the field of view of a sensor. The position of the light spot is detected by a planar PIN diode and then correlated with the mirror position to compute distance and eventually the shape of a weld seam [2].

The proximity sensor that we have developed is configured around a planar PIN diode. Multiple conical arrays of light sources focus beams of light into the center of the field of view of the sensor where the distortion and nonlinearity of the sensor chip are smallest. The proximity sensor uses 18 light sources configured into 5 conical arrays. The sensor has a range of 10 cm where it can measure the distance of approximately 200 discrete points per second with a precision of 0.1 mm, and then compute surface orientation with a precision of 1 °.

This paper focuses on the design and performance of the new proximity sensor. How well this type of sensor can measure the orientation of a target surface depends on the number of projected light spots, and how they are distributed. Hence, the new sensor was designed with the aid of a statistical argument that takes into account the geometry of the sensor to minimize measurement uncertainty. The performance of the sensor was tested by measuring the distance and shape of various target surfaces.

2. Measurement of Distance and Shape

2.1. Overview

The proximity sensor is based on the principle of active illumination and triangulation. Figure 1 shows the basic configuration of the proximity The sensor head in this drawing consists of a ring of light sources, a lens, and a light spot position sensor chip.

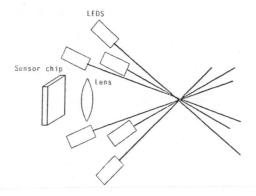

Figure 1: Configuration of the Multi-Light Source Proximity Sensor

The principle of operation is as follows. A beam of light emitted from the sensor head is interrupted by a target surface. The resultant light spot on the target surface is focused by the lens onto the analog light sensor that is sensitive to the intensity and position of a light spot in its field of view. Since the direction of the incident beam of light is known, and the light sensor measures the location of the light spot in the field of view, the three dimensional (3-D) coordinates of the light spot on the target surface can be computed by triangulation. The proximity sensor has multiple light sources, and as a result the coordinates of several discrete points on the surface can be measured. The average distance, orientation, and curvature of the target surface are computed by fitting a plane and then a quadric surface to this set of points.

Three features of the analog sensor chip make it an attractive device for use in the proximity sensor: its position linearity, its intrinsic ability to measure the centroid of a light spot on its surface, and its speed of response. The chip, a planar PIN diode, measures both the intensity and the position of the light spot on its surface. If the spot of light on a target surface is distorted because of the curvature or orientation of the surface, the sensor chip responds to the stimulus by measuring the coordinates of the centroid of the image of the light spot. To minimize error when computing the coordinates of the light spot, the spot size must be kept small. Therefore, light emitting diodes (LEDs) are used that have a light source diameter of about 150 μm.

2.2. Measuring the Distance of a Surface

The coordinates of a light spot on a target surface are computed by triangulation. The trajectory of each light beam that illuminates a surface is fixed and is measured in advance. The line of sight to the light spot on the target surface is obtained from the sensor chip. The calculation of the coordinates of the light spot is based on a simple optical model: first, the light source beam is a line, and thus the projected light spot is a point; second, the camera lens forms a flat, undistorted image; and finally the sensor chip is linear in finding the light spot position and is insensitive to defocus. Since the sensor is subject to small errors because of deviations from the simple model, they are treated as perturbations from the model and error corrections are calculated beforehand.

The position and orientation of a line in 3-D space can be defined by the

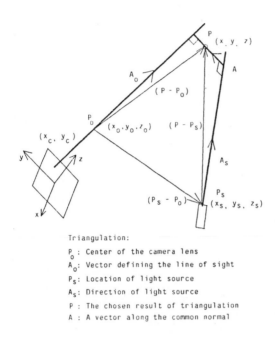

Triangulation:

P_o : Center of the camera lens
A_o: Vector defining the line of sight
P_s: Location of light source
A_s: Direction of light source
P : The chosen result of triangulation
A : A vector along the common normal

Figure 2:   Calculation of Point Coordinates by Triangulation

coordinates of a point on the line and the direction cosines of the line. In Figure 2 light beam trajectory is defined by the light source origin $P_s = (x_s, y_s, z_s)$ and the light beam direction vector $A_s = (a_s, b_s, c_s)$. The line of sight has an orientation $A_o = (a_o, b_o, c_o)$ that is defined by the camera lens center $P_o = (x_o, y_o, z_o)$ and the centroid of the spot of light $P_c = (x_c, y_c)$ on the sensor chip. Let us denote the coordinate of the spot position we want to locate as $P = (x, y, z)$. In principle, the two three-dimensional lines (the light source beam and the line of sight) should intersect at the location of the spot on the surface. In practice, however, the two spatial lines thus obtained only come very close due to noise and the simplified optical model being used. We therefore determine the spot location P to be located along the line of closest approach (i.e., along the common normal) between the two skew lines. A set of equations from which the coordinates of the light spot can be calculated is found by observing that the four vectors consisting of the baseline from the center of the camera lens to the light source $(\mathbf{P}_s - \mathbf{P}_o)$, the light beam from the light source to the common normal $(A_s\mathbf{A}_s)$, the common normal AA, and the line of sight from the common normal to the center of the camera lens $(-A_o\mathbf{A}_o)$ sum to zero, that is,

$$(\mathbf{P}_s - \mathbf{P}_o) + (A_s\mathbf{A}_s) + AA + (-A_o\mathbf{A}_o) = 0. \tag{1}$$

Let the baseline have a length B and a direction specified by the unit vector $\mathbf{B} = (\mathbf{P}_s - \mathbf{P}_o)/|\mathbf{P}_s - \mathbf{P}_o|$. Then Equation (1) can be rewritten as

$$B\mathbf{B} + A_s\mathbf{A}_s + A\mathbf{A} - A_o\mathbf{A}_o = 0 \tag{2}$$

Taking the scalar product of Equation (2) first with the direction vector of the line of sight $\mathbf{A}_o$ and then with the direction vector of the light beam $\mathbf{A}_s$ leads to a set of two simultaneous equations. Noting that $\mathbf{A}_s \cdot \mathbf{A} = 0$ and $\mathbf{A}_o \cdot \mathbf{A} = 0$ yields

$$A_s\mathbf{A}_o \cdot \mathbf{A}_s - A_o = -B\mathbf{A}_o \cdot \mathbf{B} \tag{3}$$

$$A_s - A_o\mathbf{A}_s \cdot \mathbf{A}_o = -B\mathbf{A}_s \cdot \mathbf{B} \tag{4}$$

After solving for A_s and A_o, the coordinates of the light spot can be chosen to be on the light beam or on the line of sight, or at any point along the common normal. Since the output of the sensor chip is subject to noise, and the position and orientation of each beam of light are stable quantities, we choose the computed coordinates of a light spot to lie on the light beam.

2.3. Surface Fitting

To estimate the position and orientation of a surface, a plane is fit to the set of three dimensional data points obtained by the proximity sensor. To estimate the curvature of a surface, a second order equation is fit. The accuracy of the calculation of position, orientation and curvature depends, therefore, on the accuracy of the individual data points, and the number and distribution of the data points on the surface. For example, if the points are too closely spaced, the orientation of the plane that is fit to the points will be very uncertain. These relationships were analyzed, and the results were used to design a sensor whose expected uncertainties are less than the design specifications.

Suppose there are n measured points on a surface, where each point (x_i, y_i, z_i) has an estimated total uncertainty in position of σ_i. A surface can be fit to this set of points using the method of least squares to determine the coefficients a_j in the equation

$$z(x,y) = \Sigma a_j X_j(x,y) \tag{5}$$

where $X_j(x,y)$ is a polynomial of the form $\Sigma c_{qr} x^q y^r$, as for example, when fitting a plane, $q + r \le 1$, and when fitting a quadric surface, $q + r \le 2$. We are concerned with how the uncertainty in the measured points and their distribution on the surface affects the uncertainty of the computed coefficients.

The values of the coefficients a_j are chosen so that they minimize the aggregate error E:

$$E = \sum_{i=1}^{n} \{ \frac{1}{\sigma_i^2}(z_i - z(x_i,y_i))^2 \} = \sum_{i=1}^{n} \{ \frac{1}{\sigma_i^2}(z_i - \Sigma a_j X_j(x_i,y_i))^2 \} \tag{6}$$

By minimizing E, each a_j can be computed by:

$$a_j = \sum_k \epsilon_{jk}\beta_k \tag{7}$$

where

$$[\epsilon_{jk}] = [\alpha_{jk}]^{-1} \tag{8}$$

$$\alpha_{jk} = \sum_{i=1}^{n} \{\frac{1}{\sigma_i^2}X_j(x_i,y_i)X_k(x_i,y_i)\} = \frac{1}{2}\frac{\partial^2 E}{\partial a_j \partial a_k} \tag{9}$$

$$\beta_k = \sum_{i=1}^{n} \{\frac{1}{\sigma_i^2}z_iX_k(x_i,y_i)\} \tag{10}$$

The symmetric matrix $[\alpha_{jk}]$ is known as the curvature matrix because of its relationship to the curvature of E in the coefficient space [4].

We can estimate the uncertainties of the coefficients, and determine their dependence on the geometry of the proximity sensor if we assume that fluctuations in the coordinates of the individual data points are uncorrelated, and assign the total uncertainty in the measurement of position to the variable z. The uncertainty of the coefficient a_j is then

$$\sigma_{a_j}^2 = \sum_{i=1}^{n} \sigma_i^2 \left.\frac{\partial a_j}{\partial z_i}\right.^2 \tag{11}$$

By calculating the partial derivatives of Equation (7), it follows that the uncertainties of the coefficients are proportional to the diagonal elements of the inverse of the curvature matrix [4], that is

$$\sigma_{a_j}^2 = \epsilon_{jj} \quad . \tag{12}$$

Let us first examine the case where the plane

$$z = Gx + Hy + I \qquad (13)$$

is fit to a set of data points. Here we have set $a_1 = G$, $a_2 = H$, $a_3 = I$, $X_1 = x$, $X_2 = y$, and $X_3 = 1$ in Equation (5). The local surface normal is $N = (G, H, -1)$. Since all of the measured points on a surface are at about the same distance from the sensor in any one measurement cycle, the total uncertainty σ_i can be taken as a constant σ for all of the data points. More accurately, since the uncertainty in the location of a light spot depends on where it is in the field of view and the intensity of the measured light, σ is considered an upper limit on the uncertainty in the location of a light spot. The curvature matrix for this case is

$$\alpha = \frac{1}{\sigma^2} \begin{bmatrix} \Sigma x_i^2 & \Sigma x_i y_i & \Sigma x_i \\ \Sigma x_i y_i & \Sigma y_i^2 & \Sigma y_i \\ \Sigma x_i & \Sigma y_i & n \end{bmatrix} \qquad (14)$$

In many applications the proximity sensor will be used to maintain an orientation normal to a surface. The light sources can be arranged symmetrically so that the resulting spots of light on a flat surface facing the sensor perpendicularly satisfy the condition

$$\Sigma x_i = \Sigma y_i = \Sigma x_i y_i = 0 \qquad (15)$$

that is, the α matrix is diagonal. The uncertainties in the coefficients of Equation (13) are then simply

$$\sigma_G = \frac{\sigma}{\sqrt{\Sigma x_i^2}} \tag{16}$$

$$\sigma_H = \frac{\sigma}{\sqrt{\Sigma y_i^2}} \tag{17}$$

$$\sigma_I = \frac{\sigma}{\sqrt{n}} \tag{18}$$

The angle between the optical axis of the proximity sensor and the surface normal is $\phi = \tan^{-1}\sqrt{G^2+H^2}$. The uncertainty in the angular orientation of a surface perpendicular to the sensor can be estimated by

$$\sigma_\phi^2 \leq \sigma_G^2 + \sigma_H^2. \tag{19}$$

Equations (16) through (19) relate the uncertainty in the measured position and orientation of a surface with the number of data points, the uncertainty in their measurement, and their distribution on the surface.

2.4. Normal Curvature

To compute the curvature of a surface and to compute a more accurate value for the distance of a curved target surface from the proximity sensor, the second order equation

$$z = Ax^2 + Bxy + Cy^2 + Dx + Ey + F \tag{20}$$

of a surface is fit to the set of data points. The curvature of the target surface can be computed at any point from the coefficients of this equation. In particular, the normal curvature, κ_N, and the principal normal curvatures, κ_1 and κ_2, can be computed where the surface intersects the axis of the sensor.

Figure 3 shows the proximity sensor over a target surface. The projected

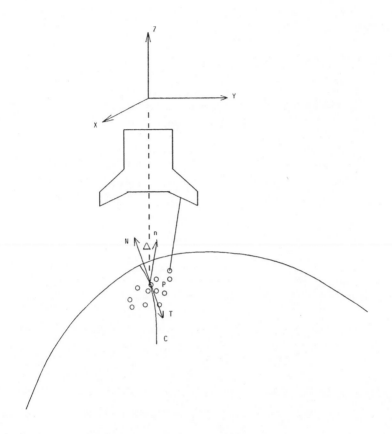

Figure 3: Measuring the Curvature of a Surface

light spots are shown as small circles on the surface. The intersection of the axis of the sensor with the target surface is located at P. The curve C on the surface is the intersection of the target surface with a plane (not shown) containing the axis of the sensor. The vector **n** is the principal normal of the curve C at P; it lies in the intersecting plane. The vector **T** is tangent to the surface, and to the curve C at P. The vector **N** is normal to the surface at P. The angle between the vectors **n** and **N** is Δ.

The equation of the fitted surface can be expressed as

$$R = xi + yj + (Ax^2 + Bxy + Cy^2 + Dx + Ey + F)k. \tag{21}$$

The surface normal, N, is given by

$$N = \frac{R_x \times R_y}{|R_x \times R_y|} = \frac{((2Ax+By+D), \, -(Bx+2Cy+E), \, 1)}{\sqrt{(2Ax+By+D)^2 + (Bx+2Cy+E)^2 + 1}}, \tag{22}$$

where the subscripts x, y denote partial derivatives. The vector, T, tangent to the surface, and to the curve C at P, is given by

$$T = \frac{dR}{ds}, \tag{23}$$

where $ds = |dR|$. The curvature vector, K, whose magnitude is the curvature of C at P is then given by

$$K = \frac{dT}{ds}. \tag{24}$$

The normal curvature vector, K_N, is the projection of K onto the surface normal N. The component of K_N in the direction of N, κ_N, is called the normal curvature of C at P:

$$K_N = (K \cdot N)N = \kappa_N N. \tag{25}$$

The magnitude of K_N is the curvature of a curve on the surface, in the plane of N, with tangent vector T at P.

To compute the curvature of a surface, one first computes the first and second fundamental coefficients of the surface [10]. The first fundamental coefficients of the surface are

$$\mathcal{E} = \mathbf{R}_x{}^2 = 1 + (2Ax + By + D)^2 , \tag{26}$$

$$\mathcal{F} = \mathbf{R}_x \cdot \mathbf{R}_y = (2Ax + By + D)(Bx + 2Cy + E) , \tag{27}$$

$$\mathcal{G} = \mathbf{R}_y{}^2 = 1 + (Bx + 2Cy + E)^2 . \tag{28}$$

The second fundamental coefficients of the surface are

$$\mathcal{L} = \mathbf{R}_{xx} \cdot \mathbf{N} = 2A , \tag{29}$$

$$\mathcal{M} = \mathbf{R}_{xy} \cdot \mathbf{N} = B , \tag{30}$$

$$\mathcal{N} = \mathbf{R}_{yy} \cdot \mathbf{N} = 2C . \tag{31}$$

In terms of these coefficients, the normal curvature of the surface is given by

$$\kappa_N = \frac{\mathcal{L} \, dx^2 + 2\mathcal{M} \, dx \, dy + \mathcal{N} \, dy^2}{\mathcal{E} \, dx^2 + 2\mathcal{F} \, dx \, dy + \mathcal{G} \, dy^2} , \tag{32}$$

where the ratio $dx{:}dy$ specifies the direction of a line tangent to the surface at point P. In terms of the coefficients of the second order surface defined by Equation (20), the normal curvature at $x = y = 0$ is given by

$$\kappa_N = \frac{1}{\sqrt{1 + E^2 + D^2}} \frac{2[A\cos^2\theta + B\sin\theta\cos\theta + C\sin^2\theta]}{[(1 + D^2)\cos^2\theta + 2DE\sin\theta\cos\theta + (1 + E^2)\sin^2\theta]} . \tag{33}$$

Here, θ defines the orientation of the intersecting plane with respect to the x–y axis.

The principal normal curvatures of the target surface where the axis of the sensor intersects the surface, κ_1 and κ_2, are the extreme values of Equation (32). In terms of the fundamental coefficients of the surface, κ_1 and κ_2 are the roots of the equation

$$(\mathcal{E}\mathcal{G}-\mathcal{F}^2)\kappa^2-(\mathcal{E}\mathcal{N}-2\mathcal{F}M+\mathcal{G}L)\kappa+L\mathcal{N}-M^2 = 0 , \tag{34}$$

The directions of the principal axes are given by the roots of the equation

$$(\mathcal{E}M-\mathcal{F}L)\tan{}^2\theta+(\mathcal{E}\mathcal{N}-\mathcal{G}L)\tan\theta+\mathcal{F}\mathcal{N}-\mathcal{G}M = 0 . \tag{35}$$

The principal curvature, κ_n, of the curve C at P is given by

$$\kappa_n = \frac{\kappa_N}{\cos\varDelta} = \kappa_N \frac{\sqrt{1+D^2+E^2}}{\sqrt{1+D^2}} \tag{36}$$

where κ_N is computed in a direction tangent to C at P, and $\varDelta$ is the angle between N and n.

3. Measurement Error

The computed coordinates of a light spot in the field of view of the sensor is subject to error. The affects of various sources of error were taken into account in designing the sensor. Features of the sensor include: the beams of light from the sensor are focused into the center of the field of view where non-linearity and distortion are smallest; the light sources are modulated and the resultant signals from the sensor chip are synchronously detected; the resultant light spots on a target surface are small when compared to the area of the field of view of the sensor; and a means for adjusting the light source intensity has been provided so that the measured light intensity is within a small range.

One of the features of the sensor chip is it intrinsically measures the center of gravity of a light spot on its surface. If the shape of a target surface doesn't vary rapidly, the location of the light spot in the field of view of the sensor can be measured accurately. However, if the surface is concave, light can be reflected about the surface. The center of gravity of the reflected light in the field of view will not coincide with the center of gravity of the projected light spot.

3.1. Distortion

To measure and then compensate for the distortion of the lens and the nonlinearity of the sensor chip, a square array of light spots was generated in the field of view of the sensor. The coordinates output by the sensor chip were then compared with the results computed according to the geometrical model of the sensor. A two dimensional transformation, $\{x_c, y_c\} \rightarrow \{x_c', y_c'\}$, maps the measured sensor chip coordinates into their expected values [7]. This transformation is applied to measured sensor chip coordinates each time a light source is turned on when the proximity sensor is in use.

3.2. Light Source Trajectories

The sensor chip was used to measure the trajectories of the beams of light that are emitted from the sensor head. To do this, a flat surface was mounted perpendicular to the optical axis of the sensor head. The z coordinate of a spot of light on the surface is simply the z position of the surface. As each light source was turned on, the sensor chip coordinates were measured and the x and y coordinates of each light spot were computed. By moving the plane along the z axis of the sensor, a set of points $\{x, y, z\}$ for each light source was acquired.

For each light source, a line was fit through a subset of the acquired data points, $\{x, y, z\}$, near the center of the field of view where distortion and nonlinearity are smallest. These lines were chosen as best estimates of the light beam trajectories. The intersection of a line with the target plane yields the coordinates of a set of points $\{\bar{x}, \bar{y}, \bar{z}\}$. These points are considered to be the best estimate of points along the trajectory of a light beam.

3.3. Error Correction

Just as the set of points $\{\bar{x}, \bar{y}, \bar{z}\}$ were computed, now a set of points $\{x, y, z\}$ are computed for each LED by triangulation. The error is taken to be the difference between the coordinates $\{x, y, z\}$ computed by triangulation and the coordinates $\{\bar{x}, \bar{y}, \bar{z}\}$ computed according to the geometrical model.

Assuming that $(\bar{x}, \bar{y}, \bar{z})$ is the correct location of a light spot, let us set

$$\bar{x} = x + \epsilon_x \,, \tag{37}$$

$$\bar{y} = y + \epsilon_y \,, \tag{38}$$

$$\bar{z} = z + \epsilon_z \,. \tag{39}$$

It would be difficult to model the effect of the departures from the simplified model precisely to determine the terms ϵ_x, ϵ_y, and ϵ_z. However, since the errors are small and usually repeatable with respect to the measured spot position (x_c, y_c) on the sensor chip, we model them as a whole by means of polynomials, for example, by third order correction polynomials of the form

$$\epsilon_t = f_t(x_c, y_c) = a_{t1}x_c^{\,3} + a_{t2}x_c^{\,2}y_c + a_{t3}x_c y_c^{\,2} + a_{t4}y_c^{\,3} \tag{40}$$

$$+ a_{t5}x_c^{\,2} + a_{t6}x_c y_c + a_{t7}y_c^{\,2}$$

$$+ a_{t8}x_c + a_{t9}y_c + a_{t10}$$

where

$$t = x, y, z \,.$$

When the proximity sensor is calibrated, these polynomials are designed for each LED using a calibrated data set. During operation, the values

ϵ_x, ϵ_y, and ϵ_z computed by Equations (40) can be added to the spatial coordinates (x,y,z) computed by triangulation to obtain the corrected location of the spot.

4. Multiple Conical Arrays of Light Sources

Uncertainty in the measurement of surface orientation ultimately depends upon the uncertainty with which the sensor chip measures light spot position. Although the resolution of the sensor chip is potentially high (about 1/5000 across its surface), the precision of a single measurement is limited by several factors. The intensity of the light that reaches the sensor chip depends upon the distance, orientation and reflectivity of the target surface. The shape of a surface also affects the intensity distribution of the light spot on the sensor chip, causing a small error in measurement. However, because the sensor chip is very fast we can use redundant multiple light sources and choose the orientations and positions of the light sources so as to increase the accuracy of the sensor.

When measuring the distance of a surface, the region of greatest interest is generally the center of the field of view of the sensor. The effects of the various distortion and non-linearity are smallest near the optical axis. In addition, even if a spot of light is out of focus or grossly distorted by a surface, errors that are the result of a portion of the light spot falling beyond the edge of the sensor chip will be minimized. For these reasons the proximity sensor is configured using multiple conical arrays of light sources. In this way, several light sources are always focused near the optical axis over the range of the sensor.

Since the sensor chip measures the centroid of the intensity distribution on its surface, the distortion of the spot of light on a flat surface does not lead to an appreciable error. However, a large light spot size does decrease the linear range of the proximity sensor, and certainly will affect the measurement of curved surfaces. To minimize the size of the light spots, the sensor uses light emitting diodes that were intended for fiber optic communication. These LEDs have a light source diameter of 150 microns. As a result the projected light spots on a target surface are on the order of 350 microns in diameter.

4.1. Placement and Orientation of Conical Arrays

Having decided that all of the light sources are to be arranged in multiple conical arrays, the angles of orientation for these light sources, and then the number of light sources comprising each cone, the number of cones, and the placement of the cones are determined by means of the statistical analysis of errors.

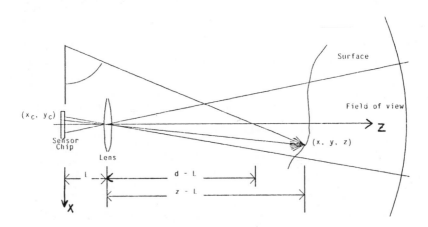

Figure 4:   Camera Model of the Proximity Sensor

Figure 4 shows the locations of the sensor chip, the lens, and one light source focused on a point along the optical axis. The angle of the light beam relative to the x axis is θ. A line is drawn from the point on a surface interrupting the light beam, through the effective center of the lens, to the sensor chip. The equation of the beam of light in the x-z plane is

$$z = x \tan \theta + d \, . \tag{41}$$

The x coordinate of the spot of light on the sensor chip is

$$x_c = -\frac{L}{z-L} x .$$ (42)

A small change in the z coordinate of the target surface by Δz corresponds to a shift in the spot of light on the sensor chip by Δx_c:

$$S = \frac{\Delta x_c}{\Delta z} = -\frac{L}{(z-L)\tan\theta} + \frac{L(z-d)}{(z-L)^2 \tan\theta} .$$ (43)

The larger the magnitude of S, the more sensitive the measurement. The orientation of the light beam θ can be chosen so any measurement of distance using a single light source will have a nominal sensitivity S_N in the center of the field of view, *i.e.*, at $z = d$. This condition places an upper bound on the value of θ:

$$\theta < \tan^{-1} \frac{L}{(d-L)} \frac{1}{S_N} .$$ (44)

Suppose one particular cone of light is generated using n light sources equally spaced in a circle. When a flat surface intersects the cone of light beams normal to the axis of the cone, the spots of light fall on a circle. The radius R of the circle is given by

$$R = |z - d| \cot\theta .$$ (45)

To determine the uncertainty in the orientation of a plane that is fit to these points, the mean-square values of the coordinates are first calculated. For $n \geq 3$:

$$\sum_{i=1}^{n} x_i^{\,2} = R^2 \sum_{i=1}^{n} \cos^2 \frac{2\pi i}{n} = \frac{nR^2}{2}, \tag{46}$$

$$\sum_{i=1}^{n} y_i^{\,2} = R^2 \sum_{i=1}^{n} \sin^2 \frac{2\pi i}{n} = \frac{nR^2}{2}. \tag{47}$$

Using Equations (16), (17), (18), and (19), the uncertainty in measured orientation when using a single cone of light can be calculated in terms of the uncertainty σ in the measurement of spot position, the radius of the circle of points, and the number of light sources:

$$\sigma_\phi \leq \frac{\sigma}{R} \frac{2}{\sqrt{n}}. \tag{48}$$

When using multiple cones of light, where the j th cone consists of n_j light sources focused towards a point d_j, the radius of the j th cone at z is

$$R_j = |z - d_j| \cot \theta_j. \tag{49}$$

The mean square value of the x coordinates of the data points is then

$$\sum_i x_i^{\,2} = \frac{1}{2} \sum_j n_j \cot^2 \theta_j (z - d_j)^2. \tag{50}$$

The uncertainty in measured orientation becomes

$$\sigma_\phi \leq \frac{2\,\sigma}{\sqrt{\Sigma n_j \cot^2 \theta_j (z - d_j)^2}}. \tag{51}$$

This equation guides the determination of the number of light sources n_j in each cone, their orientation θ_j, and their placement d_j.

4.2. Design of the Proximity Sensor

Several goals guided the design of the proximity sensor: the sensor should be compact, with dimensions on the order of 10 cm, in order that it can be used by a robotic manipulator; the sensor should have a useful range of about 10 cm; the incident angle of the light beams on a target surface should be small to minimize light spot distortion; there should be at least three separate conical rings of light sources; and each individual light source should measure position with a sensitivity of at least 0.25 mm.

The primary constraint on the design of the proximity sensor was the light source intensity. The intensity of the light that is incident on a target surface depends on the intensity of the light emitter, the distance of the emitter from its focusing lens, and the diameter and focal length of the focusing lens. These last two quantities determine the size of the sensor and the number of light sources that can be mounted on the sensor head. The intensity of the reflected light that reaches the sensor chip depends on the distance of the camera lens from the target surface and the diameter of the lens.

In a tradeoff between sensitivity, range, and the size of the field of view, the distance from the camera lens to the plane that is in best focus was chosen to be 3.42 times the distance from the sensor chip to the lens. As a result, the area of the field of view that is in best focus is approximately 4.0 cm square. In the version of the sensor that has been built we use a 16mm, f/1.6 camera lens. If a camera lens with a shorter focal length were used, a smaller sensor could have been built. This sensor would focus over a closer range and have a wider field of view. However, the sensitivity of the sensor would then be decreased at longer distances as the field of view of the sensor rapidly expanded. If we used a longer focal length camera lens the angular field of view would be smaller, and the sensitivity greater, but it is difficult to focus small spots of light with sufficient intensity at the greater distance that would then be in focus.

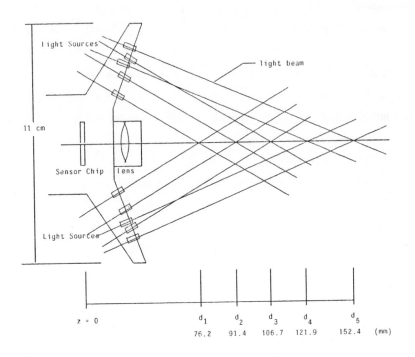

Figure 5: Geometry of the New Proximity Sensor

Figure 5 shows the configuration of the proximity sensor that has been built. To begin the analysis, we choose three cones of light each consisting of a three light sources. Equation (44) determined the choice of the angle of orientation of the light sources. With this angle chosen, we use Equation (51) to calculate the minimum spacing between the vertices d_1, d_2 and d_3 of the cones of light, where $z = d_2$ is in best focus.

First we have to estimate the uncertainty σ in the measurement of spot position at $z = d_2$. Referring to Equation (43) and Equation (44), we first set $\theta = 60°$, and $(d_2 - L) / L = 3.42$ and find $S_N = 6.67$. A smaller angle would increase the sensitivity of the sensor, but the light spot on a surface would then be further elongated. We assume that when a flat surface is measured, a signal to noise ratio can be achieved so that the effective resolution of the sensor chip is at least 1/500 its linear dimensions. This corresponds to $\Delta x_c = 0.026$ mm , and a shift in the position of the target surface by $\Delta z = 0.15$ mm . This value of Δz is used to estimate σ_z. The total uncertainty σ is estimated by observing that when the position of a point is calculated by triangulation, the solution is constrained to lie on the line determined by the light beam. Since $\sigma^2 \approx \sigma_x^2 + \sigma_y^2 + \sigma_z^2$, and $\tan 60^o = \sqrt{3}$,

$$\sigma_z^2 \approx 3(\sigma_x^2 + \sigma_y^2) \tag{52}$$

and therefore,

$$\sigma^2 \approx 1.33 \sigma_z^2. \tag{53}$$

The maximum value of the uncertainty in the measurement of surface orientation σ_ϕ occurs near the vertex of the middle cone. We use Equation (51) to calculate the minimum spacing between the vertices d_1, d_2 and d_3 of the cones of light by insisting the uncertainty in any single measurement of orientation at $z = d_2$ be less than 1.0^o (17.4 milliradians). Assuming these three cones are equally spaced, and consist of three light sources, we find

$$14.0 \text{ mm} \;\leq\; d_2 - d_1 = d_3 - d_2 \tag{54}$$

Figure 6 shows how the uncertainty in orientation σ_ϕ calculated by Equation (51) decreases as cones of light are added to the proximity sensor head. $\sigma_\phi(1)$ is graphed assuming a single cone of light consisting of three light sources is focused toward d_2. Then $\sigma_\phi(2)$ and $\sigma_\phi(3)$ are graphed as successive cones of light are added: first the cone with a vertex at d_3, then the cone with a vertex at d_1.

In order to further increase the accuracy of the sensor a second cone of light is focused at d_1 to augment the original light sources. In order to extend the range of the sensor, two more cones of light with an orientation of

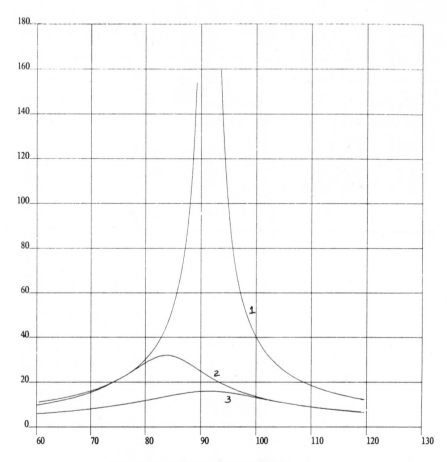

Vertical axis: angular uncertainty. (milliradians)
Horizontal axis: distance from sensor. (mm)

Figure 6: These graphs display the uncertainty when measuring the orientation of a flat target surface. Curves 1, 2, and 3 show the uncertainty when using a single, then two, and then three cones of light, each having three light sources

70 $^\circ$ are focused towards points further along the optical axis. As a result, the range of the proximity sensor consists of two overlapping regions: first an inner region with a small field of view, a depth of 6 cm, and many redundant light sources where the sensor accurately measures the distance and shape of a surface, and then an outer region that extends the range of

the sensor by another 4 cm . The proximity sensor actually has an extended range of 20 cm. At this distance several of the light beams leave the field of view. However, the center of the field of view is devoid of light spots.

The computed distance and orientation of a target surface are estimated to be the values at the intersection of the surface with the axis of the sensor. If there are data points in this region, the estimate is more reliable than otherwise. The spacing of the vertices of the cones of light, and the orientations of the light sources determine how localized the data points are near the axis of the sensor over the range of the sensor. In the sensor that has been built, a light spot on the target surface is always within 5mm of the axis of the sensor.

The distances where the beams of light enter and leave the field of view of the sensor were computed for each light source. The angular field of view of the sensor depends on the active area of the sensor chip. Table 1 lists the entrance and exit distances for each light source assuming the active area of the sensor chip has a radius of 4 mm. In addition, the table lists the distances where each cone of light intersects the optical axis. From the table, one can see over what distances the cones of light overlap.

	Entrance (mm)	Axis Intercept (mm)	Exit (mm)
Cone 1	62.26	76.2	104.18
Cone 2	73.67	91.4	127.11
Cone 3	85.09	106.7	150.03
Cone 4	86.78	121.9	236.79
Cone 5	106.68	152.4	301.85

Table 1: Entrance and Exit Distances for Each Light Source.

This proximity sensor uses a total of 18 light sources arranged in 5 conical arrays. The diameter of the sensor is 11 cm. Assuming a sufficiently high signal to noise ratio, the overall accuracy in the measurement of position when using the innermost 12 light sources is expected to be on the order of .1 mm. The accuracy in the measurement of surface orientation is expected to be better than 1°.

Figure 7:   Photograph of the new proximity sensor

The proximity sensor head shown in Figure 7 is composed of a barrel that houses the analog sensor chip, and into which the camera lens and eighteen light sources are fixed. The barrel was machined from a single piece of aluminum to insure the concentricity of the cones of light sources with the optical axis of the camera lens and with the center of the sensor chip. Any suitable C-mount lens can be mounted in front of the sensor chip. The light sources screw into the flange of the barrel.

To summarize, the final parameters of the proximity sensor are:

Camera lens focal length: $f = 16$ mm
Sensor chip to lens distance: $L = 20.67$ mm
Distance from sensor chip to plane in best focus: $D = 91.4$ mm
Orientation of inner rings of light sources: $\theta_{1,2,3} = 60°$
Orientation of outer rings light sources: $\theta_{4,5} = 70°$
Number of light sources in ring 1: $n_1 = 6$
Number of light sources in outer four rings: $n_{2,3,4,5} = 3$
Distance between inner light source target points along
    the optical axis: 15.24 mm

This particular sensor head weighs approximately one pound. Future versions of the sensor can be constructed of composite materials to decrease the sensor's weight.

5. Performance

5.1. Measuring Distance and Orientation

The proximity sensor was used to measure the distance and orientation of flat target surfaces to determine the accuracy of the sensor. The target surfaces were moved by high resolution linear and rotary platforms in the field of view of the sensor. Each surface was rotated in $2°$ increments from $-16°$ to $+16°$, and moved in 2 mm steps in front of the sensor. The distance and orientation of the surfaces were measured and compared with their expected values.

The results for a uniform white target surface are summarized in Figures 8 and 9. The graphs show the error in the measured distance and orientation of the surface as a function of the actual distance and orientation of the surface. The error in orientation is typically less than $2°$, and the error in the measurement of distance is less than 0.2 mm.

The amplifier gains of the proximity sensor were set so that the sensor could measure the distance of a matte surface over a range of 10 cm. The

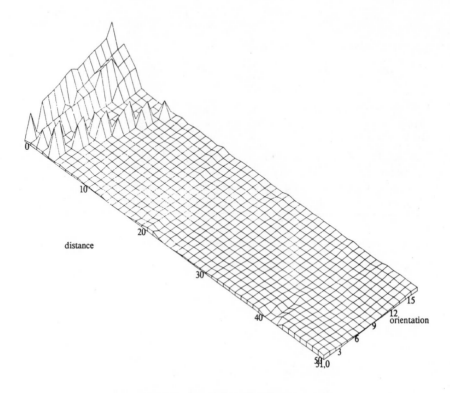

distance

Orientation axis: -16° to 16° in 2° steps.
Distance axis: 66.4 mm (0) to 166.4 mm (50) in 2 mm steps.

Figure 8: White surface: error in measured distance versus orientation and distance. The error in measured distance is typically less than 0.2mm

light intensity measured by the sensor agreed with the expected result when the white surface was used as a target. However, the reflectivity of an unpainted aluminum surface, for example, is less than that of a white surface, and in addition the reflected light has a large specular component. As a result, the range of the sensor was limited to 5 cm when measuring the distance of this type of surface. Either an insufficient amount of light is scattered by the surface towards the sensor, or alternatively, light is specularly reflected directly towards the sensor and overflows the amplifiers.

The measured coordinates of a light spot on a target surface also has a

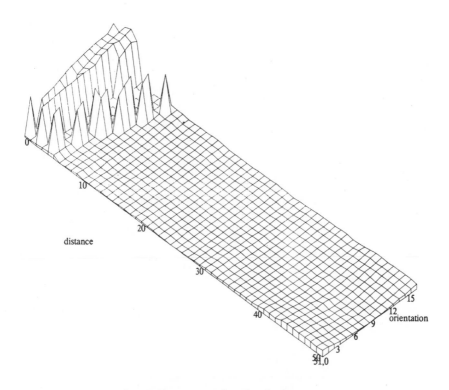

Orientation axis: -16° to 16° in 2° steps.
Distance axis: 66.4 mm (0) to 166.4 mm (50) in 2 mm steps.

Figure 9: White surface: error in measured orientation versus orientation and distance. The error in measured orientation is typically less than 2°

small dependence on the intensity of the light reaching the sensor chip. As a result, when the sensor was calibrated for the white target surface and then used to measure the distance and orientation of the other target surfaces the results were in error. However, by recalibrating the proximity sensor for each of these surfaces, (that is, recomputing error correction polynomials), the performance of the sensor was improved so that the measurement error at normal incidence was less than 0.5 mm. Figures 10 through 13 show the errors computed for two target surfaces, a brushed and a tarnished aluminum surface, after recalibrating the sensor for each of these surfaces.

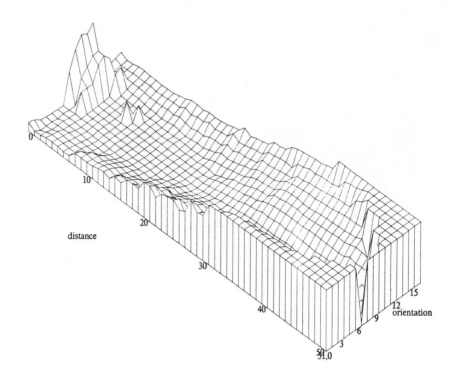

distance

Orientation axis: -16° to 16° in 2° steps.
Distance axis: 66.4 mm (0) to 166.4 mm (50) in 2 mm steps.

Figure 10: Aluminum surface: error in measured distance versus orientation and distance after recalibrating the sensor.

5.2. Curvature

To measure the contour of a surface, a surface was moved through the field of view of the proximity sensor. At each position of the surface, the distance of the surface from the proximity sensor was measured, and the orientation and curvature of the surface were computed. The results were compared with the actual shape of each surface, and with shape of each surface that was computed assuming the proximity sensor behaved ideally.

The measurement of surface shape was simulated to evaluate the expected results. The coordinates of the resultant light spots on a simulated target surface were computed for each position of the surface. The distance,

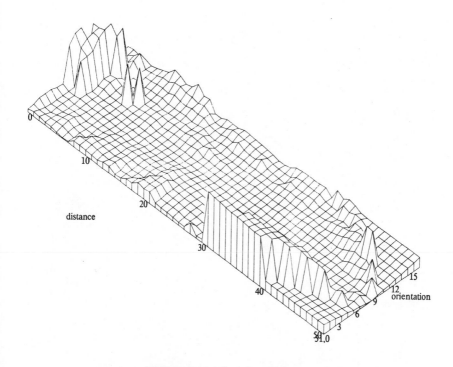

Orientation axis: -16° to 16° in 2° steps.
Distance axis: 66.4 mm (0) to 166.4 mm (50) in 2 mm steps.

Figure 11: Aluminum surface: error in measured orientation versus orientation and distance after recalibrating the sensor.

orientation and curvature of the simulated target surface were then computed, and the results were compared with the actual measurements.

Since measuring the contour of a surface uses only a small interval of the range of the sensor, relative measurement errors are generally small. For example, when a flat surface with uniform reflectivity was moved across the field of view of the sensor, the error in measured distance was much less than the errors that were reported when the distance and orientation of the target surface were varied over a range of 10 cm. In many applications of the proximity sensor, a target surface will be nearly flat. In addition, the sensor can be moved to maintain a fixed distance and orientation above a target

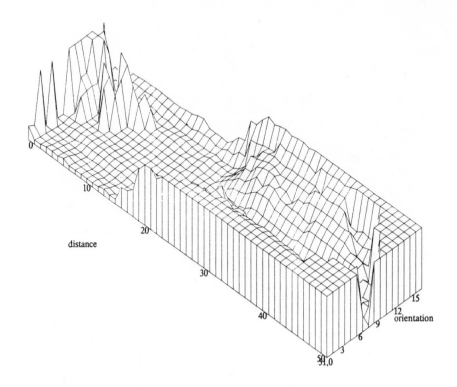

distance

Orientation axis: -16° to 16° in 2° steps.
Distance axis: 66.4 mm (0) to 166.4 mm (50) in 2 mm steps.

Figure 12: Tarnished surface: error in measured distance versus orientation and distance after recalibrating the sensor.

surface to achieve a better measurement.

The shapes of a cylinder and a cone were measured by moving each surface linearly across the field of view of the sensor a distance of 5 cm in 2 mm steps. The cylinder was in focus when closest to the sensor; the cone was positioned 6 mm further away. Both the cylinder and the cone were covered with a sheet of matte white paper. The cylinder had a 57.15 mm (2.25 inch) radius. In Figure 14 the contour of the cylinder is graphed as the sensor was moved across its surface.

The cone that was used as a target had a base angle of 81°. At the height at which the axis of the sensor intersected the surface of the cone, the cone

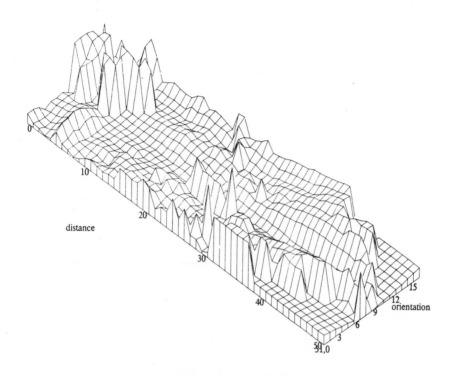

Orientation axis: -16° to 16° in 2° steps.
Distance axis: 66.4 mm (0) to 166.4 mm (50) in 2 mm steps.

Figure 13: Tarnished surface: error in measured orientation versus orientation and distance after recalibrating the sensor.

had a radius of 45 mm. The distance between the sensor and the cone was 97.4 mm at their closest position. This is 6 mm further away from the sensor than the distance of best focus. In Figure 15 the contour of the cone is graphed as the sensor was moved across its surface. Curve '1' shows the measurement results. Curve '2' shows the expected results that were computed according to the geometrical model of the sensor which assumes, for example, there were no extraneous reflections. Curve '3' shows the actual contour of the cone.

Figures 14 through 17 show that the proximity sensor can accurately measure the distance and orientation of a curved surface when the surface is

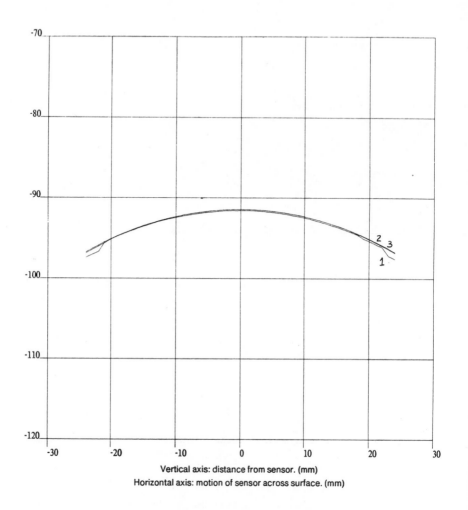

Vertical axis: distance from sensor. (mm)
Horizontal axis: motion of sensor across surface. (mm)

Figure 14:   Contour of the surface of a cylinder: the measured 1 and
the simulated 2 results, along with the actual 3 values

near normal to the axis of the sensor. When the orientation of each of these
surfaces was computed from the coefficients of the second order equation,
the results were not accurate. Fitting a quadric surface to the data points on
a curved surface provides a better estimate of the range of a surface, but the
gradient and the curvature of the surface are not stable.   The local
orientation of the target surfaces was better represented by fitting a plane to
the data points. Figures 16 and 17 show the measurement results and the

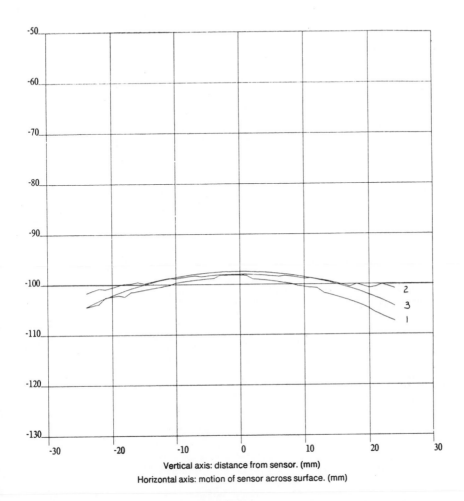

Figure 15:   Contour of the surface of a cone: the measured 1 and the simulated 2 results, along with the actual 3 values

simulation results.

The principal normal curvatures, κ_1 and κ_2, of the cylinder and the cone were also computed. Figures 18 and 19 respectively show the smaller value for the computed principal radius of curvature of the cylinder and of the cone. The curves labeled '1' are the measured results and the curves labeled '2' show the simulation results. The solution for the curvature of the cylinder and the cone can be chosen where the other principal radius of

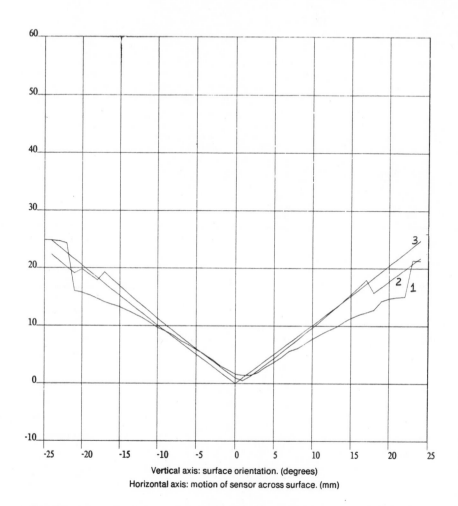

Vertical axis: surface orientation. (degrees)

Horizontal axis: motion of sensor across surface. (mm)

Figure 16: Surface orientation of the cylinder: measured 1 and simulated 2 results, and the actual value 3. The orientation is measured with respect to the axis of the proximity sensor

curvature has a maximum. For the cylinder this occurs where $x = -6$. For the cone this occurs where $x = 2$.

The proximity sensor can accurately measure the distance and orientation of a smooth surface when maintained at a near normal position relative to the surface. According to the simulation results, the proximity sensor has the potential to accurately measure the curvature of a surface.

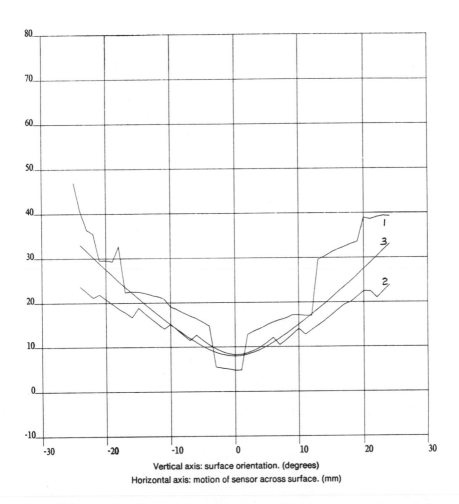

Vertical axis: surface orientation. (degrees)

Horizontal axis: motion of sensor across surface. (mm)

Figure 17: Surface orientation of the cone: measured 1 and simulated 2 results, and the actual value 3. The orientation is measured with respect to the axis of the proximity sensor

The simulated measurement of curvature is stable over a range of orientation and distance relative to a surface. However, the simulation did not take into account specular reflection of light toward the sensor head, finite light spot size and other perturbations. Either increasing the number of light sources, focusing smaller light spots onto the target surface, or using data points from earlier measurements in the computation, should increase

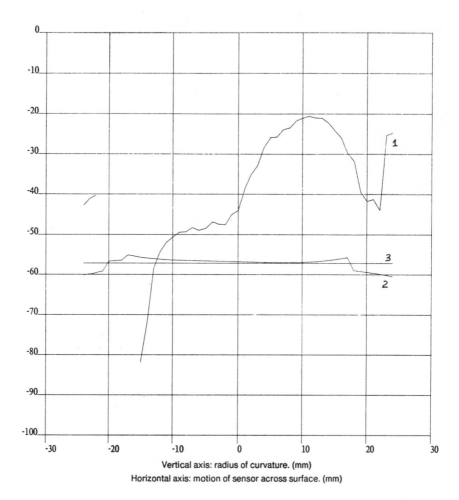

Vertical axis: radius of curvature. (mm)
Horizontal axis: motion of sensor across surface. (mm)

Figure 18: The smaller principal radius of curvature of a cylinder: measured 1 and simulated 2 results, and the actual value 3.

the accuracy of the computed curvature.

6. Conclusion

A new compact multi-light source proximity sensor has been developed. The sensor is based on the principle of active illumination and triangulation: each time a light source is pulsed, the coordinates of the resulting spot of the light on a surface are calculated. The proximity sensor that has been built

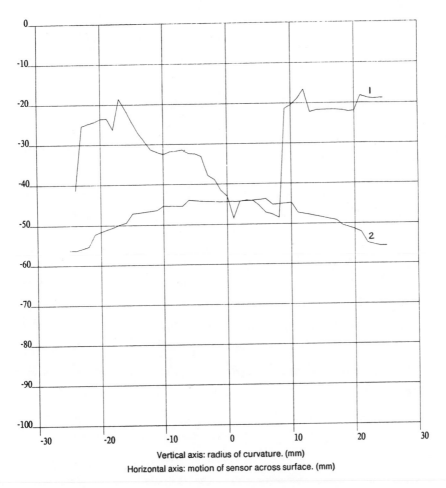

Vertical axis: radius of curvature. (mm)

Horizontal axis: motion of sensor across surface. (mm)

Figure 19: The smaller principal radius of curvature of a cone: measured 1 and simulated 2 results, and the actual value 3.

has a range of 10 cm where it can measure distance with a precision of 0.1 mm and surface orientation with a precision of 1.0°. The sensor can acquire data at a rate exceeding 200 points per second.

The design of the proximity sensor was guided by a statistical analysis of the uncertainty in the measurement of position using a single light source and the measurement of surface orientation using several light sources. The analysis related the number of light sources needed to measure orientation within a specified accuracy to the distribution of light spots on a target surface. The new design is also intended to be a compact sensor for use on a robotic manipulator.

The proximity sensor is subject to error because of various types of distortion, nonlinearity, and noise. To compensate for the distortion of the sensor chip and imaging optics, a square array of light spots was generated in the field of view. A two dimensional transform was computed that maps the measured image of this array on the sensor chip into its ideal undistorted image. This transform was applied to the sensor chip coordinates during the operation of the sensor.

To further compensate for errors in the computation of distance, error correcting polynomials were computed that map the distance of a surface computed by triangulation into its actual distance. These polynomials were used during the operation of the sensor to improve measurement results.

The features of this proximity sensor include its simple principle of operation and its fast speed. The sensor chip which detects the spot of light on the target surface is an analog device which outputs the position of the centroid of a spot of light on its surface. Although the accuracy of any individual measurement of light spot position is limited by the distortion of a spot of light by a surface, and the noise and sensitivity of the sensor, we take advantage of the speed of the sensor and use multiple light sources to increase the overall accuracy of measurements. The geometrical arrangement of the light sources was guided by the statistical analysis to achieve the design specification for accuracy. Measurement of distance, surface orientation and surface curvature can exploit the geometrical redundancy of the device.

Acknowledgments

We thank Regis Hoffman and Donald Schmitz for useful discussions and help in software and hardware development. This research was partially supported by the Office of Naval Research (ONR) Grant No. N00014-81-K-0503, and by the National Science Foundation Grant No. ECS-8320364.

References

[1]   Bamba, T.,Maruyama, H., Kodaira, N. and Tsuda, E.
        A Visual Seam Tracking System for Arc Welding Robots.
        In *Fourteenth International Symposium on Industrial Robots.*
            1984.

[2]   Bamba, T.,Maruyama, H., Ohno, E. and Shiga, Y.
        A Visual Sensor for Arc Welding Robots.
        In *Eleventh International Symposium on Industrial Robots.* 1982.

[3]   Bejczy, A.
        Smart Sensors for Smart Hands.
        Nasa Report Number 78-1714 , 1978.

[4]   Bevington, Philip R.
        Data Reduction and Analysis for the Physical Sciences.
        McGraw-Hill Book Company, 1969.

[5]   Diffracto Limited.
        Diffracto Robotics Vision Sensors.

[6]   Fuhrman, M. and Kanade, T.
        *Design of an Optical Proximity Sensor using Multiple Cones of
            Light for Measuring Surface Shape.*
        Technical Report TR-84-17, Carnegie Mellon University Robotics
            Institute, 1984.

[7]   Hayes, J.G.
        Numerical Aproximation to Functions and Data.
        Athlone Press, University of London, 1970.

[8]   Kanade, T. and Sommer, T.
        *An Optical Proximity Sensor for Measuring Surface Position and
            Orientation for Robot Manipulation.*
        Technical Report TR-83-15, Carnegie Mellon University Robotics
            Institute, 1983.

[9]   Kanade, T. and Asada, H.
        Noncontact visual three-dimensional ranging devices.
        In B. R. Altschuler (editor), *3-D Machine Perception*, pages 48-53.
            SPIE, 1981.

[10]   Lipschutz, M. M.
        *Schaum's Outline of Theory and Problems of Differential
            Geometry.*
        McGraw Hill Book Co., 1969.

[11]   Morander, E.
        The Optocator. A High Precision, NonContacting System for
            Dimension and Surface Measurement and Control.
        In *Proc. Fifth Intern. Conf. on Automated Inspection and Product
            Control*, pages 393-396. , 1980.

[12]   Nakamura, Y., Hanafusa, H.
        A New Optical Proximity Sensor for Three Dimensional Autonomous
            Trajectory Control of Robot Manipulators.
        In *Proceedings: '83 International Conference on Advanced
            Robotics*. 1983.

[13]   Okada, T.
        A Short Rangefinding Sensor for Manipulators.
        Bulletin of Electrotechnical Laboratory 42, 1978.

[14]   Thring, M.W.
        Robots and Telechirs.
        Ellis Horwood Limited, 1983.

PART II: 3-D FEATURE EXTRACTIONS

Toward a Surface Primal Sketch

Jean Ponce
Michael Brady

Massachusetts Institute of Technology
Artificial Intelligence Laboratory
545 Technology Square
Cambridge, MA 02139

Abstract

This paper reports progress toward the development of a representation of significant surface changes in dense depth maps. We call the representation the Surface Primal Sketch by analogy with representations of intensity changes, image structure, and changes in curvature of planar curves. We describe an implemented program that detects, localizes, and symbolically describes: steps, where the surface height function is discontinuous; roofs, where the surface is continuous but the surface normal is discontinuous; smooth joins, where the surface normal is continuous but a principal curvature is discontinuous and changes sign; and shoulders, which consist of two roofs and correspond to a step viewed obliquely. We illustrate the performance of the program on range maps of objects of varying complexity.

1. Introduction

This paper describes an implemented program that detects, localizes, and symbolically describes surface changes in dense depth maps:

- *steps*, where the surface height function is discontinuous;
- *roofs*, where the surface is continuous but the surface normal is discontinuous;
- *smooth joins*, where the surface normal is continuous but a principal curvature is discontinuous and changes sign; and
- *shoulders*, which consist of two roofs and correspond to a *step* viewed obliquely.

Figures 4, 7, 9, and 10 show the idealized instances of these surface changes that are the basis of the mathematical models used by the program.

The work reported here continues our investigation [3] of surface descriptions based on the concepts of differential geometry. Section 2 summarizes our ideas and shows the kind of geometric (CAD) description we are aiming at. An important component of our work is the identification and isolation of a set of *critical surface curves,* including significant surface changes. To this end, we report progress on the development of a representation we call the *Surface Primal Sketch* by analogy with:

- Marr's *Primal Sketch* [22] representation of significant intensity changes;
- Asada and Brady's *Curvature Primal Sketch* [1] representation of significant curvature changes along planar contours; and
- Haralick, Watson, and Laffey's *Topographic Primal Sketch* [14] representation of image structure.

In each case, there are three distinct problems: to *detect* significant changes; to *localize* those changes as accurately as possible; and to *symbolically describe* those changes. We follow the approach of Asada and Brady [1], as sketched in Section 3. A key component of that approach is scale space filtering, pioneered by Witkin [28]. Yuille and Poggio [30], [31] have proved that, in principle, scale space filtering enables a discontinuity to be accurately localized. Canny [7] uses the smallest scale at which a given intensity change can be detected to most accurately localize it.

Brady, Ponce, Yuille, and Asada [3] report initial experiments that adapt Asada and Brady's algorithm [1] to find surface changes. We describe a number of problems, both mathematical and implementational, with that approach.

Section 4 describes a robust algorithm to find *roofs, steps, smooth joins,* and *shoulders. Roofs* are found from extrema of curvature (positive maxima and negative minima), whereas *steps, shoulders,* and *smooth joins* are found from parabolic points: zero crossings of the Gaussian curvature. We use scale space behavior to discriminate *steps, shoulders,* and *smooth joins.* Section 6 shows the algorithm at work.

2. Background

In this section, we recall some of the main features of our work on representing visible surfaces. We work with dense depth maps that are the output of "shape-from" processes such as stereo or, more usually, direct ranging systems. There are three principal problems to be addressed:

1. Finding surface intersections. These enable the *description* of the depth map to be partitioned into a set of smooth surface patch descriptions. This is the problem addressed in the present paper. Surface intersections do not, in general, partition the depth map. Consider, for example, a bulbous end of an American telephone handset (Figure 1). The surface intersection marked on the figure peters out by the time the cylindrical portion is reached. Each surface intersection has an associated description that includes its type (*step, roof, smooth join*, etc.). In general, the type of surface intersection may vary along its length [16], [27]. If a surface intersection has a special property, such as being planar, that property is included in the description.

2. Generating descriptions for the smooth surface patches that result from the partitioning in (1). This is the problem addressed by Brady, Ponce, Yuille, and Asada [3], who introduce a representation called *Intrinsic Patches*. This is discussed further below.

3. Matching surface descriptions to a database of object models that integrate multiple viewpoints of a surface. We have not yet addressed this problem. Grimson and Lozano-Pérez [13], Faugeras, Hebert, Pauchon, and Ponce [12], and Faugeras and Hebert [10] have made a solid start on the problem, though they restrict attention to the case of polyhedral approximations to surfaces. However, Faugeras and Hebert [10], [11], illustrate the advantages of representations based on sculptured surfaces. Brou [4], Little [21], and Ikeuchi and Horn [17] have developed the Extended Gaussian Image (EGI) representation for recognition and attitude determination. The EGI is an information-preserving representation only for complete maps of convex objects, a rare situation in practice. Not much has been done to extend the representation to handle non-convex objects.

The *Intrinsic Patch* representation that we are developing is based on concepts of differential geometry, principally because it provides a hierarchy of increasingly stringent surface descriptions. A surface may simply be (doubly) curved, but, in some cases, it may be ruled, even developable, even

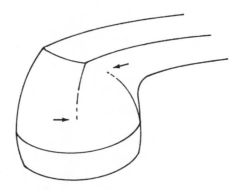

Figure 1:   A telephone handset illustrates that surface intersections on   curved surfaces do not, in general, partition the surface into a patchwork of smooth components.

conical.   Our aim is to find the most appropriate and most stringent descriptors for portions of a surface. If, for example, there is a connected region of umbilic points, indicating that part of the surface is spherical, then it is made explicit, as is the center of the corresponding sphere (Figure 2). If there is a portion of the surface that is determined to be part of a surface of revolution, it is described as such, and the axis is determined (see Figures 2 and 16).

Similarly, if there is a line of curvature or an asymptote that is planar or whose associated curvature (principal curvature or geodesic curvature respectively) is constant, then it is made explicit.   For example, the asymptote (which in this particular case is also a parabolic line) that marks the smooth join of the bulb and the stem of the lightbulb in Figure 2, as well as the surface intersections marked on the oil bottle in Figure 16, are noted in the representation. The program described in Section 3 cannot compute the asymptote on the lightbulb; but that described in Section 4 can. We may associate a description with a curve that is a surface intersection; but only if it has an important property such as being planar. For example, a slice of a cylinder taken oblique to the axis of the cylinder produces a planar

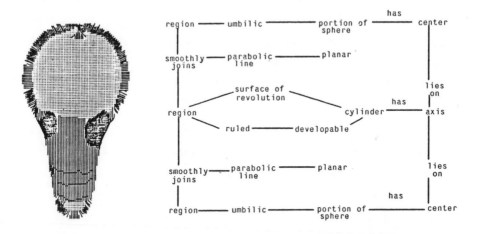

Figure 2: The representation of a light bulb. (a) The dotted region consists of umbilic points, indicating that the bulb is spherical. The parallel lines are the meridians of the cylindrical stem. The parallels, which are also rulings, are not shown. (b) The representation that we are working towards for the lightbulb. All save the rightmost column can be automatically computed by existing programs.

curve of intersection. Machining operations such as filleting tend to produce planar curves. Similarly, the intersection of a finger of a dextrous robot hand [25], [18], [19] and an object surface is planar. On the other hand, the intersection of two cylinders is not a planar curve.

Figure 2(b) illustrates the representation we are aiming at. The stem of the lightbulb is determined to be cylindrical, because it is ruled and because it is a surface of revolution. We can compute the axis of the stem. The bulb is determined to be a portion of a sphere, because it is a connected region of umbilic points. The center of the sphere can be computed. Similarly, the center of the spherical portion that forms the threaded end can be determined. The stem is smoothly joined to the bulb. Moreover, the axis of the cylindrical stem passes through the centers of the spheres defined by the bulb and threaded end. This distinguishes the diameters of each sphere that are collinear with the stem axis, showing that the lightbulb is a surface of

revolution. All of Figure 2(b) can be computed by the algorithms described
in this paper and in Brady, Ponce, Yuille, and Asada [3], except for the
rightmost column, which relates to the inferences that derive from attaching
the spherical portions to the cylindrical stem. Currently, we are working on
the inference engine (see also Kapur, Mundy, Musser, and Narendran [20]).

3. Surface intersections from lines of curvature

Asada and Brady [1] introduce a representation, called the *Curvature
Primal Sketch*, of the significant changes of curvature along a planar curve.
We review that work here because our extension to surfaces follows an
analogous development. Asada and Brady describe an algorithm that not
only detects and localizes significant changes, but describes those changes
symbolically. The simplest descriptor is *corner*, where two arcs meet
continuously but where the tangent is discontinuous. Other descriptors are
composed of two or more instances of the corner model. The curve that is
input to the algorithm is represented by its tangent $\theta(s)$ where s is the
intrinsic arclength coordinate. The algorithm is based on a mathematical
analysis of a set of models that are idealized instances of the descriptors.
For example, the corner model is formed by the intersection of two circles.
Note that this is intended as a *local* approximation to a corner to facilitate
analysis. It does not prejudice the subsequent approximation of the contour
to be piecewise circular. Rather, it suggests a set of knot points for any
appropriate spline approximation.

Asada and Brady derive a number of salient features of the curvature of
the models as they vary with the scale of the smoothing (Gaussian) filter.
For example, a corner generates a curvature maximum, equivalently a
positive maximum flanking a negative minimum in the first derivative of
curvature. The height and separation of these peaks varies in a
characteristic fashion over scale. The salient features are the basis of the
tree matching algorithm that locates a curvature change and assigns it a

descriptor. Note that the distance between peaks varies approximately linearly (in arclength) with scale.

In the next section we develop an analogous mathematical framework for significant surface changes. Our surface analysis is local and based upon smoothing (locally) cylindrical functions with a Gaussian distribution. This is because of the following theorem, proved in Brady, Ponce, Yuille, and Asada [3].

The Line of Curvature Theorem: The convolution of a cylindrical surface with a Gaussian distribution is cylindrical. In more detail, let $f(x,y,z)$ be a surface that is the cross product of a planar curve and a straight line. The lines of curvature of the convolution of f with a Gaussian distribution are in the plane of the curve and parallel to the generating line.

In vector notation, a cylindrical surface has the form $r(x,y)=x\mathbf{i}+y\mathbf{j}+f(x)\mathbf{k}$, and consists of parallel instances of a curve $f(x)$ in the $x-z$ plane. Our models for *roof*, *step*, *smooth join*, and *shoulder* correspond to different choices for the function $z=f(x)$.

The curvature of the smoothed curve is given by the non-linear expression

$$\kappa_{smooth}(x) = \frac{z''_{smooth}}{(1+ z'^2_{smooth})^{3/2}} . \tag{1}$$

Since Asada and Brady [1] could work with tangent directions $\theta(s)$ along a planar curve, the curvature was the linear expression $d\theta(s)/ds$, so that the curvature of a smoothed contour is simply equal to the smoothed curvature of the original contour. This is *not* the case for surfaces represented as height functions $z(x,y)$. For example, the (constant) curvature of the parallels of a surface of revolution are modified (see Figure 15(a)). The non-linearity of curvature considerably complicates the analysis of surface change models presented in Section 4 relative to those used by Asada and Brady. Non-linearity affects smoothing too, as we discuss in Section 5.

Brady, Ponce, Yuille, and Asada [3] used the *Line of Curvature*

Theorem directly in a *two-step process* to detect, localize, and symbolically describe surface intersections, as follows:

1. Compute the lines of curvature on the surface;
2. Compute significant changes of curvature along the lines of curvature found in the first step.

The lines of curvature are computed using a best-first region growing algorithm [3]. A *good continuation* function is defined between neighboring points of the surface. The function involves the Cartesian distance between the points and the inner product of the tangent vectors corresponding to the curvature principal directions at the two points. The region growing algorithm joins the point pair whose good continuation function is globally maximum, and incorporates the new link into the developing set of lines of curvature. Brady, Ponce, Yuille, and Asada [3] show several illustrations of the algorithm's performance. In the second step of finding surface intersections, Asada and Brady's algorithm for computing the Curvature Primal Sketch, described in the previous section, is applied to the lines of curvature in turn.

The two step process has been tested on the objects shown in Brady, Ponce, Yuille, and Asada [3]: a lightbulb, a styrofoam cup, and a telephone receiver. It is robust and gives good results, suggesting that the method has competence. Nevertheless, there are several problems with the method:

- **The method is inefficient.** Typical running times on a lisp machine for a smoothed depth map that is 128 points square are of the order of one hour. Much of the time is spent on further smoothing each of the (typically hundreds of) lines of curvature at multiple scales, as required by the Curvature Primal Sketch algorithm.

- **Multiple multiple-smoothing is mathematically confused.** The raw surface data are smoothed at multiple scales σ_i, giving a set of surfaces z_i. The Curvature Primal Sketch algorithm further smooths the lines of curvature of z_i at multiple scales σ_j yielding a set of smoothed lines of curvature $r_{ij}(s_{ij})$. There is no obvious relation between the scales σ_i and σ_j.

- **Discretization makes implementation difficult.** The lines of curvature of an analytic surface form a dense orthogonal web. The (smoothed) depth maps we work with are discrete approximations to analytic surfaces. In practice, the lines of curvature found by the two step process are sometimes broken. The lines of curvature near the perceptual join of the stem and bulb of the lightbulb shown in Figure 2 illustrates this problem. This is due in part to quantization effects, but is also because the principal directions change rapidly near surface discontinuities. This is why the *smooth join* between the bulb and the stem of the lightbulb is not found by the two step process.

- **The** *Line of Curvature Theorem* **only applies locally.** In practice, few surfaces are cylindrical in the sense of the *Line of Curvature Theorem*. The Theorem is only approximately true in general, and then only locally. The application of the Curvature Primal Sketch algorithm in the second step does not respect this.
- **Lines of curvature on smoothed surfaces are not planar curves.** The models that are embodied in the Curvature Primal Sketch algorithm are not a complete set for surface intersections.

The success of the two step process suggests that the method is on the right track. The problems just enumerated suggest that reducing the problem to apply an existing algorithm developed for planar curves, though expedient, is wrong. Together, these observations suggest that a real two-dimensional extension of the Curvature Primal Sketch should be developed along analogous lines. The next section reports our progress toward such an extension.

4. Toward a surface primal sketch

4.1. A three-step process

In this section we develop a method for finding certain types of changes in the height of a surface that overcomes the difficulties of the two-step process described in the previous section. The types of changes we have analyzed and implemented are as follows: *steps*, where the surface height function is discontinuous; *roofs*, where the surface is continuous but the surface normal is discontinuous; *smooth joins*, where the surface normal is continuous but a principal curvature is discontinuous and changes sign; and *shoulders*, which consist of two roofs and correspond to a *step* viewed obliquely. It turns out that *roofs* consist of extrema of the dominant curvature; that is, maxima of the positive maximum curvature or minima of the negative minimum curvature. On the other hand, *steps, smooth joins*, and *shoulders* consist of parabolic points, that is zero crossings of the Gaussian curvature. They are distinguished by their scale space behavior.

We have implemented the following *three-step process* (Figure 3) that is illustrated in the examples presented in Section 6:

1. Smooth the surface with Gaussian distributions at a set of scales σ_i, yielding surfaces z_i. Compute the principal directions and curvatures everywhere;

2. In each smoothed surface z_i, mark the zero-crossings of the Gaussian curvature and the (directional) extrema of the dominant curvatures.

3. Match the descriptions of the surfaces z_i to find points that lie on *roof, step, smooth join*, and *shoulder* surface discontinuities.

The computation of principal curvatures is described by Brady, Ponce, Yuille, and Asada [3]. They investigate parabolic lines and lines of curvature as *global* descriptors of surfaces, and suggest that such lines need additional global properties, such as planarity, to be perceptually important. In this paper, we are interested in parabolic points and curvature extrema as *local cues* for significant surface intersections.

We discuss smoothing in more detail in the next section. The next four subsections analyze *steps, roofs, smooth joins*, and *shoulders*. Finally, we discuss the matching algorithm and further work that is needed to elaborate the model set.

Is it necessary to use multiple scales to find surface intersections? Arguments supporting multiple scales for edge finding in images have been advanced elsewhere (see [23]). However, it might be supposed that it would be sufficient to smooth depth maps with a single coarse filter. Figures 12(b) and 15(a) show that this is not so. Even after thresholding, there is still a large curvature extremum in the neck of the bottle running parallel to the

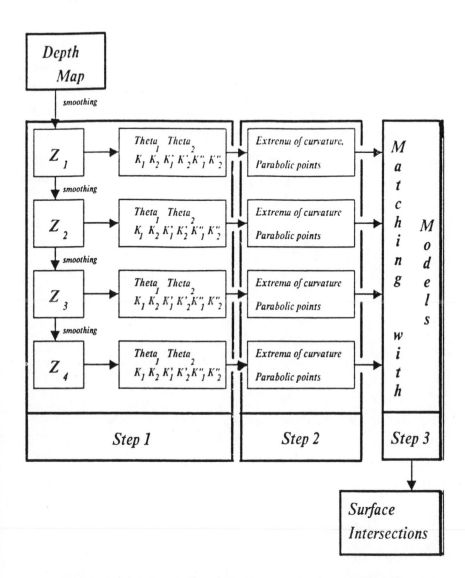

Figure 3: Schematic of the three-step process described in this paper and in Brady, Ponce, Yuille, and Asada [3]. The analysis of the set of models and the matching algorithm are the topic of the present paper. Results of running the process are shown in Section 6.

axis. This extremum is an artifact of non-linear smoothing, and it cannot be eliminated at a single scale. Instead, we reject it because it does not change over scale in the characteristic manner of a *roof*.

4.2. Step discontinuities

A *step* occurs when the surface itself is discontinuous. The model we use consists of two slanted half planes whose normals lie in the $x-z$ plane. They are separated by a height h at the origin (Figure 4).

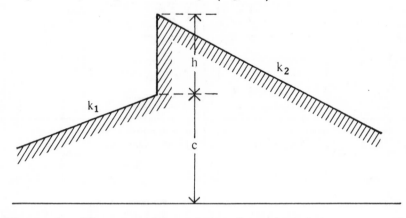

Figure 4: The *step* model, consisting of two slanted planes separated by a height h at the origin. The *roof* model corresponds to the case $h{=}0$ and $k_1 \neq k_2$.

Using the line of curvature theorem, we study the one dimensional formulation of this model.

Let the curve $z = f(x)$ be defined by

$$z = \begin{cases} k_1 x + c, & x < 0; \\ k_2 x + c + h, & x > 0; \end{cases} \tag{2}$$

where h and c are constants. In this expression, h is the height of the *step*. We now derive the result of smoothing this function with a Gaussian distribution at a given scale σ. To obtain a symmetric form for this smoothed version, we introduce the two following parameters:

$$k = (k_1 + k_2)/2;$$

$$\delta = k_2 - k_1 \, .$$

If we denote the smoothed curve $G_\sigma {*} z_\sigma$ by z_σ, we then obtain

$$z_\sigma = c + \frac{h}{\sigma\sqrt{2\pi}}\int_{-\infty}^{x} exp(-\frac{t^2}{2\sigma^2})dt \qquad (3)$$

$$+ kx + \frac{\delta x}{\sigma\sqrt{2\pi}}\int_{0}^{x} exp(-\frac{t^2}{2\sigma^2})dt$$

$$+ \frac{\delta\sigma}{\sqrt{2\pi}} exp(-\frac{x^2}{2\sigma^2}).$$

The first and second derivatives of z_σ are given by

$$z'_\sigma = k + \frac{\delta}{\sigma\sqrt{2\pi}}\int_{0}^{x} exp(-\frac{t^2}{2\sigma^2})dt + \frac{h}{\sigma\sqrt{2\pi}} exp(-\frac{x^2}{2\sigma^2}); \qquad (4)$$

$$z''_\sigma = \frac{1}{\sigma\sqrt{2\pi}}(\delta - \frac{hx}{\sigma^2}) exp(-\frac{x^2}{2\sigma^2}). \qquad (5)$$

In particular, the curvature κ, given by Equation (1), has a zero crossing at the point $x_\sigma = \sigma^2\delta/h$. This is at the origin if and only if $k_1 = k_2$; otherwise, the distance from x_σ to the origin is proportional to σ^2. This is illustrated in Figure 5 for the *step* between the cylindrical body and the cylindrical base of the oil bottle shown in Figure 16. From Figure 5, we calculate δ/h to be 0.105. The actual height of the step is about 1.5 millimeters. By the way, the position of the zero crossing shown in Figure 5 moves by about 3 pixels over one octave.

Using the fact that the second derivative of z_σ is zero at x_σ, it is easy to show that

$$\frac{\kappa''}{\kappa'}(x_\sigma) = \frac{z'''_\sigma}{z'''_\sigma}(x_\sigma) = -\frac{2\delta}{h} \qquad (6)$$

So the ratio of the second and first derivatives of the curvature at the zero crossing is constant over the scales. Calculating δ/h this way gives 0.11,

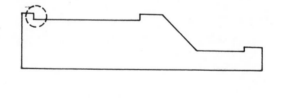

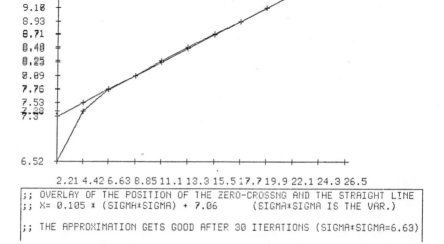

;; OVERLAY OF THE POSITION OF THE ZERO-CROSSNG AND THE STRAIGHT LINE
;; X= 0.105 * (SIGMA*SIGMA) + 7.06   (SIGMA*SIGMA IS THE VAR.)

;; THE APPROXIMATION GETS GOOD AFTER 30 ITERATIONS (SIGMA*SIGMA=6.63)

Figure 5:   Variation over scale of the position of the zero crossing of
the curvature of the smoothed *step* between the cylindrical body and the
cylindrical base of the oil bottle shown in Figure 16. The abscissa is σ^2,
and the ordinate is the position of the zero crossing. The height of the
step is about 1.5 millimeters. The slope is $\delta/h = 0.105$.

which is close to the value given by the slope in Figure 5. This suggests that
one ought not be overly coy about computing first and second derivatives of
curvature of appropriately smoothed versions of a surface, even though they
correspond to third and fourth derivatives.

4.3. Roof discontinuities

A *roof* occurs when the surface is continuous, but the surface normal is discontinuous. Specializing Equations (2) to (5) to the case $h=0$, we obtain

$$\kappa = \frac{1}{\sigma\sqrt{2\pi}} \frac{\delta e^{-\frac{x^2}{2\sigma^2}}}{\left[1 + (k + \frac{\delta}{\sigma\sqrt{2\pi}}\int_0^x exp(-\frac{t^2}{2\sigma^2})dt)^2\right]^{3/2}} \tag{7}$$

$$= \frac{1}{\sigma\sqrt{2\pi}} \frac{.\delta e^{-\frac{x^2}{2\sigma^2}}}{\left[1 + (k + \frac{\delta}{\sqrt{2\pi}}\int_0^{x/\sigma} exp(-\frac{u^2}{2})du)^2\right]^{3/2}}$$

From Equation (7), we deduce that for a roof, we have $\kappa(x,\sigma)=(1/\sigma)\kappa(x/\sigma,1)$. In particular, this implies that the extremum value of κ is proportional to $1/\sigma$, and that its distance from the origin is proportional to σ. This is illustrated in Figure 6 for the *roof* discontinuity between the cylindrical neck and the conical shoulder of the oil bottle shown in Figure 16. Figure 6(a) shows the variation in the position of the negative minimum of curvature as a function of scale. Figure 6(b) shows that curvature is directly proportional to $1/\sigma$. It is also easy to show that the second derivative of the curvature, κ'', is proportional to $1/\sigma^3$. However, we do not use this property in the current implementation of the program, relying instead on the the variation of the extremum height over scale.

We first look for points that are local maxima (minima) of the maximum (minimum) curvature in the corresponding direction. The curvature directions can be estimated accurately. Then we use non-maximum suppression [7] to reject local extrema. The location of the peak, its height, its type (maximum or minimum), and its orientation, are the features we use

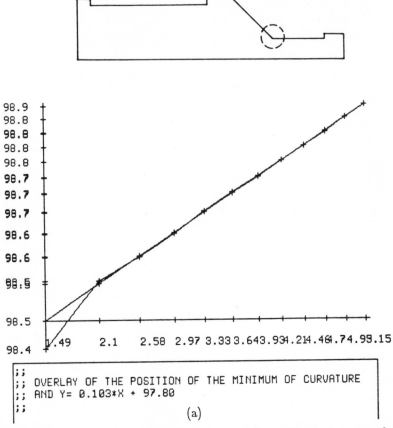

(a)

Figure 6:   Scale space behavior of the *roof* discontinuity between the cylindrical neck and the conical shoulder of the oil bottle shown in Figure 16. (a) The position of the negative minimum of curvature varies linearly as a function of scale as predicted by the analysis.

for the subsequent matching over scales.

4.4. Smooth join discontinuities

In certain circumstances, one can perceive surface changes where both the surface and its normal are continuous, but where the curvature is discontinuous. We call such a surface change a *smooth join* discontinuity. If the curvature changes sign at a *smooth join*, the surface has a parabolic point. As we shall see, such changes can be found from zero crossings of a principal curvature. It is well-known (see [1] for discussion and references)

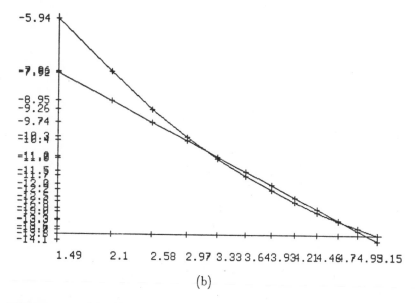

(b)

Figure 6: (b) The curvature is directly proportional to $1/\sigma$, as predicted.

that a *smooth join* where the curvature does not change sign is perceptible only when the (discontinuous) jump in curvature is "sufficiently large". In such a case, there is not a zero crossing of curvature; rather there is a level crossing, and the curvature typically inflects. We do not yet have a complete analysis of that case.

Our model of a *smooth join* consists of two parabolas that meet smoothly at the origin (the curve is differentiable). Figure 7 shows the two distinct cases of the model. Though the two cases appear to be perceptually distinct and lead to different matching criteria, they are governed by the same equation, so it is convenient to analyze them together at first.

Consider the curve $z(x)$ defined by

$$z = \begin{cases} \dfrac{1}{2}c_l x^2 + b_l x + a_l, & x < 0; \\ \dfrac{1}{2}c_r x^2 + b_r x + a_r, & x > 0. \end{cases} \tag{8}$$

The continuity and differentiability of the curve at $x=0$ imply that $b_l = b_r = b$, say, and $a_l = a_r = a$, say. As in the case of the *step*, we introduce

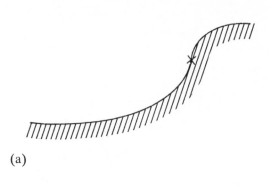

(a)

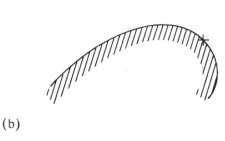

(b)

Figure 7:   The model for a *smooth join* consists of two parabolas meeting smoothly at the origin.   (a) The curvature changes sign generating a parabolic point on the surface;   (b) The curvature does not change sign.   Such *smooth joins* are typically perceivable only when there is a large, discontinuous jump in curvature.

the parameters

$$c = (c_l + c_r)/2,$$

$$\delta = c_r - c_l.$$

We can express the surface, smoothed at the scale σ, as

$$z_\sigma = \frac{1}{2}(c + \frac{\delta}{\sigma\sqrt{2\pi}} \int_0^x exp(-\frac{t^2}{2\sigma^2})dt \ )\ x^2 \tag{9}$$

$$+ (b + \frac{\delta\sigma}{2\sqrt{2\pi}} exp(-\frac{x^2}{2\sigma^2}))\ x$$

$$+ (a + \frac{c\sigma^2}{2} + \frac{\delta\sigma}{2\sqrt{2\pi}} \int_0^x exp(-\frac{t^2}{2\sigma^2})dt \ )$$

The first and second derivatives of z_σ are now given by:

$$z'_\sigma = (c + \frac{\delta}{\sigma\sqrt{2\pi}} \int_0^x exp(-\frac{t^2}{2\sigma^2})dt \ )\ x \tag{10}$$

$$+ (b + \frac{\delta\sigma}{\sqrt{2\pi}} exp(-\frac{x^2}{2\sigma^2})\ );$$

$$z''_\sigma = c + \frac{\delta}{\sigma\sqrt{2\pi}} \int_0^x exp(-\frac{t^2}{2\sigma^2})dt \ . \tag{11}$$

In particular, we deduce from Equation (11) that the curvature has a zero crossing if and only if

$$\frac{1}{\sqrt{2\pi}} \int_0^{x/\sigma} exp(-\frac{u^2}{2})du = -\frac{c}{\delta}. \tag{12}$$

This equation has a solution if and only if the absolute value of c/δ is less than $1/2$, which simply corresponds to c_l and c_r having opposite signs. It follows that *smooth joins* of the sort shown in Figure 7(a) generate a zero crossing in curvature, hence a parabolic point on the surface. Those shown in Figure 7(b) do not.

For example, the parabolic lines found on the light bulb shown in Figure 8 are *smooth joins*. The two-step process utilizing the Curvature Primal

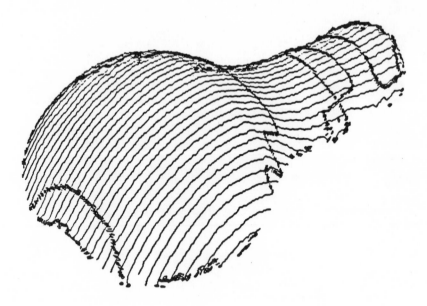

Figure 8:   Parabolic lines found by the program described in Section 4 for the lightbulb shown in Figure 2. The smooth join between the stem and bulb were not found by the program described in Section 3.

Sketch algorithm, discussed in the previous section, failed to find the *smooth joins* (see [3] Figure 1).

Equation (12) implies that the distance from the zero crossing location x_σ to the origin is proportional to σ. Using this property and the fact that z'' is zero at x_σ, it is then easy to use the Implicit Function Theorem to show that

$$\frac{\kappa''}{\kappa'}(x_\sigma) = \frac{z_\sigma''''}{z_\sigma'''}(x_\sigma) = \frac{\gamma}{\sigma} \tag{13}$$

for some constant γ. It follows that the ratio of the second and first derivatives of the curvature in x_σ is inversely proportional to σ. This scale space behavior allows us to discriminate zero crossings due to steps from those due to smooth joins.

4.5. Shoulder discontinuities

A *step* discontinuity confounds information both about the geometry of the surface and the viewpoint. Shifting the viewpoint to the half space defined by the outward normal of the "riser" of the step typically changes the depth discontinuity to a pair of *roofs* of opposite sign whose separation again confounds geometry and viewpoint. We introduce the *shoulder* discontinuity to cater for this situation (Figure 9).

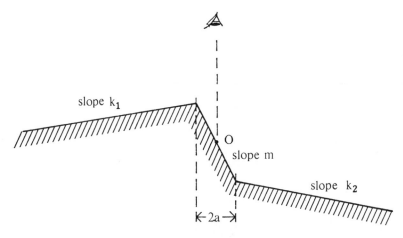

Figure 9: The *shoulder* discontinuity consists of two *roofs* of opposite sign. The shoulder appears as a *step* when viewed from the half-space defined by the inward normal of the "riser".

We may expect the scale space behavior of the *shoulder* to closely resemble that of the *step* when the projected separation $2a$ of the *roofs* is small compared to the filter size σ, perhaps becoming more like a pair of *roofs* as the viewpoint shifts. This is what we find.

We model the *shoulder* by the function

$$z = \begin{cases} k_1 x + (k_1 - m)a, & x < -a; \\ mx & x \in [-a, +a]; \\ k_2 x + (m - k_2)a, & x > +a. \end{cases} \tag{14}$$

If we denote $k_1 - m$ by δ_1 and $k_2 - m$ by δ_2, then $\delta_i \neq 0$, and δ_2 / δ_1 is positive (otherwise the curve is always convex, or always concave). It is easy to show that the second derivative of the shoulder, smoothed at scale σ is

$$z''_\sigma = \frac{\delta_2}{\sigma\sqrt{2\pi}} exp(-\frac{(x-a)^2}{2\sigma^2}) - \frac{\delta_1}{\sigma\sqrt{2\pi}} exp(-\frac{(x+a)^2}{2\sigma^2}). \tag{15}$$

Since δ_2/δ_1 is assumed positive, we deduce from Equation (15) that the curvature has a zero crossing. The location of the zero crossing is given by

$$x_\sigma = \frac{\sigma^2}{2a} log(\frac{\delta_1}{\delta_2}). \tag{16}$$

Using the Implicit Function Theorem as in the case of the *roof*, it is then straightforward to show that

$$\frac{\kappa''}{\kappa'}(x_\sigma) = \frac{z'''''_\sigma}{z'''_\sigma}(x_\sigma) = -\frac{1}{a} log(\frac{\delta_1}{\delta_2}), \tag{17}$$

so the ratio of the first and second derivatives at the zero crossing is constant over scales.

4.6. Thin bar and other compound discontinuities

The models considered so far involve *isolated* surface changes. Even though a *shoulder* may appear as a pair of *roofs* if it is viewed close to the normal to the riser and if the riser subtends a sufficient visual angle, its more typical behavior is like that of a *step*. As two (or more) surface changes are brought more closely together, so that the filter width σ approaches half the separation of the changes, the filter responses due to the individual changes interfere with each other. Since certain kinds of compound surface discontinuity are important for recognition and use of objects, they must be modeled and matched by the program.

This observation raises two questions: (i) which compound surface changes should be modeled and matched; and (ii) how shall instances be found by the program? Ultimately, the answer to (i) is application-dependent, though the *thin bar*, consisting of a *step* up closely followed by a *step* down, presses for inclusion (Figure 10). *Thin bars* occur as ribs on many

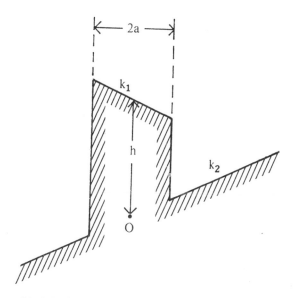

Figure 10: Model of a *thin bar* compound surface discontinuity. It consists of a plateau of height h, width $2a$, and slope k_2, resting on flat ground of slope k_1.

surfaces, for example along the sides of the neck of a connecting rod. Also, it seems [22], [24] that the mammalian visual system is sensitive to thin intensity stripes. In the case of curvature changes along a planar curve, the *crank* [1] is analogous to a thin bar since it consists of a corner followed closely by one of opposite sign. Other compound surface changes that might be important are a rounded corner and a moulding that is like a thin bar but with one of its risers smooth and concave. We have not studied such configurations.

Restricting attention to *thin bars* raises question (ii): how shall instances be recognized? First, let us conjecture what the curvature response to a *thin bar* might look like. We may base our conjecture on Asada and Brady's analysis [1] of a *crank*, though we need to be cautious because their

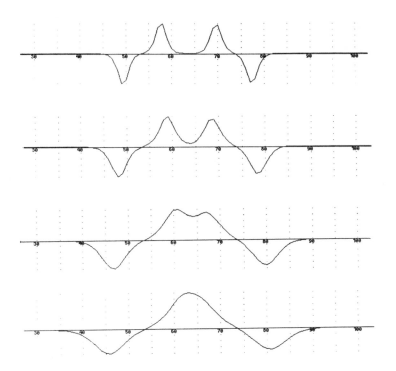

Figure 11: Expected curvature response of a *thin bar* to Gaussian filters. At fine scales, the *thin bar* signals two separate *step*s; at coarser scales it resembles a difference-of-Gaussians. The *step* responses begin to interfere when σ equals half the separation of the risers.

operators were linear. Figure 11 shows the response that might be expected, indeed the response that is generated in one special case (see below). Unfortunately, the response becomes substantially more complex in the general case.

Note that the *thin bar* in Figure 11 generates as many as five curvature peaks at fine scales, reducing to three at coarser scales. Note also that there appears to be a curvature peak at the origin. Asada and Brady extracted peaks at all scales and developed a matcher that linked peaks across scales. The *crank* model explicitly checked for three peaks splitting to five in the way shown in Figure 11. Matching such compound (planar curve curvature) changes was the source of the complexity of their program. In view of the non-linearity of surface curvature, and the two-dimensionality of surfaces, we are reluctant to implement an analogous peak matching program. In practice, the peaks from a *thin bar* may cover as many as fifteen pixels, suggesting error-prone and inefficient search. In this paper, we have sought *local statements* that apply to a single zero crossing or curvature extremum and studied its scale space behavior in isolation. As we shall see, however, the analysis is quite difficult for a thin bar, even in simple cases.

We analyze a model of a *thin bar* consisting of a plateau of height h, width $2a$, and slope k_2, resting on flat ground of slope k_1 (Figure 10). We model the *thin bar* by the function

$$z = \begin{cases} k_1 x, & x < -a; \\ h + k_2 x, & x \in [-a, +a]; \\ k_1 x, & x > +a. \end{cases} \tag{18}$$

We denote $k_2 - k_1$ by δ. The first and second derivatives of the smoothed *thin bar* are given by

$$z'_\sigma = k_1 + \frac{\delta}{\sigma\sqrt{2\pi}} \int_{x-a}^{x+a} exp(-\frac{t^2}{2\sigma^2}) dt \tag{19}$$

$$+ \frac{(h-\delta a)}{\sigma\sqrt{2\pi}} exp(-\frac{(x+a)^2}{2\sigma^2}) - \frac{(h+\delta a)}{\sigma\sqrt{2\pi}} exp(-\frac{(x-a)^2}{2\sigma^2});$$

$$z''_\sigma = \frac{1}{\sigma\sqrt{2\pi}} (\delta - \frac{(h-\delta a)(x+a)}{\sigma^2}) exp(-\frac{(x+a)^2}{2\sigma^2}) \tag{20}$$

$$- \frac{1}{\sigma\sqrt{2\pi}} (\delta - \frac{(h+\delta a)(x-a)}{\sigma^2}) exp(-\frac{(x-a)^2}{2\sigma^2}).$$

These expressions simplify considerably in the case that the plateau surface is parallel to the ground, that is $\delta = 0$. In particular, in that case $z'''_\sigma = 0$. However, the curvature attains an extremum at the origin (equivalently, $\kappa' = 0$) only when $k_1 = 0$. This is the case depicted in Figure 11, but it is too restrictive since it is too sensitive to changes in viewpoint. Further work is needed here. The special case $k_1 = k_2 = 0$ is that typically studied in psychophysical studies of intensity *thin bars* (e.g., [24]). It would be interesting to know what is the response to *thin bars* of intensity superimposed on a linear intensity ramp.

4.7. The matching algorithm

We now use the models introduced in the previous sections to *detect* and *localize* the surface intersections. We use a coarse to fine tracking of the extrema of curvature and parabolic points found at each scale, and direct it by using the particular features associated to each model.

We first smooth the original depth map with a Gaussian distribution at a variety of scales σ (see Figure 3). We then compute, for each smoothed version of the surface, the principal curvatures and their directions (using the method described in [3]). We also compute the first and second derivatives of the principal curvatures in their associated directions. All parabolic points, as well as directional maxima of the maximum curvature and minima of the minimum curvature, are then marked (Figure 12(a)). The marked points are thresholded, according to their type: all extrema whose curvature value is less than a given value are removed; all zero crossings whose slope is less than another value are also removed (Figure 12(b)).

The values of the thresholds vary according to the scale. The extremum threshold varies proportionally to $1/\sigma$, as suggested by the *roof* model. There is unfortunately no such clue for the zero crossing threshold. This is not a major problem however, as the thresholding step is used for selecting a set of candidates for the subsequent matching process, rather than finding the surface intersections themselves. For example, the curvature extrema parallel to the axis of the oil bottle, that are due to the non-linear smoothing (Figure 15), cannot be eliminated by thresholding, but are rejected by the

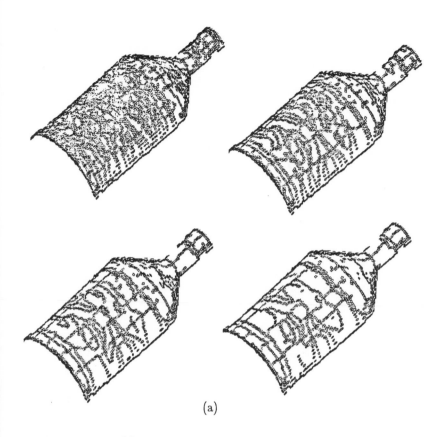

(a)

Figure 12: (a) The extrema and zero crossings of curvature marked on the oil bottle at four different increasing scales, from left to right.

matching algorithm (Figure 16).

The basic matching algorithm is a simple 2D extension of the Asada and Brady method. We track the thresholded extrema and zero crossings across the scales, from coarse to fine. We obtain a forest of points, equivalent to a "fingerprint" [29]. We then use the models developed in the previous sections to parse this forest. Paths in the forest are assigned a type, *roof,* *step, smooth join,* or *shoulder,* according to the behavior of the associated curvature and its first and second derivatives across the scales. This provides us with a *symbolic description* of the models instances that are found. Finally, we use a stability criterion (see Witkin [28]) to find the significant surface intersections (*detection phase*); they correspond to paths in the

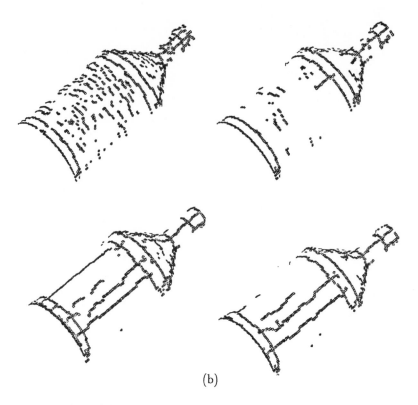

(b)

Figure 12: (b) The thresholded points. Note the curvature extrema parallel to the bottle axis are due to the non-linear smoothing, and the numerous parabolic points not thresholded at the finest scale. These non-significant points will be eliminated by the matching algorithm (Figures 13 and 16).

forest that go from the coarsest scale to the finest one. Points at the finest scale that have no ancestor at the coarsest scale are eliminated. The remaining points at the finest scale ensure the best *localization* of the surface intersections.

There are some problems with a straightforward implementation of this method. In order to stress the first one, we momentarily restrict ourselves to the case of the *roof*, which is the only one of our models to be characterized by a curvature extremum. In the *Curvature Primal Sketch*, the extrema of curvature are isolated points on a one-dimensional curve, and this makes the construction of the trees relatively simple. However, in the two-dimensional case, the surface intersections themselves form continuous curves. This makes it difficult, for each scale, to associate to each point a single ancestor, and to build an explicit representation of the forest. This, in turn, makes difficult an *a posteriori* interpretation of this forest. Instead, we associate to each couple of marked points belonging to two successive images a compatibility function. This function involves the Cartesian distance between the points and the angle between their associated principal directions. It also takes into account the *roof* model by comparing the ratio of the curvatures to the inverse of the ratio of the associated scales. At each scale, and for each thresholded extremum, we look for an ancestor inside a square window of the previous scale image. If an ancestor with a good enough score is found, then the point is kept as a potential ancestor for the next scale. Otherwise it is removed. This way, the forest is never explicitly built, and the interpretation is done during the tracking itself, as the only extrema tracked are those which correspond to potential *roofs*. In particular, this is how the artifacts due to non-linear smoothing are removed.

The case of the parabolic points is slightly more complicated, as they may correspond to different types of discontinuities. In fact, a point may be at the same time a zero crossing of the Gaussian curvature and an extremum of a principal curvature. We again define compatibility functions for *steps*, *shoulders*, and *smooth joins* that take into account their mathematical models. The behavior of the ratio of the second and first derivatives of the curvatures is the basis for these compatibility functions. At each scale, a point may be a candidate for several different types of intersections, and be associated to an ancestor of each of these types. The use of the succession of scales is usually sufficient to disambiguate these cases. However, if several interpretations subsist until the finest level, the interpretation with the best cumulated compatibility score is chosen.

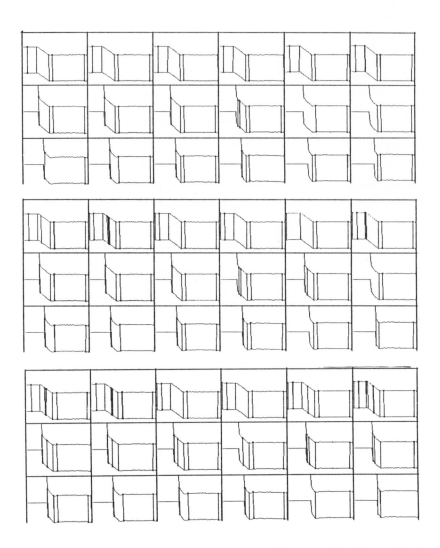

Figure 13: Matching between the different scales. The three parts of the figure show, for each couple of scales, points that have been matched on different slices of the oil bottle. These points are linked by a vertical line. Note that the matching is in fact made between points that are in a square window, not necessarily in the same slice. The display is done slice by slice for the sake of readability.

Figure 13 shows the matching algorithm at work, and Figure 16 shows the final result on the oil bottle. Note that, although we have analyzed the behavior of the position of the characteristic point of each of our models, and have found it simple and reliable, we do not use it in the compatibility functions. The reason is that, for each intersection found, the real origin on the surface is unknown. This implies that at least three matched points are necessary to estimate the parameters of the motion of the extremum or zero crossing across the scales, and makes the measure of this movement unsuitable for a scale-to-scale tracking. Note however that the estimation of the movement parameters could be used for an *a posteriori* verification of the surface intersections found.

5. Smoothing a surface with a Gaussian distribution

Brady, Ponce, Yuille, and Asada [3] discuss techniques for smoothing a depth map with a Gaussian distribution. The main difficulty stems from bounding contours, where the surface normal turns smoothly away from the viewer, and where there is typically a substantial depth change between points on the surface of the object and the background. In general, the bounding contour is easy to find with a simple edge operator or by thresholding depth values. The problem is how to take the boundary into account when smoothing the surface.

Brady, Ponce, Yuille, and Asada observed that if the smoothing filter is applied everywhere, the surface "melts" into the background and changes substantially. Figure 14 is reproduced from [3] and shows this. They suggested instead using repeated averaging [5], [6] as well as adapting Terzopoulos' technique [26] of computational molecules to prevent leakage across depth boundaries. This smooths the surface without substantially altering it (see Figure 14(c)).

Here we point out a slight difficulty in smoothing surfaces using the technique illustrated in Figure 14(c), and suggest a refinement. Although the smoothed surface appears to be close to the original, small orientation-dependent errors are introduced. These errors are magnified in computing the curvature (Figure 15(a)), to produce "false" curvature extrema near the

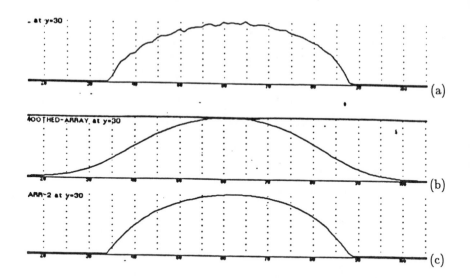

Figure 14:   (a) Raw data from a cross section of an oil bottle after scanning using the INRIA system;   (b)   Smoothing across surface boundaries with a Gaussian mask that is applied everywhere;   (c) Gaussian smoothing using repeated averaging and computational molecules. (Reproduced from [3], Figure 12)

boundary (compare the overshoot phenomenon in Terzopoulos' work [26] on detecting surface discontinuities). The overshoots do not exemplify Gibbs' ringing as we originally thought. Instead, the phenomenon has two causes:

- **The coordinate frame is not intrinsic.** The smoothing filter is applied in the $x-y$ plane, and since this is not intrinsic to the surface, the result is orientation-dependent. For example, the difference between a cylinder and its smoothed version monotonically increases toward the boundary from a value of zero where the normal faces the viewer.
- **Points near the boundary are not smoothed as much.** Such points are relatively unsmoothed as several of their computational molecules are continually inhibited. The result is that the difference between the smoothed and original surfaces decreases at a certain distance from the boundary. This creates an inflection point, which in turn creates an extremum of curvature.

It is possible to substantially reduce the effect of this problem by smoothing in intrinsic coordinates. At each point of the surface, the normal

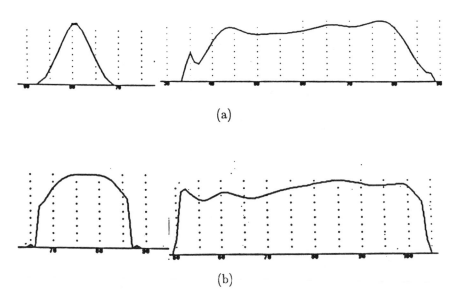

(a)

(b)

Figure 15:   (a) The curvature computed from the smoothed surface shown in Figure 14(c). Small orientation-dependent errors in smoothing are magnified. The first figure is the neck, the second part of the body; (b) The curvature on the slices shown in (a) computed using the intrinsic coordinate method described in the text

is estimated. Instead of smoothing z, the surface point is moved along its normal a distance that depends upon the projected distances of the point's neighbors from the tangent plane. In the case that the normal faces the viewer, this is equivalent to the previous technique. However, the result is no longer orientation-dependent. Figure 15(b) illustrates the computation of curvature after smoothing the oil bottle by this method. The drawback with the technique is the computation time; it requires one to compute the tangent plane at every point.

The technique described in Brady, Ponce, Yuille, and Asada [3] has been used in all the examples presented here, as it represents a good tradeoff between computational efficiency and faithful rendering of the smoothed surface.

6. Examples

In this section, we present a number of examples of the surface discontinuities found on simple objects by our algorithm. In all the examples, we use four different scales corresponding to 20, 40, 60, and 80 iterations of the smoothing filter described in Brady, Ponce, Yuille, and Asada [3]. Viewing the resulting centrally-limiting Gaussian distributions as approximately bandpass filters, they span one octave. Figure 16 shows the final output of the algorithm for the oil bottle. The points detected during the matching step are linked together using a connected components exploration algorithm. The smallest components (less than 3 or 4 pixels) are then removed. Conversely, points may have been missed during the previous phases, creating gaps in the lines that are found. These gaps are filled by adding points that have characteristics compatible with the detected points. The bottle is finally segmented into six parts, separated by three step edges and two roofs.

Brady, Ponce, Yuille, and Asada [3] showed that the coffee cup shown in Figure 17 is best represented as the join of a cylindrical body and a tube surface that corresponds to the handle. Here we show that the handle can be separated from the body using the algorithms described in this paper. Note that the surface intersections are of type *roof*.

The third example shows the surface intersections found on a telephone handset, Figure 18. All the major intersections have been found. The representation is not symmetric because the handset was not quite perpendicular to the scanner, causing part of the surface to be occluded. Note that the surface intersections are more reliably detected at the coarsest scale, but are more accurately localized at the finest scale.

The surface intersections found on a few simple tools, namely a hammer, a drill, and the head of a screwdriver are shown in Figure 19. Figure 20 shows the surface intersections found on an automobile part that has featured in several papers by the group at INRIA. On this complicated

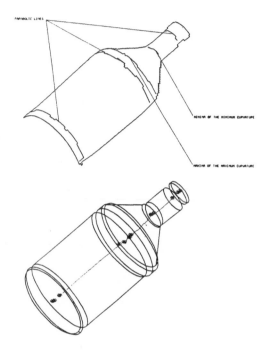

Figure 16: The oil bottle is segmented into six parts. Three step and two roof intersections are found by the algorithm described in this paper. The algorithms described in Brady, Ponce, Yuille, and Asada [3] determine the lines of curvature of the parts of the oil bottle, fit circles to the parallels, and fit axes to the centers of those circles.

object, global lines of curvature have no significance, so the Curvature Primal Sketch would not perform well. Notice in particular the circular step edge found on the left "head" of the part: it corresponds to a shallow depression whose depth is about one millimeter. This is approximately at the resolution limit of the laser scanner, and underlines the practical significance of the algorithms described here.

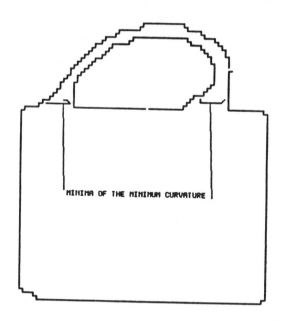

MINIMA OF THE MINIMUM CURVATURE

Figure 17: The joins of the handle to the body of the coffee mug are computed by the algorithms described in this paper. They are determined to be of type *roof*.

Figure 21 shows the surface intersections found on the head of a connecting rod. The current state of our algorithm cannot deal with the *thin bars* located on the sides of the neck of the connecting rod, but performs well for the other intersections.

The last example is the mask of a human face (Figure 22). The program finds face features as the nose, the eyes, and the mouth. This shows its ability to deal with arbitrary curved surfaces, usually not found in man-made objects.

Although our primary concern in this paper has been with the intrinsic geometry of a surface as found by a three-dimensional vision system, one might suppose that the methods described in this paper could be applied straightforwardly to *extract* and *interpret* significant intensity changes in images, considered as surfaces. To interpret intensity changes, it is necessary to take irradiate effects into account, since intensity changes do not always correspond to surface changes. Rather, they may signify

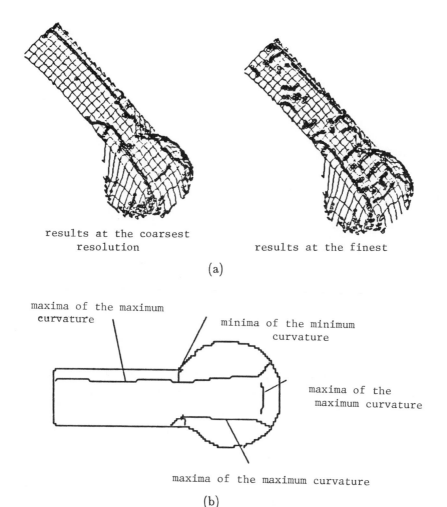

results at the coarsest
resolution

results at the finest

(a)

maxima of the maximum
curvature

minima of the minimum
curvature

maxima of the
maximum curvature

maxima of the maximum curvature

(b)

Figure 18: (a) The surface intersections found on a telephone handset at the coarsest and finest scales, approximately an octave apart. The intersections are more reliably detected at the coarsest scale; they are more accurately localized at the finest scale. (b) The results of matching the changes across scales.

reflectance or illumination changes. Extraction and interpretation was in fact the intent of Marr's original work [22] on the Primal Sketch, though considerably more attention was paid to extraction than interpretation. More recently, Haralick, Watson, and Laffey [14] have advocated a representation of image surface geometry that involves concave and convex

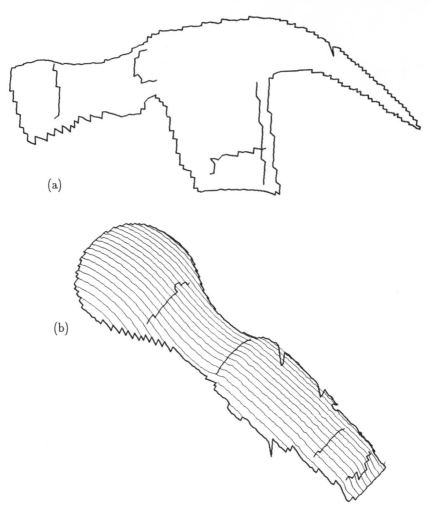

Figure 19:   The intersections found by the program on simple tools.
(a) a hammer; (b) a drill

hills, planar regions, and saddles. These geometrical aspects of the image
surface are not interpreted in terms of the intrinsic geometry of the surface,
or of illumination or reflectance changes.

Some preliminary work exists on interpretation of intensity changes. An
early edge finder developed by Binford and Horn [2] included filters for step,
roof, and "edge effect" changes.   Horn's study of intensity changes
[15] included a suggestion that occluding boundaries and reflectance
changes correspond to *step* intensity changes, while concave surface

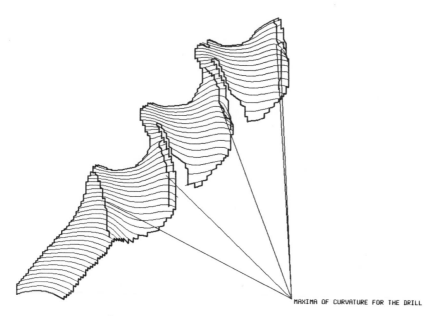

MAXIMA OF CURVATURE FOR THE DRILL

Figure 19: (c) a screwdriver

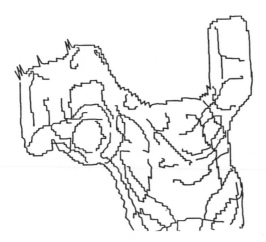

Figure 20: Surface intersections found on an automobile part (see, for example, [8], [9]). The circular step edge found on the left "head" of the part corresponds to a shallow depression whose depth is about one millimeter. This is approximately at the resolution limit of the laser scanner.

intersections generate *roof* intensity changes (because of mutual illumination). Finally, Yuille [29] suggests that certain points along lines of

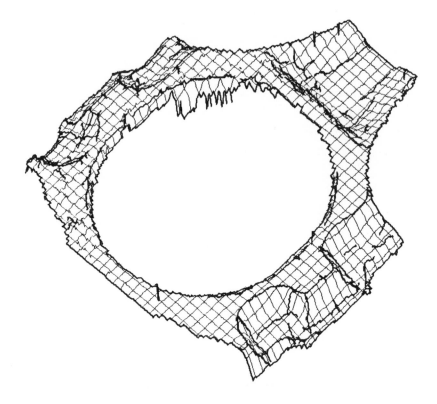

Figure 21:   Surface intersections found on a connecting rod. Only the head of the part is shown. The inclusion of the *thin bar* models in the algorithm would allow a fine description of the neck of this object.

curvature of a surface can be extracted directly from an image. There is much scope for additional work along these lines.

Figure 23 shows an initial experiment we have carried out on applying the methods developed in this paper to image surfaces. The join of the wing to the fuselage of the airplane is determined to be *roof* changes, consistent with Horn's suggestion.

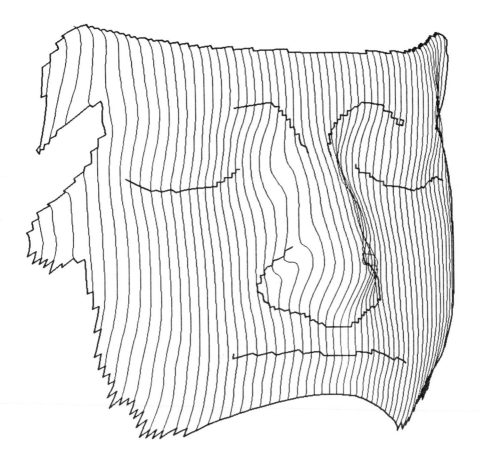

Figure 22: Surface intersections found on a mask. The mouth, the nose and the eyes are found as corners by the program.

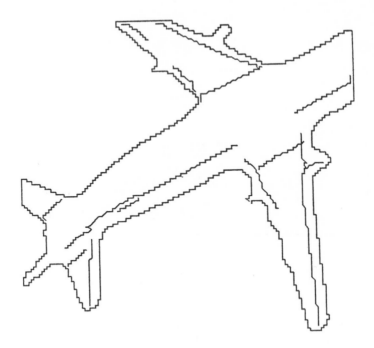

Figure 23: Application of the methods of the paper to an intensity surface. The interest is in the type of the internal intensity changes. The join between the wings and fuselage is a roof, suggesting a concave surface intersection.

Acknowledgments

    This report describes research done at the Artificial Intelligence Laboratory of the Massachusetts Institute of Technology.   Support for the laboratory's Artificial Intelligence research is provided in part by the Advanced Research Projects Agency of the Department of Defense under Office of Naval Research contract N00014-80-C-0505, the Office of Naval Research under contract number N00014-77-C-0389, and the System Development Foundation.   This work was done while Jean Ponce was a visiting scientist on leave from INRIA, Paris, France. We thank several people who have commented on the ideas presented in this paper, particularly Ruzena Bajcsy, Bob Bolles, Jean-Daniel Boissonat, Bernard and Hilary Buxton, Olivier Faugeras, Eric Grimson, Berthold Horn, Ramesh Jain, Takeo Kanade, Christopher Longuet-Higgins, David Mumford, Joe Mundy, Tommy Poggio, Robin Stanton, Demetri Terzopoulos, and Alan Yuille. We are indebted to Olivier Faugeras and the INRIA group for giving us access to their ranging system, and to Philippe Brou for bringing his ranging system back to life.

References

[1] Asada, Haruo and Brady, Michael.
The curvature primal sketch.
Technical Report AIM-758, MIT, Artificial Intelligence Lab, 1984.

[2] Binford, Thomas O.
Inferring surfaces from images.
Artificial Intelligence 17:205-244, 1981.

[3] Brady, M., Ponce, J., Yuille, A. and Asada, H.
Describing surfaces.
In *Proc. Seond Int. Symp. on Robotics Research*, pages 434-445.
Kyoto, Japan, 1984.

[4] Brou, P.
Finding the orientation of objects in vector maps.
Int. J. Rob. Res. 3(4):89-175, 1984.

[5] Burt, Peter J.
Fast filter transforms for image processing.
Comp. Graph. and Im. Proc. 16:20-51, 1981.

[6] Burt, Peter J.
Fast algorithms for estimating local image properties.
Comp. Graph. and Im. Proc. 21:368-382, 1983.

[7] Canny, J. F.
Finding edges and lines in images.
Technical Report TR-720, MIT, Artificial Intelligence Lab, 1983.

[8] Faugeras, O.D., et. al.
Towards a flexible vision system.
In A. Pugh (editor), *Robot Vision*. IPS, UK, 1982.

[9] Faugeras, O. D.
New steps toward a flexible 3-D vision system for robotics.
*Preprints of the Second International Symposium of Robotics
Research. Note: also see In Proc. 7th Int. Conf. Pattern
Recogn., pages 796-805; Montreal, Canada, July* :222-230, 1984.

[10] Faugeras O.D., Hebert, M.
A 3-D recognition and positioning algorithm using geometrical
matching between primitive surfaces.
In *Proc. Eighth Int. Joint Conf. On Artificial Intelligence*, pages
996-1002. Los Altos: William Kaufmann, August, 1983.

[11]   Faugeras, O. D. and Hebert, M.
        The representation, recognitin, and positioning of 3-D shapes from
            range data.
        to appear in the Int. Journal of Robotics Research , 1985.

[12]   Faugeras, O. D., Hebert, M., Pauchon, E., and Ponce, J.
        Object representation, identification, and positioning from range
            data.
        First International Symposium on Robotics Research.
        MIT Press, 1984, pages 425-446.

[13]   Grimson, W. E.
        Computational experiments with a feature based stereo algorithm.
        Technical Report 762, MIT, Artificial Intelligence Lab, 1984.

[14]   Haralick, Robert M., Watson, Layne T., Laffey, Thomas J.
        The topographic primal sketch.
        Int. J. Rob. Res. 2(1):50-72, 1983.

[15]   Horn, B. K. P.
        Understanding image intensities.
        Artif. Intell. 8, 1977.

[16]   Huffman, D. A.
        Impossible objects as nonsense sentences.
        Machine Intelligence 6.
        Edinburgh University Press, Edinburgh, 1971, pages 295-323.

[17]   Ikeuchi, K., Horn, B. K. P., et al.
        Picking up an object from a pile of objects.
        First International Symposium on Robotics Research.
        MIT Press, 1984, pages 139-162.

[18]   Jacobsen, S. C., Wood, J. E., Knutti, D. F., and Biggers, K. B.
        The Utah/MIT dextrous hand: work in progress.
        Int. J. Rob. Res. 3(4), 1984.

[19]   Jacobsen, S. C., Wood, J. E., Knutti, D. F., Biggers, K. B., and
        Iversen, E. K.
        The verstion I UTAH/MIT dextrous hand.
        Proc. of the 2nd International Symposium on Robotics Research.
        MIT Press, Cambridge, 1985.

[20]   Kapur, D., Mundy, J. L., Musser, D., and Narendran, P.
       Reasoning about three-dimensional space.
       In *IEEE Int. Conf. on Robotics*, pages 405-410. St. Louis, MO,
          March, 1985.

[21]   Little, J. J.
       Recovering shape and determining attitude from extended
          Gaussian images.
       Technical Report, Univ. of British Columbia, Dept. of CS, 1985.

[22]   Marr, D.
       Early processing of visual information.
       Phil. Trans. Roy. Soc. B275:843-524, 1976.

[23]   Marr, D. and Hildreth, E. C.
       Theory of edge detection.
       Proc. R. Soc. London B 207:187-217, 1980.

[24]   Richter, J. and Ullman, S.
       A model for the spatio-temporal organisation of X and Y type
          ganglion cells in the primate retina.
       Biol. Cyb. 43:127-145, 1982.

[25]   Salisbury, J. Kenneth and Craig, John J.
       Articulated hands: force control and kinematic issues.
       Int. J. Rob. Res. 1(1):4-17, 1982.

[26]   Terzopoulos, D.
       The role of constraints and discontinuities in visible-surface
          reconstruction.
       Proc. 7th Int. Jt. Conf. Artif. Intell., Karlsruehe :1073-1077, 1983.

[27]   Turner, K.
       Computer perception of curved objects using a television camera.
       Technical Report, Edinburgh University, 1974.

[28]   Witkin, A.
       Scale-space filtering.
       In *7th Int. Jt. Conf. Artificial Intelligence*, pages 1019-1021.
          Karlsruhe, 1983.

[29]   Yuille, A. L.
       Zero-crossings on lines of curvature.
       In *Proc. Conf. Amer. Assoc. Artif. Intell..* Washington, August,
          1983.

[30]   Yuille, A. L. and Poggio, T.
        Fingerprints theorems for zero-crossings.
        Technical Report AIM-730, MIT, AI Lab, 1983.

[31]   Yuille, A. L. and Poggio, T.
        Scaling theorems for zero crossings.
        Technical Report AIM-722, MIT, AI Lab, 1983.

3-D Object Representation from Range Data Using Intrinsic Surface Properties

B. C. Vemuri, A. Mitiche, and J.K. Aggarwal

Laboratory for Image and Signal Analysis
College of Engineering
The University of Texas at Austin
Austin, TX 78712

Abstract

A representation of three-dimensional object surfaces by regions homogeneous in intrinsic surface properties is presented. This representation has applications in both recognition and display of objects, and is not restricted to a particular type of approximating surface.

Surface patches are fitted to windows of continuous range data with tension splines as basis functions. Principal curvatures are computed from these local surface approximations, and maximal regions formed by coalescing patches with similar curvature properties.

1. Introduction

Range (depth) data provide an important source of 3-D information about the shape of object surfaces. Range data may be derived from either intensity images or through direct measurement sensors.

In recent years, considerable attention has been devoted to the problem of analyzing intensity images of 3-D scenes to obtain depth information [18], [2], [14]. Much of the work has been focussed on overcoming the difficult problems caused by projection, occlusion, and illumination effects such as shading, shadows, and reflections. Many monocular depth cues can be exploited to produce three-dimensional interpretations of two-dimensional images. These cues include texture gradient [15], size perspective (diminution of size with distance), motion parallax [25], occlusion effects [17], depth from optic flow, surface orientation from surface contours [17], and surface shading variations [12].

According to many researchers the problem of deriving the 3-D structure of an object from its intensity image is equivalent to inferring the orientations of the surfaces of the object. The word inference is used because there is insufficient information in the intensity image to allow the visual system to merely recover or extract the surface orientation. Rather, the visual system must make explicit assumptions about the nature of the surfaces it sees. Although depth cues in a single intensity image are weak, there are generally strong cues to surface orientation; these were exploited by Stevens [27]. To determine 3-D structure from intensity information obtained from a single view of an object would require several restrictive constraints in the form of additional information about the objects that populate the environment of interest. Fundamentally this is because of the many-to-one nature of perspective projections; an infinite number of 3-D objects can correspond to any single view. Stereo-scopic vision [19] and structure from motion [30] may be used to effect three-dimensional reconstruction of objects. Both these methods may be considered as correspondence techniques, since they rely on establishing a correspondence between identical items in different images; the difference in projection of these items is used to determine the surface shape. If this correspondence is to be determined from the image data there must be sufficient visual information at the matching points to establish a unique pairing. Two basic problems arise in relation to this requirement. The first arises at parts of the image where uniformity of intensity or color makes matching impossible, the second when the image of some part of the scene appears in only one view of a stereo pair because of occlusion (the missing part problem) or a limited field of view. The further apart the two camera positions, the potentially more accurate the disparity depth calculation, but the more prevalent the missing part problem and the smaller the field of view. Another limitation of correspondence-based methods stems from the large number of computations required to establish the correspondence.

The use of direct range measurements can eliminate some of the problems associated in deriving 3-D structure from intensity images. Capturing the third dimension through direct range measurement is of great utility in 3-D scene analysis, since many of the ambiguities of interpretation arising from occasional lack of correspondence between object boundaries and from inhomogeneities of intensity, texture, and color can be trivially resolved [13]. One of the disadvantages of direct range measurement is that it is somewhat time consuming. Several range finding techniques have been proposed in literature; a good survey of those in use is presented in Jarvis [13]. Two widely used ranging methods are based on the principle of triangulation [26], [1], [24] and time of flight [22].

There have been several studies on scene description using range data. Will and Pennington [32] describe a method by which the locations and orientations of planar areas of polyhedral solids are extracted through linear frequency domain filtering of images of scenes illuminated by a high contrast rectangular grid of lines. Shirai and Suwa [26] use a simple lighting scheme with a rotating slit projector and TV camera to recognize polyhedral objects. The facts that projected lines are equally spaced, parallel and straight on planar surfaces are exploited in reducing the computational complexity; only the end points of each line segment are found, and line grouping procedures are used to identify distinct planar surfaces. Agin and Binford [1] describe a laser ranging system capable of moving a sheet of light of controllable orientation across a scene; their aim was to derive descriptions of curved objects based on a generalized cylinder model. Their method is interactive and is useful in describing objects with cylinder like or elongated parts. Mitiche and Aggarwal [20] discuss an edge detection technique for noisy range images. Their motivation for detecting edges was to delineate regions on the surface of objects, and the technique is limited to finding edges in prespecified directions. Duda, Nitzan and Barret [8] use intensity as a guide for finding planes, with plane equations being determined from the registered range images. Their technique assumes underlying surfaces to be planar. Faugeras, et al. [9] are quite successful in segmenting range data from approximately planar and quadratic surfaces. Their segmentation process partitions data points into subsets in which approximation by a polynomial of order at most two yields an error less than a given threshold. However, their edge detection scheme is prone to large errors due to the fact that planes are used to detect surface creases. Also, region merging requires explicit updating of region equations and could be quite time consuming. Grimson [10] presents a theory of visual surface interpolation, where the source of range data is a pair of stereo images. The theory deals with determining the best-fit surface to a sparse set of depth values obtained from the Marr and Poggio [19] stereo algorithm. Grimson's algorithm assumes that the entire scene is a single surface although this is not a valid assumption along occluding contours. Also, the iterative algorithm suggested converges slowly. Oshima and Shirai [23] describe scenes by surfaces and relations between them. Their intent was to develop a program which described a scene containing both planar and curved objects in a consistent manner. The primitives used in their algorithm are planar surface elements which are not adaptable to the shape of the surface. Following a region growing process, which requires explicit updating of region equations, a region refinement procedure is described which involves fitting quadrics to curved regions. This process is computationally expensive. Terzopolous [29] presents a technique for visible surface computation from multiresolution images. His surface reconstruction process treats surfaces of objects as thin plate patches bounded by depth discontinuities and joined by membrane strips along loci of orientation discontinuities. He suggests the use of intrinsic properties of surfaces, computed post reconstruction, to describe shape. His motivation for computing these intrinsic properties was to demonstrate the robustness of his surface representation scheme.

In this paper a representation of visible 3-D object surfaces by regions that are homogeneous in certain intrinsic surface properties is presented. First smooth patches are fitted to the object surfaces; principal curvatures are then computed and surface points classified accordingly. The surface fitting technique is general, efficient and uses existing public domain software for implementation. An algorithm is presented for computing object descriptions. The algorithm divides the range data array into windows and fits approximating surfaces to those windows that do not contain discontinuities in range. The algorithm is not restricted to polyhedral objects nor is it committed to a particular type of approximating surface. It uses tension splines, which make the fitting patches locally adaptable to the shape of object surfaces. Computations are local and can be implemented on a parallel architecture . Our ultimate goal is to recognize objects independent of the viewpoint; therefore, this necessitates a representation which is invariant to rigid motion.

In Section 2 we describe the data acquisition mechanism, some relevant concepts in differential geometry are recalled, and the surface fitting technique used in building object descriptions is introduced. Section 3 describes in detail the algorithm for building object descriptions. In section 4 some experimental results using real range images are presented. Section 5 contains a summary.

2. Surface Fitting and Intrinsic Features Computation

The range information used in this study is obtained from a single view of the scene by the White scanner [28], a laser ranging system shown in Figure 1. This system works on the principle of light sheet triangulation. A sheet of light produced with a laser and a cylindrical lens is cast on a scene, resulting in a bright curve when viewed through a video camera. A rotating mirror causes this sheet of light to move across the scene. The intersection of

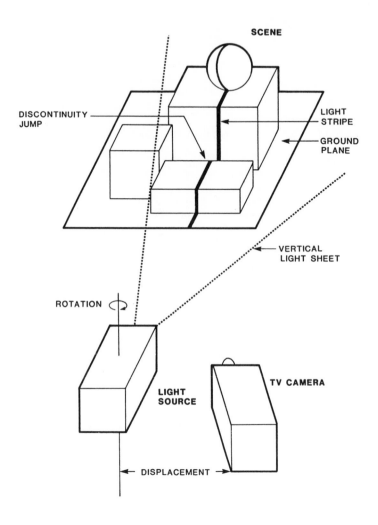

Figure 1:   Laser ranging system [13]

this plane of laser light with projecting rays through the video image (bright curve) determines the position of points in space. The actual output of the device is a set of (x,y,z) coordinates of points in the scene. These coordinates are stored in three two-dimensional arrays of size $m \times n$.

2.1. Surface Fitting

In this section we discuss the details of the surface fitting technique used in computing object descriptions.

The surface fitting problem can be stated as follows: given $m \times n$ arrays of $x, y,$ and z coordinates, we want to determine a surface that approximates the data in the least-squares sense and is smooth. Smoothness is interpreted as the twice differentiability of the surface at all points. Fitting a surface to an $L \times L$ window of data is equivalent to computing a surface fit for a roughly rectangular grid of data values. Intuitively, a roughly rectangular grid is one that has been obtained from a rectangular mesh that has been deformed to fit the surface. The surfaces need not be single valued; i.e., one coordinate need not be a function of the other two, as in $z = f(x,y)$. In fact, the surfaces are represented in the parametric form

$$x = f(s,t),$$

$$y = g(s,t), \tag{1}$$

$$z = h(s,t),$$

where s and t are the parameters and the functions f, g, and h are tensor products of splines in tension. This fitting is a mapping from a two-dimensional parameter space to a three-dimensional Cartesian space.

Note that points (x,y,z) and (x,y,z') could both occur within the grid with $z \neq z'$. It is assumed however that no two adjacent rows and no two adjacent columns are identical.

More precisely, let $p_{ij} = (x_{i,j}, y_{i,j}, z_{i,j})$. We require

$$p_{i,j} \neq p_{i+1,j}, \tag{2}$$

$$p_{i,j} \neq p_{i,j+1} \tag{3}$$

for all $i=1,\ldots,m$ and $j=1,\ldots,n$. If these conditions are not satisfied then the parameterization fails.

The first step towards a solution to the surface fitting problem is to obtain the parametric grids $\{s_i\}_{i=1}^m$ and $\{t_j\}_{j=1}^n$ so that for each $i=1,\ldots,m$ and each $j=1,\ldots,n$ there exist s_i and t_j such that

$$(x(s_i,t_j), y(s_i,t_j), z(s_i,t_j)) = (x_{i,j}, y_{i,j}, z_{i,j}) \qquad (4)$$

where x, y and z are twice differentiable functions of s and t. It is required to have $s_1=t_1=0$ and $s_m=t_n=1$. The quantity $s_{i+1}-s_i$ is made equal to the average $(\text{over } j=1,\ldots,n)$ normalized distance between $p_{i+1,j}$ and $p_{i,j}$. Similarly, $t_{j+1}-t_j$ is made equal to the average $(\text{over } i=1,\ldots,m)$ normalized distance between $p_{i,j+1}$ and $p_{i,j}$. Precisely,

$$s_{i+1} = s_i + \frac{1}{n}\sum_{j=1}^n \left(\frac{\|p_{i+1,j}-p_{i,j}\|}{\sum_{l=1}^{m-1}\|p_{l+1,j}-p_{l,j}\|} \right), \qquad (5)$$

$$t_{j+1} = t_j + \frac{1}{m}\sum_{i=1}^m \left(\frac{\|p_{i,j+1}-p_{i,j}\|}{\sum_{l=1}^{n-1}\|p_{i,l+1}-p_{i,l}\|} \right), \qquad (6)$$

where $\|p_{i+1,j}-p_{i,j}\|$ is the Euclidean distance. It is easy to show, given these definitions and $s_1=t_1=0$, that indeed $s_m=t_n=1$.

Having determined the parametric grids, the surface approximation problem is now reduced to three standard surface fitting problems over rectangular grids, namely

- $x(s_i, t_j) = x_{i,j}$ *for* $i=1, \ldots, m$ *and* $j=1, \ldots, n$

- $y(s_i, t_j) = y_{i,j}$ *for* $i=1, \ldots, m$ *and* $j=1, \ldots, n$

- $z(s_i, t_j) = z_{i,j}$ *for* $i=1, \ldots, m$ *and* $j=1, \ldots, n$

These problems can be solved using the following formulation by Cline [3], [4], [5]. We are given a grid defined by two strictly increasing abscissa sets, $\{s_i\}_{i=1}^m$ and $\{t_j\}_{j=1}^n$, an ordinate set $\{\{z_{ij}\}_{i=1}^m\}_{j=1}^n$, positive weights $\{\delta s_i\}_{i=1}^m$ and $\{\delta t_j\}_{j=1}^n$, a non-negative tolerance S, and a non-negative tension factor σ. We wish to determine a function g which minimizes

$$\int_{s_1}^{s_m} \int_{t_1}^{t_n} \{g_{sstt}(s,t)\}^2 dsdt + \sigma^2 \sum_{i=1}^{m-1} \sum_{j=1}^{n-1} \int_{s_i}^{s_{i+1}} \int_{t_j}^{t_{j+1}} \{Q_{ij}(s,t)\}^2 dsdt \quad (7)$$

over all functions $g \in C^2[s_1, s_m] \times [t_1, t_n]$ such that

$$\sum_{i=1}^m \sum_{j=1}^n \{\frac{(g(s_i, t_j) - z_{ij})}{\delta s_i \delta t_j}\}^2 \le S. \quad (8)$$

where

$$Q_{ij}(s,t) = g_{st}(s,t) - \{\frac{(g(s_{i+1}, t_{j+1}) - g(s_i, t_{j+1}) - g(s_{i+1}, t_j) + g(s_i, t_j))}{(s_{i+1} - s_i)(t_{j+1} - t_j)}\}$$

It can be shown that the solution to this problem is a tensor product of splines under tension (i.e., g is a one dimensional spline under tension when restricted to any value s or to any value of t). Also, g can reproduce local monotonicity, local positivity and local convexity as the one dimensional smoothing spline under tension can [5]. Because all three subproblems have several components in common it is possible to solve them concurrently.

The surface fitted to the roughly rectangular grid can be represented by three tensor products of splines under tension:

$$x(s,t) = \sum_i \sum_j \alpha_{ij} \, \phi_i(s) \psi_j(t) \, ,$$

$$y(s,t) = \sum_i \sum_j \beta_{ij} \, \theta_i(s) \nu_j(t) \, , \tag{9}$$

$$z(s,t) = \sum_i \sum_j \gamma_{ij} \, \rho_i(s) \kappa_j(t) \, ,$$

where ϕ, ψ, θ, ν, ρ and κ are tension splines; and α, β, and γ are coefficients of the tensor products.

2.2. Curvature Computation

In this section we discuss the computation of principal curvatures at a point on the parameterized, fitted, surface. Principal curvatures at a point on a surface indicate how fast the surface is pulling away from its tangent plane at that point. In the presence of an edge the principal curvatures will achieve a local maximum. Therefore it is appropriate to declare as edges those points at which the curvature is above a given threshold. The partial derivative information required to compute the principal curvatures are provided by the surface fitting software [5].

In general, when a plane passing through the normal to the surface at a point P is rotated about this normal, the radius of curvature of the section changes; it will be at a maximum r_1 for a normal section s_1 and a minimum r_2 for another normal section s_2 [11]. The reciprocals $k_1 = 1/r_1$ and $k_2 = 1/r_2$ are called the principal curvatures; the directions of tangents to s_1 and s_2 at P are called the principal directions of the surface at P (see Figure 2). The Gaussian curvature at point P is defined as the product of the two principal curvatures. The Gaussian curvature is an intrinsic property of the surface,

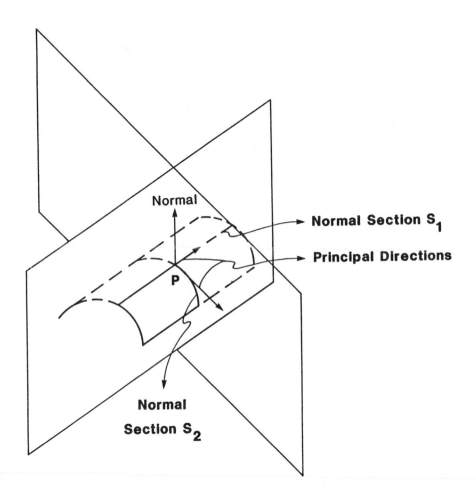

Figure 2: A surface with normal sections along the principal directions

since it depends only on the coefficients of the first fundamental form and their derivatives, whereas the two principal curvatures are extrinsic properties. The coefficients of the first and second fundamental forms (E, F, G and e, f, g as given below) determine the surface uniquely up to a rigid body transformation [7]. The principal curvatures at a point on a surface are computed in terms of the parameters s and t as follows. Let $\vec{X}(s,t)$ represent the surface.

$$\vec{X}(s,t) = (x(s,t),\, y(s,t),\, z(s,t)),\tag{10}$$

$$\vec{X}_s = dX/ds = [x_s\, y_s\, z_s],\tag{11}$$

$$\vec{X}_t = dX/dt = [x_t\, y_t\, z_t],\tag{12}$$

$$\vec{X}_{st} = d^2X/dsdt = [x_{st}\, y_{st}\, z_{st}],\tag{13}$$

$$\vec{X}_{ss} = d^2X/ds^2 = [x_{ss}\, y_{ss}\, z_{ss}],\tag{14}$$

$$\vec{X}_{tt} = d^2X/dt^2 = [x_{tt}\, y_{tt}\, z_{tt}].\tag{15}$$

Let

$$E = |\vec{X}_s|^2 = (x_s^2 + y_s^2 + z_s^2),\tag{16}$$

$$F = \vec{X}_s \cdot \vec{X}_t = (x_s x_t + y_s y_t + z_s z_t).\tag{17}$$

$$G = |\vec{X}_t|^2 = (x_t^2 + y_t^2 + z_t^2).\tag{18}$$

The unit normal is then given by

$$\vec{X} = \frac{\vec{X}_s \times \vec{X}_t}{\sqrt{EG - F^2}},\tag{19}$$

The Gaussian curvature K at any point on a surface is defined as the product of two principal curvatures k_1 and k_2, and is given by

$$K = k_1 k_2 = \frac{eg - f^2}{EG - F^2},\qquad(20)$$

where e, f and g are

$$e = \vec{N} \cdot \vec{X}_{ss},$$

$$f = \vec{N} \cdot \vec{X}_{st},\qquad(21)$$

$$g = \vec{N} \cdot \vec{X}_{tt}.$$

The mean curvature H is given by

$$H = \frac{(k_1 + k_2)}{2} = \frac{eG - 2fF + gE}{2(EG - F^2)},\qquad(22)$$

and the principal curvatures are

$$k_1 = H - \sqrt{H^2 - K},\qquad(23)$$

$$k_2 = H + \sqrt{H^2 - K}.\qquad(24)$$

The above formulas are used to compute Gaussian curvature, and points on a surface are classified according to its sign. Points for which $K > 0$ are called elliptic, $K < 0$ are called hyperbolic, $K = 0$ are called parabolic, $K =$ constant (i.e., $k_1 = k_2$) are called umbilic, and $k_1 = k_2 = 0$ are called planar umbilic [7]. This classification will be used subsequently to group object points into regions.

3. Computation of Object Description

In this section we present an algorithm to compute the object description in terms of jump boundaries, internal edges, and regions homogeneous in sign of Gaussian curvature. Our algorithm is:

1. Divide the range image into overlapping windows.

2. Detect jump boundaries and fit patches to windows not containing jump boundaries.

3. Compute the principal curvatures and extract edge points.

4. Classify each non-edge point in a patch into one of parabolic, elliptic, hyperbolic, umbilic or planar umbilic, and group all points of the same type in a patch and its neighboring patches into a region.

The details of each step in the algorithm follow. In step 1 the division of the input arrays of x, y, and z coordinates into $L \times L$ windows should be overlapped in order to ensure that an internal edge is always contained in a patch.

In step 2 the standard deviation of the Euclidean distance between adjacent data points in the $L \times L$ window gives an indication of the scatter in the data. In the vicinity of the jump boundaries of an object the Euclidean distance between adjacent points will be large. This will cause the standard deviation of the windows of data containing such jump boundaries to be very high. By setting a threshold on the standard deviation, jump boundaries can be detected. After the jump boundaries have been detected, smooth patches are fitted to the data in the windows not containing a jump discontinuity. In the presence of a jump boundary, the window size is adapted so that the window no longer contains the jump boundary. The sizes of the objects in the scene are assumed to be much larger than the $L \times L$ window, allowing us to detect fine detail such as internal edges in an object. Fitting a surface to an $L \times L$ window of data is equivalent to computing a surface fit to a roughly rectangular grid of data values, as mentioned in the previous section. The surfaces are represented parametrically, and this fitting is a mapping from the two-dimensional parameter space to three-dimensional space.

In step 3 of the algorithm we detect all those points that belong to or fall on internal edges of objects in the scene. Intuitively it is reasonable to expect high curvature to occur in the vicinity of an edge, and we define a point P in a patch as an edge point if the absolute value of one of the principal curvatures becomes greater than a threshold [31].

Due to the presence of noise in the data and inappropriate choice of thresholds for detecting edges, clusters of edge points may appear in the vicinity of a true edge position. In order to eliminate these clusters of computed edge points, non-maxima suppression is applied at every edge point in a direction perpendicular to the edge, which is the direction associated with the maximum absolute principal curvature. Non-maxima suppression will declare no edge present at points where the curvature is not a local maximum.

The next step is to group object points into homogeneous regions. All points in a patch are classified as either elliptic, hyperbolic, parabolic, planar umbilic, or umbilic. All points of the same type are grouped together into a large region. The merging of points does not require explicit fitting of a new surface to this homogeneous region. This is because we can make the second derivatives across patch boundaries match one another, thereby resulting in a C^2 surface [6]. Each homogeneous region is assigned a label depicting its type (elliptic, hyperbolic, parabolic, umbilic, or planar umbilic). The grouping of points into regions immediately follows the surface fitting process.

The above definition of a region allows for internal edges to occur within regions. The extent of regions can be delineated by (a) jump boundaries, (b) internal edges, and (c) curvature edges, which we define as places where there is a change in curvature-based classification. Figure 4(d) shows an example of a wedge, where the regions are bounded by jump boundaries and internal edges. For instance, Figure 8(c) depicts an example of a coffee jar, where the regions are bounded by jump boundaries and curvature edges.

The object representation in terms of regions and curvature based properties of regions, has the advantage of being invariant under the transformation of independent parameters of the surface or rigid body transformations and is therefore independent of viewpoint. In addition, our algorithm for computing object descriptions has the advantage of being able to operate directly on the laser range data without applying any coordinate transformations, and uses existing, tested public domain software.

4. Experimental Results

This section presents examples of the object description algorithm. Real and synthetic data are used to demonstrate the performance of the algorithm.

We consider three elementary surfaces and a composite surface in our examples. Elementary surfaces are those that contain regions made up of one of parabolic, umbilic, planar umbilic, hyperbolic or elliptic points. A composite surface is one that is made up of a combination of elementary surfaces. The three elementary surfaces are chosen because they constitute regions that occur most often in practice. The composite surface is chosen to demonstrate the performance of the algorithm on complex objects composed of different types of regions.

In order to extract maximal regions of an object from a scene in the presence of noise, preprocessing of the range data is necessary. A simple filtering scheme, neighborhood averaging, is applied to the range image to smooth out noise in the data.

Synthetic data are obtained using the approach discussed in [16]. After completing the surface fitting process to each window of data, the (x,y,z) coordinate values on the surface are converted into intensity values using

Lambert's law of direct illumination [21]; this is done to facilitate easy viewing of the reconstructed surface. Figure 3(a) shows the range image of an automobile. The image has been pseudo-colored; red indicates points far from the viewpoint and blue indicates points nearer to the viewpoint. Figure 3(b) illustrates the reconstructed object. Figure 3(c) depicts the

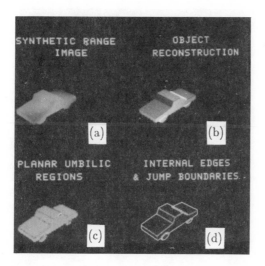

Figure 3: (a) Synthetic range image; (b) Object reconstruction; (c) Planar umbilic regions on the automobile; (d) Jump boundaries and internal edges

various regions extracted. In this case all the regions obtained are planar-umbilic regions. Regions in this and all the examples to follow are colored with the following code:

Yellow	Planar umbilic
Red	Parabolic
Green	Elliptic
Blue	Hyperbolic
Cyan Blue	Umbilic

The results are consistent with the theoretical predictions. Figure 3(d) depicts the various jump boundaries (in dark) and internal edges (in light) of the automobile.

The real data are obtained using the laser scanner described earlier. Figure 4(a) shows the pseudo-colored range image of a wedge obtained from

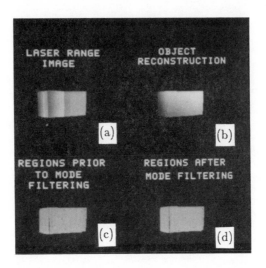

Figure 4: (a) Laser range image; (b) Reconstructed object; (c) Region classification prior to mode filtering; (d) Planar umbilic regions obtained after mode filtering

a laser range finder. A number of points in the scene have no (x,y,z) coordinate values; such points are called holes, (Figure 5). Holes are caused, when parts of the scene illuminated by the laser are not visible to the camera or when parts of the scene visible to the camera are not illuminated by the laser. Figure 4(b) depicts the reconstructed object, Figure 4(c) shows the regions extracted from the range image. Note that the regions obtained are not maximal. In order to extract maximal regions, mode filtering is applied to the initial region classification shown in Figure 4(c). Mode filtering replaces each label of a point in a region by the most dominant label in its neighborhood. Figure 4(d) shows the regions obtained after applying mode filtering to the Figure 4(c). Every region obtained is a planar umbilic region. The results obtained here are in agreement with the theoretical predictions.

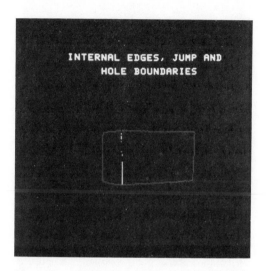

Figure 5: Jump, hole boundaries and internal edges on the wedge

Figure 5 shows the hole boundaries and internal edges. Figure 6(a) shows the pseudo-colored range image of a cylinder placed on a block, obtained from the laser scanner. Figure 6(b) shows the reconstructed objects. Figure 6(c) depicts the regions extracted (after mode filtering) from the range image. In this case there are two regions, one parabolic (on the cylinder) and the other planar umbilic (on the block). Figure 6(d) shows the hole and jump boundaries detected in the range image. Figure 7(a) shows another example of objects consisting of elementary surfaces, a balloon placed on a block. Figure 7(b) is the reconstructed objects and Figure 7(c) depicts the regions extracted after mode filtering. In this case there are two regions, one elliptic (on the balloon) and the other planar umbilic (on the block). Figure 7(d) shows the jump boundaries and hole boundaries on the balloon and the block.

Finally, Figure 8(a) shows the pseudo-colored range image of a composite

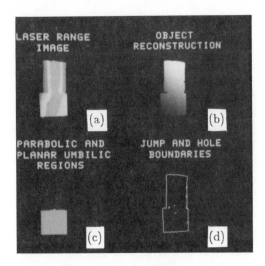

Figure 6: (a) Laser range image of a cylinder placed on a block; (b) Reconstructed objects; (c) Parabolic (cylinder) and planar umbilic (block) regions; (d) Jump and hole boundaries on the objects

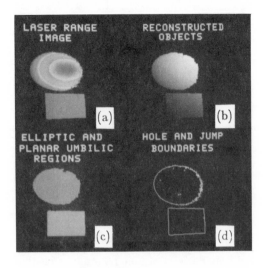

Figure 7: (a) Laser range image of a balloon placed on a block; (b) Reconstructed objects; (c) Elliptic (balloon) and planar umbilic (block) regions; (d) Jump and hole boundaries on the objects

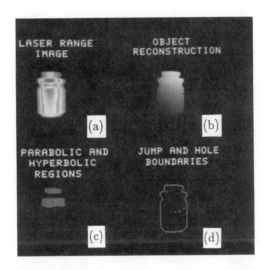

Figure 8: (a) Laser range image of a coffee jar; (b) Reconstructed object; (c) Parabolic and hyperbolic regions on a coffee jar; (d) Jump boundaries on a coffee jar

surface (a coffee jar) obtained from the laser scanner. Figure 8(b) shows the reconstructed object. Figure 8(c) depicts the regions extracted after mode filtering. In this case and as expected, there are four regions, two parabolic and two hyperbolic as shown in Figure 8(c). Figure 8(d) depicts the jump boundaries and hole boundaries on the coffee jar.

It should be noted that surface fitting is an important operation in the overall process. Indeed, refer to Figures 9(a), 9(b), 9(c), and 9(d) which show the classification results obtained when the principal curvatures are computed directly. These results confirm the need for smoothing (surface fitting) the raw data prior to computation of principal curvatures. The results of direct computation are very sensitive to noise in the data, and misclassification is high as can be seen from the examples.

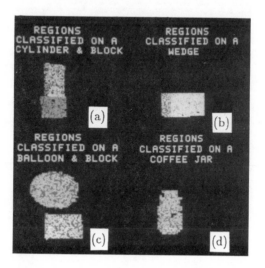

Figure 9:   (a) Region classification of a cylinder placed on a block; (b) Region classification of a wedge; (c) Region classification of a balloon placed on a block; (d) Region classification of a coffee jar

The examples chosen constitute both planar and curved surfaces of fair complexity. The results obtained are most often in agreement with the theoretical predictions.

5. Summary

This paper has presented an approach where 3-D objects are represented by regions which are defined as a collection of patches homogeneous in curvature-based surface properties.

The input to the object description algorithm consists of a collection of 3-D points which need not be represented with one coordinate as a function of the other two. First a local process fits a smooth, tension spline based surface to the data. The surface fitting algorithm is general, efficient (linear time [4]) and uses existing public domain numerical software. The principal curvatures are then computed and the surface points are classified accordingly. Regions on the surface of an object are grown on the basis of this classification. The object description so obtained is viewpoint independent and hence will prove to be valuable in the context of object recognition.

Future research will be directed towards organizing the region descriptions in a meaningful fashion (segmentation) and developing efficient algorithms for matching object descriptions with model descriptions to achieve recognition.

Acknowledgments

The work was supported in part by the Air Force Office of Scientific Research under Contract F49620-85-K-0007 and in part by National Science Foundation under contract DCR8517583.

References

[1]   Agin, G. J. and Binford, T. O.
      Computer description of curved objects.
      In *Proceedings of the Third International Joint Conference on
          Artificial Intelligence*, pages 629-640. August, 1973.
      Also appeared in IEEE Transactions on Computers, 1976, Vol. C-25,
          pages 439-449.

[2]   Barnard, S. T. and Thompson, W. B.
      Disparity analysis of images.
      IEEE Trans. Pattern Anal. Mach. Intell. PAMI-2(4):333-340, 1980.

[3]   Cline, A. K.
      Curve fitting using splines under tension.
      Atmos. Technology 3:60-65, September, 1973.

[4]   Cline, A. K.
      Smoothing by splines under tension.
      Technical Report CNA-168, Univ. of Texas at Austin, Dept. of CS,
          1981.

[5]   Cline, A. K.
      Surface smoothing by splines under tension.
      Technical Report CNA-170, Univ. of Texas at Austin, Dept. of CS,
          1981.

[6]   Coons, S. A.
        Surfaces for computer aided design of space forms.
        MIT Project MAC TR-41, MIT, June, 1967.

[7]   Do Carmo, M. P.
        Differential Geometry of Curves and Surfaces.
        Prentice-Hall, Inc., Englewood Cliffs, NJ, 1976.
        pages 134-156.

[8]   Duda, R. O., Nitzan, D., and Barrett, P.
        Use of range and reflectance data to find planar surface regions.
        IEEE Trans. Pattern Anal. Machine Intell. PAMI-1:259-271, July,
            1979.

[9]   Faugeras, O. D., Hebert, M., and Pauchon. E.
        Segmentation of range data into planar and quadratic patches.
        In *CVPR*, pages 8-13. 1983.

[10]   Grimson, W. E. L.
        An implementation of computational theory for visual surface
            interpolation.
        CVGIP 22:39-69, 1983.

[11]   Hilbert, D. and Cohn-Vossen, S.
        Geometry and the Imagination.
        Chelsea Publishing Co., New York, 1952.
        pages 183-198.

[12]   Horn, B.K.P.
        Obtaining shape from shading information.
        In P.H. Winston (editor), *The Psychology of Computer Vision*,
            pages 115-156. Mc Graw-Hill, New York, 1975.

[13]   Jarvis, R. A.
        A perspective on range finding techniques for computer vision.
        IEEE Trans. Pattern Anal. Machine Intell. PAMI-5:122-139,
            March, 1983.

[14]   Levine, M. D., O'Handley, D. A., and Yagi, G. M.
        Computer determination of depth maps.
        Computer Graphics and Image Processing 2:131-150, 1973.

[15]   Lieberman, L. I.
        Computer recognition and description of natural scenes.
        PhD thesis, Univ. of Pennsylvania, 1974.

[16] Magee, M. J. and Aggarwal, J. K.
Intensity guided range sensing recognition of three dimensional
objects.
Proc. CVPR :550-553, June, 1983.

[17] Marr, D.
Analysis of occluding contours.
Technical Report AI Memo 372, MIT, 1976.

[18] Marr, D. and Poggio, T.
Cooperative computation of stereo disparity.
Science 194:283-288, 1976.

[19] Marr, D. and Poggio, T.
A computational theory of human stereo vision.
Proc. R. Soc. Lond. (B204):301-328, 1979.

[20] Mitiche, A. and Aggarwal, J. K.
Detection of edges using range information.
In *Proc. IEEE Int. Conf. Acoust., Speech, Signal Processing*, pages
1906-1911. Paris, France, May, 1982.
Also appeared in IEEE Trans. on PAMI, Vol. 5, No. 2, March, 1983,
pp. 174-178.

[21] Newman, W. M. and Sproull, R. F.
2nd edition: Principles of Interactive Computer Graphics.
McGraw-Hill, 1979.
pages 498-501.

[22] Nitzan, D., Brain, A. E., and Duda, R. O.
The measurement and use of registered reflectance and range data in
scene analysis.
Proc. IEEE 65:206-220, Feb., 1977.

[23] Oshima, M. and Shirai, Y.
A scene description method using three-dimensional information.
Pattern Recognition 11:9-17, 1978.

[24] Rocker, F. and Kiessling, A.
Methods for analyzing three dimensional scenes.
Proc. 4th IJCAI :669-673, September, 1975.

[25] Rogers, B. and Graham, M.
Motion parallax as an independent cue for depth perception.
Perception 8:125-134, 1979.

[26]   Shirai, Y. and Suwa, M.
        Recognition of polyhedrons with a range finder.
        In *Proceedings of the Second International Joint Conference on
            Artificial Intelligence*, pages 80-87. 1971.

[27]   Stevens, K. A.
        Representing and analyzing surface orientation.
        Artificial Intelligence: An MIT Perspective.
        P. H. Winston and R. H. Brown, 1980, pages 103-125.

[28]   Technical Arts Corp.
        White Scanner 100a User's Manual
        Seattle, WA, 1984.

[29]   Terzopolous, D.
        Multiresolution computation of visible surface representation.
        PhD thesis, MIT, Dept. of EE and CS, January, 1984.

[30]   Ullman, S.
        The interpretation of visual motion.
        MIT Press, Cambridge, MA, 1979.

[31]   Vemuri, B. C. and Aggarwal, J. K.
        3-D reconstruction of objects from range data.
        In *7th International Conference on Pattern Recognition*, pages
            752-754. Montral, July, 1984.

[32]   Will, P. M. and Pennington, K. S.
        Grid coding: A preprocessing technique for robot and machine vision.
        2nd Int. Joint Conf. on AI 2:319-329, 1971.

Use of Vertex-Type Knowledge for Range Data Analysis

Kokichi Sugihara[1]

Department of Mathematical Engineering
and Instrumentation Physics
University of Tokyo
Hongo, Bunkyo-ku
Tokyo 113, Japan

Abstract

This paper describes a knowledge-guided system for range data analysis. Inputs to the system are raw range data obtained by the slit-light projection method and outputs are surface patch descriptions (i.e., organized structures of vertices, edges, and faces) of the visible part of the objects. A junction dictionary is constructed to represent the knowledge about the physical nature of the object world, and is used by the system for the extraction and organization of edges and vertices. At each step of the analysis, the system consults the dictionary to predict positions, orientations and physical types of missing edges. Those predictions guide the system to decide where to search and what kinds of edges to search for as well as how to organize the extracted features into a description.

1. Introduction

Three-dimensional range data contain rich information about object shapes. However, raw range data are simply collections of distance values from the viewer to surface points, and consequently are not directly suitable for use as object identification and object manipulation; they should be condensed into more concise descriptions.

Generalized cylinders [1], [11] and surface patches [12], [13], [5], [3] are the prevailing tools in terms of which the descriptions are constructed from

---

[1]Previously in the Department of Electrical Engineering, Nagoya University

range data. The generalized cylinders are based on natural principal axes of objects, and so enable us to construct object-centered descriptions relatively easily [9]. Descriptions based on generalized cylinders, however, are unstable or non-unique if the objects are not composed of elongated parts. The surface patches, on the other hand, can be used for non-elongated objects, but it is usually difficult to segment smooth, curved surfaces into patches in a stable manner. If the objects are limited to polyhedrons, however, the surfaces can be decomposed into natural patches, that is, planar faces. Thus, polyhedral scenes can be described naturally in terms of surface patches.

This paper presents a method for extracting surface patch descriptions of polyhedral objects from range data [18], [20]. In doing this, knowledge about vertex types is represented in terms of a junction dictionary, and used for constructing scene descriptions. For this purpose we borrow some results in the study of interpretation of line drawings.

One of the greatest milestones in interpretation of line drawings is a labelling scheme proposed by Huffman [6] and Clowes [2]. They introduced labels to represent physical properties of edges (such as convex and concave edges), and found that the number of junction types that can appear in labeled line drawings is very small. On the basis of this observation they proposed the labeling scheme in which, once the list of all possible junctions is constructed and registered, line drawings can be interpreted by finding consistent assignments of labels in the sense that all the resulting junctions belong to the list. This scheme has been extended to pictures with cracks and shadows [22], pictures with hidden lines [19], and pictures of the Origami world [8]. Thus, junction knowledge has been used for categorizing lines in line drawings.

In the case of range images, on the other hand, we can classify convex, concave, and some other types of lines directly from the data without using any additional information; the junction knowledge is not necessary for categorizing lines. Therefore, this knowledge can be used for another purpose, that is, for predicting missing edges. Comparing a line drawing extracted from the range data with the list of the possible junctions, we can tell whether the present drawing is consistent with physical reality. If it is not, we can predict the locations and physical categories of possible missing edges. The extraction of edges can be efficiently guided by those predictions. This approach is practical, since noisy raw range data can be dealt with, while the junction knowledge has been used for the analysis of "perfect" line drawings in the previous studies.

In Section 2 we introduce a local operator that can extract edges from raw range data, and in Section 3 we construct a junction dictionary, which is a condensed representation of vertex-type knowledge. Using these two tools, we construct a system for extracting surface patch descriptions of the scenes, which is described in Section 4.

2. Edge Detection from Range Data

This section describes the method for detecting edges from range data. First, the representation of range data is explained, then we define a local operator to detect edge points, and finally we present the strategy to track edge points, which can extract selectively a particular type of a line running in a particular direction.

2.1. Range Data

One way to get range information is using a rangefinder by active illumination. As shown in Figure 1, a beam of light is projected from the source, say O, onto an object, and the illuminated spot is observed from another angle by a camera. Let $P(X,Y,Z)$ be the illuminated spot on the surface, and P' be its image. Furthermore, let $D(P)$ denote the distance from the left margin of the image to the image point P'. This $D(P)$ is what we can observe directly by the method in Figure 1. Indeed, if a TV camera is used, $D(P)$ can be measured as the time interval from the time when the

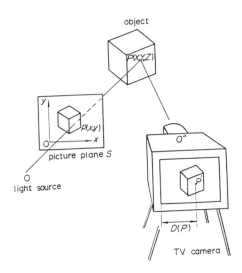

Figure 1:　Range Finding Method

horizontal scan begins to the time when the video signal shows the maximum intensity.

　　Let us place a virtual plane, say S, with a Cartesian coordinate system (x,y) between the light source O and the object, and let $p\,(x,y)$ be the point of intersection of the light beam OP and the plane S. Since the direction of the light beam OP is specified uniquely by the point $p\,(x,y)$ on S, the value $D(P)$ can be considered as a function of the point $p\,(x,y)$. Hence, we hereafter denote $D(x,y)$ instead of $D(P)$.

　　When the scene is swept by the light beam, the point $p\,(x,y)$ sweeps some region on S and we get raw data $D(x,y)$ at many points (x,y), say

$$\{D(i,j)\mid i{=}1,\ldots,m,\ \ j{=}1,\ldots,n\}\,,$$

where we use (i,j) instead of (x,y) in order to indicate that the data are obtained only at discrete points. This collection of the data is similar to a usual digital image in that the pixel values are defined at grid points (i,j) on the image plane, i.e., the plane S. The only difference is that the values $D(i,j)$'s defined above are not light intensity values at the grid points. We shall call the collection of $D(i,j)$'s a raw range picture.

If we use a spot light, $D(i,j)$ is measured only at one point for each light projection, which requires a long time to measure ranges for the whole scene [7]. For practical purposes, therefore, we project light through a narrow vertical slit so that we can obtain many data $\{D(i,j) \mid j=1, \ldots, n\}$ along a vertical line simultaneously [16] (see also [10] and [15]).

2.2. Edge Detecting Operators

To construct an operator for detecting edges from a range picture, we need to consider the relations between geometrical configurations of objects and the values of $D(i,j)$'s. Let ξ be a line on a picture plane. Let $p_\xi(s)$ denote a point which moves along the line ξ with a line parameter s (that is, s is the length along the line from a certain fixed point on ξ to $p_\xi(s)$). Further, let $P_\xi(s)$ denote the corresponding point on the three-dimensional surface of the object. Referring to Figure 2, we will examine how $D(P_\xi(s))$ varies as $P_\xi(s)$ moves across geometrical features of objects. The following observations are intuitively obvious.

When $P_\xi(s)$ moves on a flat surface, $D(P_\xi(s))$ is approximately a linear function of s as shown in Figure 2(a) (indeed, it is strictly linear if the light is projected in parallel and its image is obtained orthographically). When $P_\xi(s)$ moves across a convex edge, $D(P_\xi(s))$ changes continuously and bends upward in the $s-D$ plane as illustrated in Figure 2(b). Similarly, when $P_\xi(s)$ moves across a concave edge, $D(P_\xi(s))$ bends downward (see Figure 2(c)). Note that these upward and downward bends corresponding to convex and concave edges, respectively, are true only when the camera is to the right of the light source. If the camera and the light source exchange their locations to each other, the changes of $D(P_\xi(s))$ at a convex edge and at a concave edge are reversed. Finally, when $P_\xi(s)$ moves across an obscuring edge, $D(P_\xi(s))$ changes discontinuously as shown in Figure 2(d).

According to our observation we can devise the following operator to

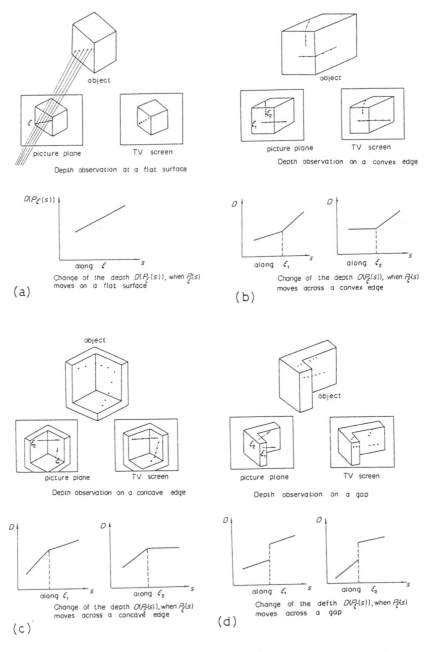

Figure 2: Changes of $D(P_\xi(s))$ as $P_\xi(s)$ Moves Across Various Features on the Object

detect edges. Consider three points $p_\xi(s)$, $p_\xi(s+\Delta s)$, and $p_\xi(s-\Delta s)$ along the line ξ on the image plane S, where Δs is a short length (see Figure 3).

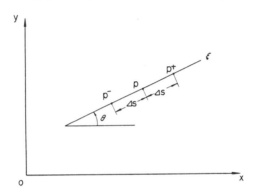

Figure 3: Three Points Used by the Edge Detecting Operator

Let θ represent the orientation of the line ξ in the image and measured counterclockwise from the x axis. For simplicity in notation, let

$$p^+ = p_\xi(s+\Delta s) \text{ and } p^- = p_\xi(s-\Delta s)$$

as illustrated in Figure 3. We will define the operator $O_\theta(p)$ by

$$O_\theta(p) = \frac{1}{2\Delta s}[D(p^+) + D(p^-) - 2D(p)]. \tag{1}$$

Figure 4 shows profiles of $O_\theta(p)$ when P moves across various types of features of the object corresponding to the cases in Figure 2. We can see that $O_\theta(p) = 0$ when P moves on a planar surface, $O_\theta(p) > 0$ when P moves across a convex edge, and $O_\theta(p) < 0$ when P moves across a concave edge.

For practical purposes we can redefine the operator on a digital range picture by restricting $\theta = 0$, $\pi/4, \pi/2$, and $3\pi/4$:

$$O_0(i,j) = [D(i+k,j) + D(i-k,j) - 2D(i,j)]/(2k), \tag{2}$$

$$O_{\pi/2}(i,j) = [D(i,j+k) + D(i,j-k) - 2d(i,j)]/(2k), \tag{3}$$

$$O_{\pi/4}(i,j) = [D(i+k,j+k) + D(i-k,j-k) - 2D(i,j)]/(2\sqrt{2}k), \tag{4}$$

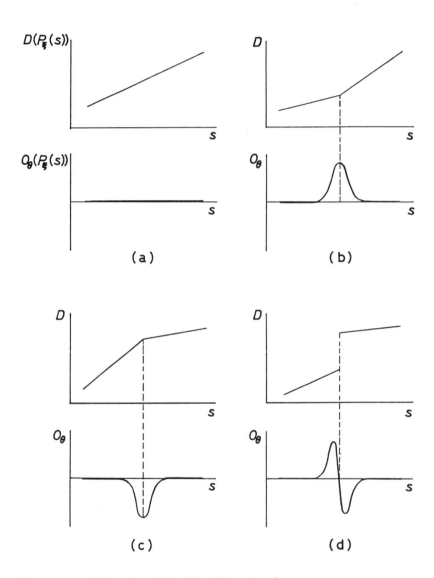

Figure 4:   Behaviors of the Edge Detecting Operator

$$O_{3\pi/4}(i,j) = [D(i+k,j-k) + D(i-k,j+k) - 2D(i,j)]/(2\sqrt{2}k). \qquad (5)$$

As we have just seen, $O_\theta(i,j)$ takes the greatest absolute value when an edge lies in the direction $\theta + \pi/2$.

In addition to the directional operators, we can also consider nondirectional operators:

$$O_{M4}(i,j) = \frac{1}{4}\sum_{\theta=0,\pi/4,\pi/2,3\pi/4} O_\theta(i,j), \tag{6}$$

$$O_{M2}(i,j) = \frac{1}{2}[O_0(i,j) + O_{\pi/2}(i,j)], \tag{7}$$

The operator in (6) or (7) will usually have a non-zero value when P is on an edge, but it does not give information about the direction of the edge.

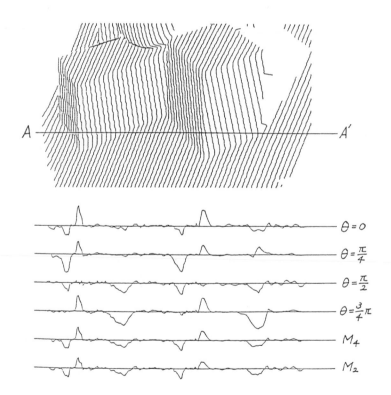

Figure 5: Output of the Edge Detecting Operators Along the Line AA' in the Range Picture

Figure 5 shows the output of the operators from (2) through (5), (6), and

(7) when the point $p(i,j)$ moves along the horizontal line AA' in the range picture, where the range picture is displayed as the image of vertical light stripes. We can see that the operators in (2) through (5) are sensitive to particular edge directions. In contrast, the operators in (6) and (7) have non-zero values at all edges. Moreover, the values of all the operators are positive at convex edges and negative at concave edges. There is not a great deal of difference between the value of O_{M4} and that of O_{M2}, and thus we mainly use only O_{M2}, which is more economical in computational time than O_{M4}.

2.3. Edge Tracking

The previous subsection presented operators to detect edge points locally. In order to extract edge lines, the following edge-tracking strategy is used. We first scan the range picture with the nondirectional operator O_{M2} or O_{M4}, and when we come across a point where the absolute value of the operator output is large enough, that is, a promising point for an edge, we then determine the direction of the line using the four directional operators from (2) through (5). We next follow the line in the given direction from that promising point. That is, we search the neighboring area for the next point along the line step by step and construct a string of points forming the edge line.

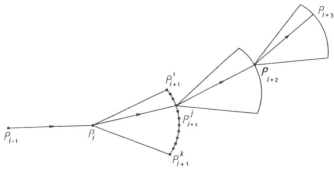

Figure 6:   Edge Following

Figure 6 illustrates these steps for edge following. Suppose that we have just found a new point p_i by searching the neighboring area of a point p_{i-1}. Then, in the next step we search the fan-shaped area spanning from the pivotal point p_i as shown in Figure 6. For practical purposes, the discrete points $p^1_{i+1}, p^2_{i+1}, \ldots, p^k_{i+1}$ on the arc of the fan-shaped area are chosen as the candidates for the next point, and the edge detecting operator is applied to those candidate points. Since the expected direction of the line is given, we need only use one of the four operations for each candidate point. Figure 7 gives the associated direction θ for each candidate point used in our system. We consider 32 points around the pivotal point and divide them into four groups, one for each directional operator. Among the candidates, we choose the point with the maximum value of the operator if the lines are convex, or the one with the minimum value (i.e., the maximum absolute value with a negative sign) if the line is concave.

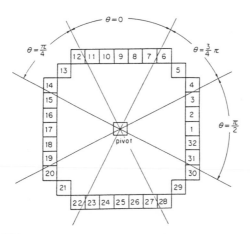

Figure 7: Thirty-two Candidate Points for the Next Point in the Edge Following

In our system we usually need not scan the whole picture to find the starting points for edge following, because the vertex-type knowledge suggests the starting point, the direction, and the type of an expected line as will be described later.

3. Junction Dictionary

Here we present our second tool for range data analysis, that is, a junction dictionary. The junction dictionary contains possible configurations at a junction, and can be considered as a kind of representation of knowledge about three-dimensional objects. It is used as a predictor of missing edges that should be extracted from range pictures.

3.1. Line Types

We consider a world consisting of solid polyhedrons, which are bounded only by planar *faces*. An *edge* is a line where two faces meet, and a *vertex* is a point where three or more faces meet. We call a collection of polyhedrons in a three-dimensional space a *scene*, and its projection on the plane a *picture*. The visible faces, edges, and vertices of the scene are associated with the polygonal *regions*, the straight *lines* (actually, line segments), and *junctions*, respectively, of the picture.

In the three-dimensional space there are only two types of edges for our planar faced objects: convex edges and concave edges. When a concave edge is visible, both of its side faces are also visible from the eye. Hence, we always call a line representing a concave edge a *concave line*, and represent it by a line segment with a minus (−) label. On the other hand, when a convex edge is visible, one of the two side faces is sometimes hidden by the other. Therefore, we distinguish between two types of lines representing convex edges: a *convex line* to represent a convex edge both of whose side faces are visible, and an *obscuring line* to represent a convex edge where one of the sides faces hides the other. A plus (+) label is used for a convex line and an arrow (→) is used for an obscuring line with the convention that, when one stands on the line facing the direction indicated by the arrow, both of the associated side faces are to the right.

The three types of labels above were introduced by Huffman [6] and have been used in various works on interpretation of line drawings. However, those three types are not enough to constitute line drawings representing range pictures. In order to introduce another line type, we have to consider the method for obtaining range information.

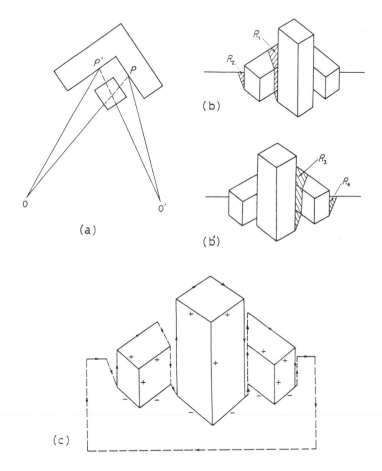

Figure 8: Line Labels for Range Pictures

A light-stripe rangefinder measures the range of a point on an object by seeing the point from two angles. As shown in Figure 8(a), we project a beam of light from a light source O and observe the illuminated spot with a camera at the point O'. This method does not measure the range of a point either when it is not illuminated (for example, see the point P in Figure 8(a)) or when it is not seen from the camera (point P'). Thus, after sweeping the scene with the beam of light, there remains some regions of points whose ranges are not determined. Examples of such regions are illustrated by hatched areas in Figure 8(b) and (b'). Figure 8(b) is the picture of the scene viewed from the light source, and hence the shaded regions represent surfaces that cannot be seen from the camera. On the other hand, Figure 8(b') is the picture seen from the camera, and the shaded regions represent surfaces that cannot be seen from the light source. We will call boundary lines of those obscured regions *obscured lines*; we will represent them by broken lines with arrows with the convention that, when we stand on the line facing the direction indicated by the arrow, the obscured region is to the left. Figure 8(c) shows the labeled line drawing of the picture in Figure 8(b) according to our conventions.

Since we have defined a range picture as that obtained by seeing the scene from the light source, unilluminated regions (for example, the regions R_3 and R_4 in Figure 8(b')) are always out of sight. Nevertheless we will represent the existence of those regions by a pair of contour lines: an obscuring line and an obscured line, parallel to each other and opposite in direction (see Figure 8(c)). Moreover, for convenience we will represent the margin of the picture plane also by obscured lines.

3.2. Possible Junctions

Now that we have introduced the four types of lines, we next enumerate possible junctions, that is, junctions that can appear in labeled pictures. For this purpose we restrict our objects to trihedral ones, that is, objects in which exactly three faces meet at each vertex.

Junctions in a picture usually correspond to vertices in a scene. However, there are junctions which are not images of actual vertices. One typical class of such junctions is found on an obscured line. An obscured line is caused when one body is in front of another. Therefore, a shape of an obscured line is sometimes due to the shape of the obscured body, sometimes due to the shape of the obscuring body, and sometimes due to both of them. Although the configurations of obscured lines may contain much information about shapes of the objects, it seems difficult to extract helpful information for our purpose. Hence, we will ignore junctions on obscured lines except in the following case.

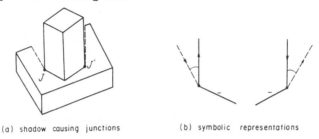

(a) shadow causing junctions   (b) symbolic representations

Figure 9: Shadow Causing Junctions and Their Representation

Figure 9(a) shows a line drawing of a scene with two blocks, one supporting the other. Junctions J and J' in the figure have the property that each of the junctions includes one obscuring line and one obscured line, and that the former causes the latter; in other words, the latter is a shadow of the former. We call a junction with this property a *shadow causing junction*. We represent a shadow causing relationship by an arc linking the obscuring line to the obscured one as illustrated in Figure 9(b). Any vertex that can generate a shadow causing junction in the picture has at least one convex edge and at least one concave edge, for only a convex edge can cast its shadow on a face associated with a concave edge.

The enumeration of possible junctions can be executed in a systematic manner. Let us consider an isolated vertex. Since the vertex is trihedral, there are exactly three faces meeting at the vertex. If the three faces are extended into unbounded planes, then they divide the surrounding space into eight parts, which we call *octants*. Some of the octants are occupied with material and the others are empty. The eye and the light source can only be in empty octants. Considering all the combinations of placing the light source and the camera in empty octants, we can exhaust possible views of the vertex.

Actually, there are only four types of trihedral vertices: a vertex composed of one nonempty octant, that of three nonempty octants, that of five nonempty octants, and that of seven nonempty octants (see [6] for the details). Hence, by the enumeration method above we eventually find that there are only fourteen possible views of trihedral vertices provided that the light source and the camera are in general positions so that neither of them is coplanar with any faces. These fourteen junctions are listed in Figure 10,

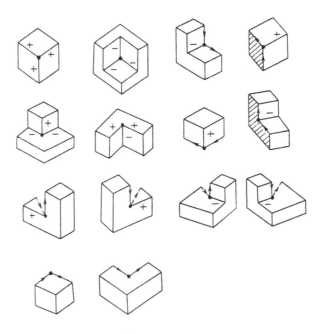

Figure 10:   All Possible Junctions Resulting from Isolated Trihedral Vertices

where the shaded regions represent the regions that can be seen from the light source but cannot be seen from the camera.

If the scene contains two or more objects, other types of junctions can also appear because contact of objects often gives rise to non-trihedral vertices. Let us assume that objects in the scene do not align accidentally, that is, neither two distinct edges nor two distinct vertices share a common position and no vertex lies on an edge of another object. Then, by the exhaustive search we can find that there are only six additional types of junctions as shown in Figure 11. After all, we have a total of twenty possible junctions, which are listed in Figure 12.

3.3. Representing a Junction Dictionary

A junction dictionary is represented in a form of a directed graph whose nodes are all subconfigurations of the possible junctions and whose arcs represent the relationships that one node can be changed into another node by an addition of one line.

Part of the junction dictionary is shown in Figure 13, where seven junctions and relationships between them are illustrated. The three junctions in rectangles (t,v, and y) are impossible ones and the four in circles (u,w,x, and z) are possible ones. The drawings beside the possible junctions show the examples of situations in which the junctions occur. An arrowed arc connecting two junctions denotes an addition of a line. If, for example, we add a convex line in the shaded area of the junction t in Figure 13, we obtain junction u. Adding a concave line to u results in w, and thus we get the path (t,u,w) from t to w. This kind of a path suggests where new lines exist and thus guides the analysis of the range data. For example, if a junction of the type u is found during the analysis, the dictionary tells us that a concave line "may" exist at that junction. If, on the other hand, a junction of the type v is found, the dictionary says that a convex line "must" exist to form w because v is impossible and w is the only possible junction that includes v as a subconfiguration. In this way, the junction dictionary can suggest what type of edges exist and in what directions they must be searched for near the vertex.

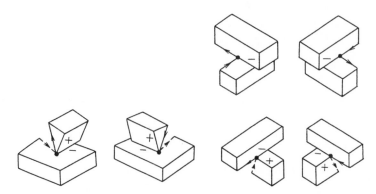

Figure 11:   All Possible Junctions Resulting From Contact of Two Objects Without Accidental Alignments

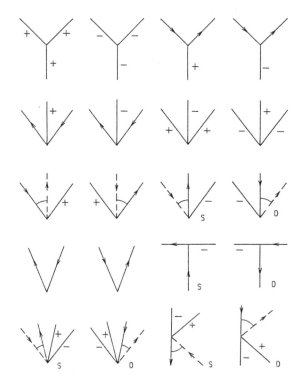

Figure 12:   Final List of Possible Junctions

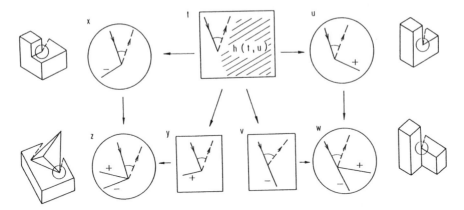

Figure 13: Part of the Junction Dictionary

3.4. Body Partitioning Rules

For object recognition it is useful to decompose a complicated scene into simple elements and to describe the spatial relationships among them. We show that the junction dictionary can also be used for the decomposition of a scene. Note that this decomposition is inherently ambiguous (see [4]). We get no information from a picture (no matter what kind of a picture it is, a light intensity picture or a range picture) to judge, for example, whether a block is merely placed on a desk or it is fixed to the desk with glue. Consequently, the rules we will propose here are necessarily empirical ones.

Our junction dictionary contains only such junctions that are found when polyhedrons in a scene are mutually in general positions; that is, without accidental alignments. However, when there are accidental alignments of objects in a scene, many new junction types can appear. If we enumerated all such accidental junctions, the junction dictionary would have become unmanageably large. Instead, the partitioning rules to be described will sometimes decompose an illegal junction (that is, a junction that is not included in the dictionary) into two legal ones. As a result, the partitioning rules help in handling cases when accidental alignments of bodies are involved.

The first partitioning rule is the following:

Partitioning Rule 1.
If there are two pairs of lines with arrows and both pairs have shadow causing relationships as shown in Figure 14(a), then partition the junction into two as shown in Figure 14(b).

The condition of this rule is usually found when a body is supported by another at a point. Figure 14(c) gives an example scene to which the rule can be applied.

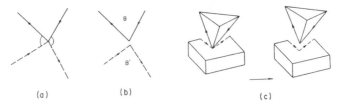

(a)   (b)   (c)

Figure 14:   Partitioning Rule 1

The second partitioning rule is based on the particular types of junctions labeled S and D in the list of possible junctions of Figure 12. These junctions occur at the point where two objects contact each other. Hence, they can be used as cues for partitioning.

Partitioning Rule 2.
If a chain of concave lines connects two S and D type junctions as shown in Figure 15(a), or if a chain of concave lines connects an S or D type junction with a junction on an obscured line as shown in Figure 15(b), then partition the body into two by changing the chain of concave lines into a pair of contour lines as shown in Figure 15(c).

Figure 15(d) shows an example scene to which the rule is applicable. The scene is divided into two along the concave lines between junctions S and D.

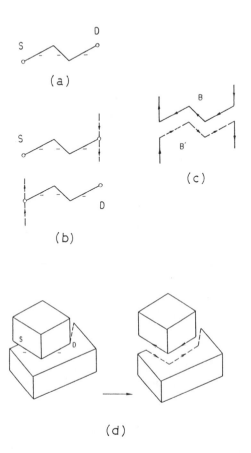

Figure 15: Partitioning Rule 2

4. Range Data Analysis Guided by a Junction Dictionary

In the preceding two sections we have developed two tools, the edge detecting operators and the junction dictionary. Here we combine them into a whole system which generates a surface patch description from a range picture and, at the same time, which partitions the scene into simple bodies. We first describe the outline of the system, and then present experimental results.

4.1. Experts in the System

The input to the system is a raw range picture of a real scene and the output is a collection of labeled line drawings of all the objects in the scene. The system is composed of the following experts: a line predictor, a prediction tester, a line tracker, a straight-line adjuster, a body partitioner, and a vertex position corrector.

When a junction is given, a *line predictor* consults the junction dictionary and predicts missing lines around the junction. New lines are predicted if and only if addition of the new lines results in possible junctions. If the present junction is an impossible one, the prediction is strong in the sense that at least one of the predicted lines must be found. If the present junction is a already a possible one, on the other hand, the prediction is weak in that the predicted lines may not exist.

A *prediction tester* judges whether the predicted lines really exist or not. The judgment is based upon two kinds of tests. One is to use the edge detecting operators to see if there are new lines to be tracked in the range picture, and the other is to examine old lines in the present line drawing if their properties meet the prediction. The latter search is necessary, because when a new line is found and is added to a junction, the end of the new line may be left unconnected though it should be joined to one of its neighboring junctions. When old lines whose properties meet the prediction are found, the prediction tester concludes that the prediction is true. Otherwise, it examines all the predictions in the range picture and concludes differently depending on the strength of the prediction. That is, if the present junction is impossible, the prediction tester returns the direction in which the edge detecting operator gives the maximum output value even if the value may not be very large. On the other hand, if the present junction is possible, the prediction tester behaves conservatively; it reports the existence of new lines only if the edge detecting operator gives very large absolute values.

When a new line is found, a *line tracker* tracks it as far as possible and extracts the whole line. The tracking terminates when the line disappears or when the physical type of the line changes (such as from concave to convex), or when it comes near to another line that has already been found.

A *straight-line adjuster* finds a sequence of straight lines which fit to the strings of points extracted by the line tracker. When a string is given, it first finds points with maximal curvatures (the curvature at a point on a string is calculated in the same way as proposed by Shirai [17]). The adjuster next divides the string at those points into substrings and finally replaces each substring with a straight line segment. If it fails in finding a well-fitting line, the substring is left for later analysis. The joints between two substrings as well as the end point of the original string are added to the drawing as new junctions.

Whenever a new line is added to the drawing, a *body partitioner* checks the conditions for the partitioning rules and, if one of the conditions is satisfied, it decomposes the line drawing into two smaller ones.

A *vertex-position corrector* revises a position of a junction whenever a new line is connected to the junction. The position is estimated as the weighted mean of intersections of all pairs of the lines incident to the junction, where the weights are chosen according to the lengths of the lines.

4.2. System Behavior

The processing of the system is divided into two stages: the preprocessing and the main processing.

In the preprocessing stage, the system first finds contour lines (i.e., obscuring lines and obscured lines) by tracking points where ranges show discontinuity, divides them into straight line segments, and locates junctions on the contours. Next it finds shadow-causing relationships and classifies the contour lines into obscuring lines and obscured lines. At this stage, shadow-causing junctions are also identified. Each connected region bounded by obscuring and obscured lines is called a body.

In the main processing stage, the experts which have been described in Section 4.1, work together to analyze the range picture using the following seven steps.

Step 1. If the descriptions of all bodies have been constructed, the system terminates the processing. Otherwise, the system picks up a body whose description has not yet been completed.

Step 2. If all the junctions of the body have been examined, the processing goes to Step 7. Otherwise, the system picks up a junction to examine next.

Step 3. The line predictor compares the junction with the dictionary, and proposes all predictions of missing lines around the junction. The prediction tester examines them to see if any lines are really there. If no lines are found, it puts the mark "examined" to the junction and the processing goes to Step 2. If old lines whose properties meet one of the predictions are found, the processing goes to Step 5. If new lines are found, the processing goes to Step 4.

Step 4. The line tracker extracts the strings of points which form the new lines. The straight-line adjuster locates new junctions on the strings, and divides them into substrings, each of which is approximated by a straight line segment.

Step 5. The lines which are extracted are connected to the present junction. The vertex-position corrector corrects the position of the junction using all the incident lines.

Step 6. The body partitioner examines the improved part of the body. If one of the conditions of the partitioning rules is satisfied, it decomposes the body into two and the processing goes to Step 1. Otherwise, it goes to Step 3 (in this case the junction is the same but its type has been changed).

Step 7. The system sweeps the body region with the edge detecting operator. If it finds a new line, it locates all junctions on it and goes to Step 2. Otherwise, it puts the mark "completed" to the body and goes to Step 1.

4.3. System Performance

The range pictures used as inputs to test the system are all from real scenes consisting of white polyhedrons. They are obtained by a rangefinder developed at Electrotechnical Laboratory [14]. Each picture consists of 200 by 240 points and the value of range $D(i,j)$ has nine bits of information.

Figure 16(a) shows a light stripe image of a scene with two blocks, one supporting the other, and Figures (b) through (f) show how the description of the scene is created progressively. The system, in its preprocessing stage, extracts the contour, finds junctions on it, and adjusts line segments to the contour. Junctions $J1$ and $J2$ are recognized as shadow causing junctions.

In the main processing stage, the system first picks the junction $J1$ and finds that $J1$ is impossible. The line predictor suggests a concave line in the left side of $J1$ (where the word "left" is used with respect to the reader) and the prediction tester actually finds the concave line. The line tracker follows the new line until it terminates at $J3$. Next the straight-line adjuster works with this line. Junction $J4$ is found on the new line (Figure 16(b)), and the resultant two portions of the string are approximated by straight line segments $J1J4$ and $J4J3$. The system next picks up the junction $J2$. It is an impossible junction and a missing line is predicted, but the old line $J3J4$ can be used to resolve the problem by merging $J2$ and $J3$. Then the condition for the partitioning rule is satisfied; that is, $J1$ belongs to the D type and $J2$ belongs to the S type, and they are connected by the concave lines $J1J4$ and $J4J2$. The body partitioner partitions the picture into two bodies, *BODY1* and *BODY2*, (Figure 16(c)). The system picks up *BODY1*. When junction $J5$ is examined, a new convex line is found. The new line terminates at $J6$ and junction $J7$ is also located on the new line (Figure 16(d)). Junction $J6$ is joined to the neighboring junction on the contour. Another new convex line $J4J8$ is found from $J4$ (Figure 16(e)). Two impossible junctions $J7$ and $J8$ are joined together. Now, since all the junctions in *BODY1* are examined and no new lines can be found, the description of *BODY1* is completed. The

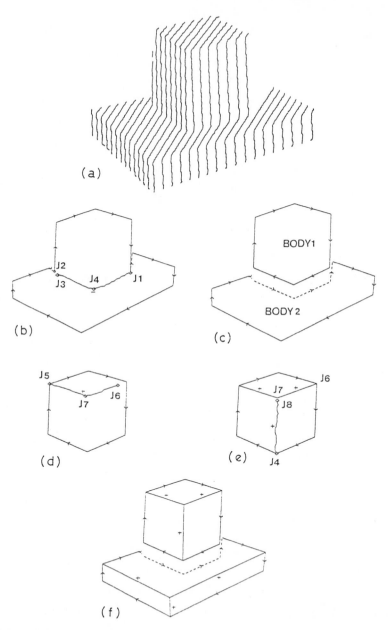

Figure 16:   Example Scene 1

system picks up the other body *BODY2* and improves its description
similarly. The final result is given in Figure 16(f).

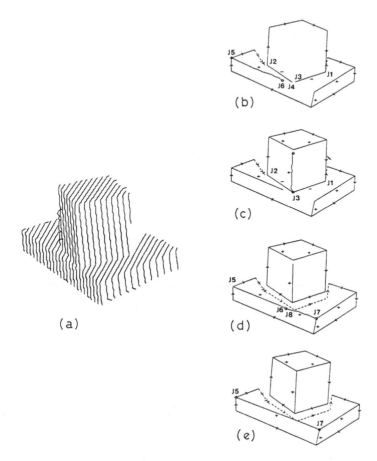

Figure 17: Example Scene 2

Figure 17(a) shows a light stripe image of another two-block scene. A vertex of one block lies accidentally on an edge of the other block, and a non-trihedral vertex occurs. The system conquers this difficulty as follows. A new concave line is found at $J1$ but it terminates at $J3$. Another concave line is found at $J2$ and it is traced to $J4$. The system finds a convex line at $J5$ and tracks it. However, the non-trihedral vertex obstructs the tracking of the new line at $J6$ (Figure 17(b)). When the system examines $J3$, $J4$ is merged to $J3$ and another convex line going upward is found at $J3$; $J3$ becomes a possible junction (Figure 17(c)). Now the partitioning rule is applied to the concave lines $J1J3$ and $J3J2$, and the scene is partitioned into two bodies. When the junction $J7$ is examined, another part of the incomplete convex line is found; $J7J8$ in Figure 17(d). Junction $J8$ is joined to $J6$ and the resultant $J6$ is eventually deleted because the two lines $J5J6$ and $J6J7$ are collinear. In this way the complete line $J5J7$ is found as is shown in Figure 17(e).

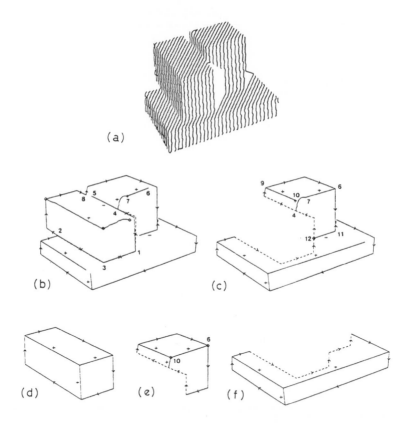

(a)

(b)

(c)

(d)

(e)

(f)

Figure 18:   Example Scene 3

The third example is shown in Figure 18. The scene contains three blocks. The system finds two contours in this scene. One is a usual clockwise contour. The other is a counterclockwise contour surrounding a vacant hole found in the gap; that is, the contour $J1J4$ and $J4J1$ in Figure 18(b). When the system finds concave lines connecting $J1$ to $J2$, the body partitioner cuts the body along the lines. However, the resulting body still remains connected. When the system examines $J4$, a new line $J4J5$ is found. Since $J4$ is still impossible, the system searches for another line and finds a convex line $J4J6$. On this line the system finds a new junction, $J7$ (though its position is not quite appropriate). Junction $J5$ is merged to $J8$ and the body is cut along the line $J4J5$ as shown in Figure 18(c) and (d). One of the resultant body is further partitioned along the line $J11J12$ into two portions shown in Figure 18(e) and (f). When the system finds a new convex line emanating from $J9$, the new line is followed to $J10$. The system next creates a new junction on the line $J4J7$, and the new junction is merged to $J10$. Then the straight-line adjuster replaces the two strings $J4J10$ and $J10J7$ with straight line segments, and since the two lines $J10J7$ and $J7J6$ are collinear, they are merged and the junction $J7$ is deleted as is shown in Figure 18(e).

5. Concluding Remarks

We have presented a range data analysis system using knowledge about vertex types. Since the junction dictionary suggests missing lines at each step of the analysis, the detection and organization of edges can be done efficiently and reliably. Moreover, the output is not merely a collection of lines but an organized structure of faces, edges, and vertices (that is, the surface patch description), which is useful for later processing such as object identification and manipulation.

The present approach can be extended in several directions. First, though we have restricted our objects to trihedral ones, our system can be applied to other objects with a slight modification. All we have to do is to replace the junction dictionary with the one containing all possible junctions in a new world. If the object world consists of a finite number of prototypes (which is often the case in industrial applications), the associated junction dictionary can be generated automatically. Further details may be found in a paper by Sugihara [21].

Secondly, our system can also be applied to curved surface objects. The greatest difficulty in dealing with curved surface objects is that curved lines appear in line drawings and consequently the number of possible junction types becomes very large. However, we can circumvent the explosion of the dictionary by replacing curved lines emanating from a junction with the straight lines tangent to them. This convention is also practical in the sense that the system need not distinguish between straight and curved lines, which is a difficult task especially when the data contain noises. Thus, curved surface objects can be dealt with by a little larger dictionary. The reader is referred to papers by Sugihara [18], [20] for further details.

References

[1]   Agin, G. J. and Binford, T. O.
        Computer description of curved objects.
        In *Proceedings of the Third International Joint Conference on
            Artificial Intelligence*, pages 629-640. August, 1973.
        Also appeared in IEEE Transactions on Computers, 1976, Vol. C-25,
            pages 439-449.

[2]   Clowes, M. B.
        On seeing things.
        Artificial Intelligence 2:79-116, 1971.

[3]   Faugeras, O. D.
        New steps toward a flexible 3-D vision system for robotics.
        *Preprints of the Second International Symposium of Robotics
            Research. Note: also see In Proc. 7th Int. Conf. Pattern
            Recogn.*, pages 796-805; Montreal, Canada, July :222-230, 1984.

[4]   Guzman, A.
        *Computer recognition of three-dimensional objects in a visual
            scene.*
        Technical Report MAC-TR-59, Ph.D. thesis, Project MAC, MIT,
            Mass. , 1968.

[5]   Herman, M. and Kanade, T.
        *The 3-D MOSAIC scene understanding system: Incremental
           reconstruction of 3-D scenes from complex images.*
        Technical Report CMU-CS-84-102, Carnegie-Mellon University,
           Computer Science Dept., 1984.

[6]   Huffman, D. A.
        Impossible objects as nonsense sentences.
        Machine Intelligence 6.
        Edinburgh University Press, Edinburgh, 1971, pages 295-323.

[7]   Ishii, M. and Nagata, T.
        Feature extraction of three-dimensional objects and visual processing
           in a hand-eye system using laser tracker.
        Pattern Recognition 8:229-237, Oct., 1976.

[8]   Kanade, T.
        A theory of Origami world.
        Artificial Intelligence 13:279-311, 1980.

[9]   Marr, D. and Nishihara, K.
        Representation and recognition of the spatial organization of three-
           dimensional shapes.
        Proceedings of the Royal Society of London B200:269-294, 1978.

[10]   Mino, M., Kanade, T., and Sakai, T.
        Parallel use of coded slit light for range finding (in Japanese).
        In *21st National Convention of the Information Processing
           Society of Japan*, pages 875-876. 1980.

[11]   Nevatia, R. and Binford, T. O.
        Description and recognition of curved objects.
        Artificial Intelligence 8:77-98, 1977.

[12]   Oshima, M. and Shirai, Y.
        A scene description method using three-dimensional information.
        Pattern Recognition 11:9-17, 1979.

[13]   Oshima, M. and Shirai, Y.
        Object recognition using three-dimensional information.
        IEEE Transactions on Pattern Analysis and Machine Intelligence
           PAMI-5(4):353-361, July, 1983.

[14]   Oshima, M. and Takano, Y.
        Special hardware for the recognition system of three-dimensional
            objects (in Japanese).
        Bulletin of the Electrotechnical Laboratory 37:493-501, 1973.

[15]   Sato, S. and Inokuchi, S.
        Range imaging by Gray coded projection (in Japanese).
        In *1984 National Convention of the Institute of Electronics and
            Communcationa Engineers of Japan, Part 6*, pages 283-284.
            1984.

[16]   Shirai, Y. and Suwa, M.
        Recognition of polyhedrons with a range finder.
        In *Proceedings of the Second International Joint Conference on
            Artificial Intelligence*, pages 80-87. 1971.

[17]   Shirai, Y.
        A context sensitive line finder for recognition of polyhedra.
        Artificial Intelligence 4:95-119, 1973.

[18]   Sugihara, K.
        *Dictionary-guided scene analysis based on depth information,
            Report on Pattern Information Processing System.*
        Technical Report 13, Electrotechnical Lab, 1977.

[19]   Sugihara, K.
        Picture language for skeletal polyhedra.
        Computer Graphics and Image Processing 8:382-405, 1978.

[20]   Sugihara, K.
        Range-data analysis guided by a junction dictionary.
        Artificial Intelligence 12:41-69, May, 1979.

[21]   Sugihara, K.
        Automatic construction of junction dictionaries and their
            exploitation for the analysis of range data.
        In *Proceedings of the Sixth International Joint Conference on
            Artificial Intelligence*, pages 859-864. 1979.

[22]   Waltz, D.
        Understanding line drawings of scenes with shadows.
        The Psychology of Computer Vision.
        McGraw-Hill, New York, 1975, pages 19-91.

PART III: 3-D RECOGNITION ALGORITHMS

The Representation, Recognition, and Positioning of 3-D Shapes from Range Data

O. D. Faugeras

INRIA
Domaine de Voluceau - Rocquencourt
B.P. 105
78150 Le Chesnay
France

M. Hebert

Carnegie-Mellon University
Robotics Institute
Schenley Park
Pittsburgh Pennsylvania 15213

Abstract

The task of recognizing and positioning rigid objects in 3-D space is important for robotics and navigation applications. In this paper we analyze the task requirements in terms of what information needs to be represented, how to represent it, what paradigms can be used to process it and how to implement them.

We describe shape surfaces by curves and patches and represent them by the linear primitives--points, lines and planes. We also describe algorithms to construct this representation from range data.

We then propose the recognizing-while-positioning paradigm and analyze the basic constraint of rigidity that can be exploited. We implement the paradigm as a prediction-and-verification scheme that makes efficient use of our representation.

Results are presented for data obtained from a laser range finder, but both the shape representation and the matching algorithm are used on other types of range data from ultrasound, stereo and tactile sensors.

1. Introduction

Recently, there has been a surge in the development of methods for dealing with range data. Recovering depth information from images has always been an important goal in computer vision and has achieved some success. Various stereo programs have been developed [3], [20], [11], [22], [21] which are capable of producing fairly accurate and dense range maps in times that may eventually become realistic if the right hardware is built. Techniques like shape from texture or shape from shadows [17], [27] have also been investigated and some of them look promising.

Other approaches to the computation of depth are based on active ranging. They basically fall into two broad categories: active triangulation and time of flight techniques. Active triangulation techniques use an extra source of light to project some pattern onto the objects to be measured, thereby reducing the complexity of the stereo matching problem [23], [8]. Time of flight techniques send some energy (ultrasonic or electromagnetic) toward the scene and measure the time it takes for some of it to return to the source after reflection. In both cases, distances to scene points are immediately available without further processing.

Much work still remains to be done to obtain data faster and more accurately, but it is important to ask what can be done with those data, what task domains can be tackled, what representations and computational paradigms are useful, what constraints they allow us to use, and how much information is actually needed, given those constraints, to solve a particular problem.

In this article we describe the task of constructing descriptions of static scenes of 3-D objects when range data and 3-D models of these objects are available, and we explore the problem of recognizing and positioning such objects in the work space. The specific range data that have been used in our examples come from a laser range finder developed at INRIA [8], but the ideas and the results should be fairly independent of the origin of the data.

Our work is related to that of Oshima and Shirai [23], who were pioneers in this area, and also to that of Grimson and Lozano-Pérez [12], Bolles and Horaud [16], and Brady, et al. [7].

2. Representing 3-D Shapes

The first step in building a vision system is to extract high-level representations from raw data. In this section we investigate the possible representations of 3-D objects. Then we present the algorithms used for extracting the basic primitives of our representation--points, boundaries, and surfaces.

2.1. Primitives for 3-D Shape Representation

The question of how to represent 3-D shapes is a very basic one. A representation is a set of data structures and algorithms that operate on them; it is meaningless to discuss a representation without reference to a class of applications. In light of this we now briefly discuss several categories of representations and relate them to our own application.

Traditionally there have been two approaches to the representation of shapes. Hierarchical representations deal explicitly with various resolutions thereby allowing the recognition program to reason about objects with different levels of precision. Homogeneous descriptions, on the contrary, deal only with one resolution. We prefer homogeneous representations because they are simpler to build and to use; it remains to be demonstrated that hierarchical representations are appropriate to our application even though some promising work is in progress [24], [7].

With volume and surface representations, we believe that the key point is that of accessibility; that is, of reliably extracting the representation from the output of the sensors. Volume representations are potentially very rich and lend themselves to the computation of many other representations. However, descriptions obtained by decomposing the object into elementary volumes such as cubes or rhombododecahedra [24] are not very well suited to our problem since they are not invariant under rigid transformations. We believe that intrinsic object representations such as the one proposed by Boissonnat [5], where the volumes of objects are decomposed into tetrahedra, will prove to be much more powerful.

Another popular type of representation derived from the volumic representation is based on the idea of a skeleton [1], [24], [6], [5] as a complete summary of the shape, as well as a way of making explicit some important symmetries. This kind of representation can also be extremely useful since it provides a compact object model, but again is not very well suited to our specific problem because it is not robust to partial occlusion.

Surface representations seem ideal for recognizing and positioning objects since object surfaces can be measured by the sensors. Many interesting ideas of the representation of surfaces can be found in the computer graphics and CAD/CAM literatures. Unfortunately, because these disciplines are primarily concerned with faithful reproduction and accurate positioning of machine tools, the representations that have been developed there do not apply to noisy data and are unsuitable for the recognition and positioning of objects.

Among all these possible representations, we propose to use a surface description in terms of points, curves and surface patches. Points are corners on the surface or centers of symmetry, curves are either occluding or internal boundaries or even symmetry axes, and surface patches are either planes or quadrics. What features are used to represent these primitives is an extremely important issue for the object recognition task.

We consider two requirements. First we want the features to be stable with respect to partial occlusion and, second, we want them to carry enough information to allow us to recover position and orientation. Standard numerical features such as elongations, length, and perimeter of the surface do not satisfy either of these requirements and should therefore be used with caution. Topological features such as connectivity, genus or number of neighbors suffer from the same handicap. In contrast, geometric features, such as equations of curves or surface patches, do not change when the region they describe is partially occluded and can be used to recover position and orientation.

An important point is that primitives are ultimately described in terms of linear entities (points, lines and planes). The reason is that most of the computations used for the control of the matching process are intractable when one uses nonlinear representations--for example, the equation of a quadric surface instead of its principal directions. Furthermore, nonlinear representations are more sensitive to noise and thus less robust. The objects we want to handle in the vision program are typically parts with a reasonably large number of curved and planar faces (Figure 1).

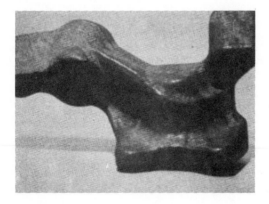

Figure 1:   Example of an industrial part

The idea of extracting and matching characteristic points has been used in image intensity analysis. The same idea may be used in the case of 3-D data. For example, one can extract spikes of an object, which are defined as local maxima of the curvature. This class of primitive is certainly not suitable for the control of the search problem but can be used for resolving ambiguities such as partial symmetries of the object.

2.2. Boundaries

2.2.1. Occluding Boundaries

Occluding boundaries are defined as the set of points where the surface presents a discontinuity in depth (see Figure 2); this is dependent on the direction of view. The main advantage of the occluding boundaries is that they are fairly easy to extract from a range image by a simple edge detector. Therefore, they can provide a quick filter for the recognition process [15] although they are generally not sufficient for deriving a unique solution of the identification/positioning problem.

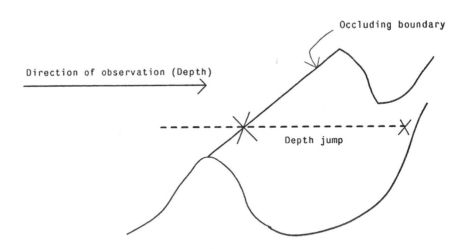

Figure 2: Definition of the occluding boundaries

One possible drawback is that these boundaries might be less robust than some other proposed primitives, mainly because the acquisition process might produce artifacts in the vicinity of a depth discontinuity.

2.2.2. Internal Boundaries

Internal boundaries such as troughs or creases contain important information about an object shape because they are localized in space and can be robustly extracted even in the presence of measurement noise. We define them as curves on the object where its surface undergoes a C^1 discontinuity (see Figure 3). In the work by Brady, et al. [7], the authors

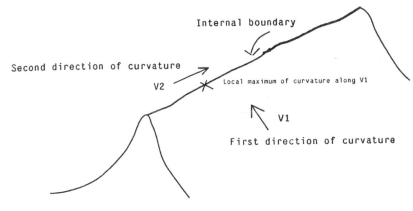

Figure 3: Definition of the internal boundaries

applied the 1-D curvature primal sketch to principal lines of curvature to detect local extrema in curvature. Linking these extremes allowed them to recover some of the internal boundaries. Their method is computationally expensive and we proceed differently.

From the definition above we can follow several roads. The first is based on the observation that when we cross an internal edge, one of the principal curvatures reaches a local maximum. This is the principal curvature in the direction perpendicular to the edge (Figure 3). Thus, method for detecting internal edges is as follows:

- Compute principal curvatures and directions at every point by fitting locally polynomials (in practice quadrics) to the set of range values. The mathematical details can be found in any book on differential geometry.

- Perform non-maxima suppression in the principal directions.

- Link the local maxima in the principal direction orthogonal to that of the maximum.

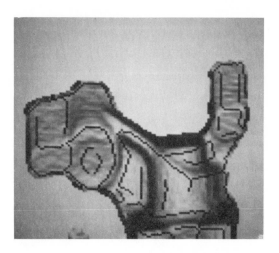

Figure 4:   Results of internal boundaries detection

This algorithm has been applied to one view of the object in Figure 1 and results are shown in Figure 4. This is the method proposed by Ponce and Brady [25].

Another method is based on the observation that when we cross an internal boundary, the normal to the surface is discontinuous. The problem is thus reduced to that of finding a local maximum of the norm of the directional derivative of the normal in some local coordinate system. Thus, the second method for detecting internal edges is as follows:

- Estimate the normal at every point either by straight-forward computation on the depth image or by fitting locally a plane.

- Compute the norm of the directional derivative of the normal in a number of directions (typically 8) at every point.

- Perform non-maxima suppression.

Mathematically the two methods are equivalent. Practically, we have found that the second method performs better because of some noise sensitivity problems attached to the estimation of principal curvatures in the first method. The results of applying the second method to one view of the object in Figure 1 are shown in Figure 4.

2.2.3. Representation of Boundaries

The outcome of the previous algorithms is a set of discrete curves in 3-D space which can yield represented in various ways. Our previous experience with 2-D curves [2] led us to believe that a polygonal approximation scheme is sufficient. Therefore, we propose to represent edges as polygonal chains. This raises the question of the ambiguity of the representation of a line in 3-D space. As shown in Figure 5, we represent a line by the pair (v,d) where d is the vector perpendicular from the origin and v a unit vector parallel to the line. Notice that the line $(-v,d)$ is equivalent to (v,d). Consequently, contrary to the case of occluding edges in which a segment has an intrinsic orientation with respect to the direction of observation, there is no intrinsic way of choosing the orientation of those line segments; this in turn increases the combinatorics of the problem.

2.3. Surface Representation

2.3.1. Type of Primitives

The type of surface primitives that can be used is constrained by the feasibility of the corresponding segmentation algorithm. It is difficult to control the segmentation algorithm when the degree of the polynomial representing the surface is high. Moreover, the robustness of the segmentation is higher when the primitives are simple. For example, the segmentation into splines can yield completely different results when applied to two different views of the same part of an object. Finally, surfaces represented by high-order polynomials cannot be as efficiently used in the matching process as the linear ones.

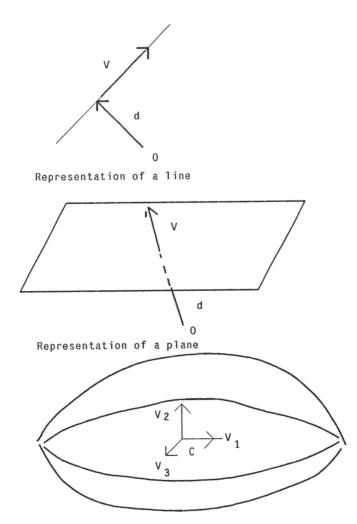

Representation of a line

Representation of a plane

Representation of a quadric

Figure 5:   Representation of the primitives

The two types of surface primitives we use are the plane and the quadric patches surface. We present our notations and conventions for representing these primitives, and then the segmentation algorithm.

- **Planes**. A plane is represented by a vector v and a scalar d. The equation of the plane is

$$\mathbf{x} \cdot \mathbf{v} - d = 0,$$

where "·" is the inner product, $\mathbf{v}$ is the normal and d the distance to the origin (Figure 5).

A plane has two different equivalent definitions $((\mathbf{v},d)$ and $(-\mathbf{v},-d))$, and the canonic orientation is defined by orienting the normal toward the outside of the object.

- **Quadrics**. The standard representation of a quadric surface is a 3x3 symmetric matrix $\mathbf{A}$, a vector $\mathbf{v}$ and a scalar d. The equation of the surface is

$$\mathbf{x}^t \mathbf{A} \mathbf{x} + \mathbf{x} \cdot \mathbf{v} + d = 0 \qquad (1)$$

As we have previously mentioned, we want to avoid using high degree polynomials. Therefore, we prefer a representation of quadric surfaces using linear features (Figure 5) which are:

The principal directions of the quadric which are the eigenvectors $\mathbf{v}_1, \mathbf{v}_2, \mathbf{v}_3$ of $\mathbf{A}$. Notice that these vectors do not have a canonic orientation.

The center of the quadric (when it exists) which is defined by

$$\mathbf{c} = -\frac{1}{2} \mathbf{A}^{-1} \mathbf{v}$$

Other information that could be used is the type of the surface (cylinder, ellipsoid, etc.) Unfortunately the type is related to the signs of the eigenvalues of $\mathbf{A}$ which are highly unstable when one eigenvalue is very near zero; this happens for useful surfaces such as cylinders and cones.

Notice that these representations are unique up to a scale factor. The way of defining a unique representation with respect to scaling is discussed in the section on surface fitting.

2.3.2. Segmentation by Region Growing

The primitive surfaces can be extracted from the original data in two ways: splitting and merging schemes. First, the whole set of points is considered as a primitive and is split into smaller regions until a reasonable approximation is reached. Second, the current set of regions, the initial set of regions being the set of original points, are merged until the approximation error becomes too large. The splitting schemes found in the literature have been used with respect to a fixed coordinate system (octrees, quadtrees). However, we think that a much better way to proceed would be to split in a way that is intrinsic to the object (prismtrees). Since we have not been able to derive such a scheme, the region growing method seems, therefore, more appropriate.

Let us assume that we are able to compute a measure $E(S)$ of the quality of the fit between region S and the generic primitive surface. (This measure is described in the next section.) The region growing algorithm merges the neighboring nodes of a current graph of surface patches, and stops when no more merges can be performed according to a maximum error value E_{max}. The initial graph can be either derived from a range image or built from a set of measurements on the object in the form of a triangulation (see Figure 6).

Several strategies can be implemented for the selection of the pair of regions to be merged at a given iteration and for the kind of control on the error. The best solution is to use a global control strategy, which means the evolution of the segmentation is determined by the quality of the overall description of the surface. The global control prevents the segmentation from being perturbed by local noisy measurements. In terms of implementation, the global control has two consequences:

- At each iteration, the regions R_i and R_j which produce the minimum error $E(R_i \cup R_j)$ among the whole set of pairs are merged.

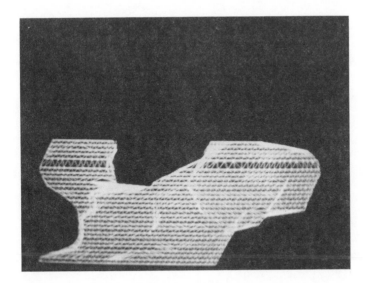

Figure 6: Triangulation of the Renault part

- The program stops when the global error $\sum_{i=1}^{N} E(R)_i$ is greater than E_{max}.

Let us define the error measure E. The cases of the planes and quadrics are treated separately.

Planes. In the case of planar patches, the error E is defined as the distance between the points of the region and the best fitting plane in the least-squares sense

$$E = Min \sum_{i=1}^{N} (\mathbf{v} \cdot \mathbf{x}_i + d)^2 = Min\ F(\mathbf{v}, d) \tag{2}$$

with constraint $\|\mathbf{v}\| = 1$. As is well known the solution of this minimization is obtained as follows:

$$\bar{x} = \sum_{i=1}^{N} x_i/N$$

$$M = \sum_{i=1}^{N} (x_i - \bar{x})(x_i - \bar{x})^t \, .$$

The direction v is the eigenvector corresponding to the smallest eigenvalue λ_{min} of M. Then

$$d = - \sum_{i=1}^{N} (\mathbf{v} \cdot x_i)/N$$

The resulting error E is the smallest eigenvalue λ_{min}.

Quadrics. The approach is the same as with the planes except that we do not have a clear definition of the distance from a point to a quadric surface. The simplest way is to define the error measure as

$$E = Min \sum_{i=1}^{N} (\mathbf{x}_i^t \mathbf{A} \mathbf{x}_i + \mathbf{x}_i \cdot \mathbf{v} + d)^2 = \mathbf{F(A,v,}d) \qquad (3)$$

where the notations of Equation (1) are used.

For the purpose of segmentation, we represent a quadric surface by 10 numbers $a_1, \ldots, a_{10}$ related to the previous description by the relations

$$\mathbf{A} = \begin{bmatrix} a_1 & a_4/\sqrt{2} & a_5/\sqrt{2} \\ a_4/\sqrt{2} & a_2 & a_6/\sqrt{2} \\ a_5/\sqrt{2} & a_6/\sqrt{2} & a_3 \end{bmatrix},$$

$$\mathbf{v} = (a_7 \ \ a_8 \ \ a_9)^t,$$

$$d = a_{10} \, .$$

The function $\mathbf{F}$ is homogeneous with respect to the parameters a_i. Therefore, we have to constrain the problem in order to avoid the trivial solution $(0, \ldots, 0)$. Several constraints can be designed since there exists no natural constraint as in the case of planes ($\| \mathbf{v} \| = 1$). The most frequently used constraints are:

$$\sum_{i=1}^{N} a_i^2 = 1$$

$$a_{10} = 1$$

$$Tr(\mathbf{AA}^t) = \sum_{i=1}^{6} a_i^2 = 1 .$$

The first two are not invariant with respect to rigid transformations, neither rotations nor translations. This implies that a surface observed from two different directions will have two different sets of parameters when expressed in the same coordinate system. So we use the third one which is invariant by translation because $\mathbf{A}$ is invariant and because the trace operator is also invariant.

Let us define the three vectors $\mathbf{P} = (a_1, \ldots, a_{10})^t$, $\mathbf{P}_1 = (a_1, \ldots, a_6)^t$, and $\mathbf{P}_2 = (a_7, \ldots, a_{10})^t$. Using these vectors, our constraint is

$$\| \mathbf{P}_1 \| = 1 . \tag{4}$$

The function $\mathbf{F}$ defined in Equation (3), being a quadratic function of the parameters, can be redefined as

$$\mathbf{F(P)} = \sum_{i=1}^{N} \mathbf{P}^t \mathbf{M}_i \mathbf{P}, \tag{5}$$

where M_i is a symmetric matrix of the form

$$M_i = \begin{bmatrix} B_i & C_i \\ C_i^t & D_i \end{bmatrix}$$

and B_i, C_i, and D_i are 6x6, 6x4, and 4x4 matrices respectively. Additionally, D_i and B_i are symmetric matrices. We shall not detail the values of the matrices that are easily computed from Equation (3).

If we define the matrices

$$B = \sum_{i=1}^{N} B_i, \quad C = \sum_{i=1}^{N} C_i, \quad D = \sum_{i=1}^{N} D_i,$$

the minimization problem (3) becomes

$$\textit{Min } P^t M P = \textit{Min } F(P), \tag{6}$$

with $\| P_1 \| = 1$,

where

$$M = \begin{bmatrix} B & C \\ C^t & D \end{bmatrix}.$$

The minimum is found by the method of the Lagrange multipliers. Using this method, solving Equation (6) is equivalent to finding the minimum λ and the corresponding vector P such that

$$\mathbf{MP} = \begin{bmatrix} \mathbf{BP}_1 & \mathbf{CP}_2 \\ \mathbf{C}^t\mathbf{P}_1 & \mathbf{DP}_2 \end{bmatrix} = \begin{bmatrix} \lambda\mathbf{P}_1 \\ 0 \end{bmatrix}. \tag{7}$$

Relation in Equation (7) gives the solution for $\mathbf{P}_2$

$$\mathbf{P}_2 = \mathbf{D}^{-1}\mathbf{C}^t\mathbf{P}_{1\ min} \tag{8}$$

Reporting this expression of $\mathbf{P}_2$ in Equation (7), the solution $\mathbf{P}_{1\ min}$ is the unit eigenvector corresponding to the smallest eigenvalue λ_{min} of the symmetric matrix

$$\mathbf{B} - \mathbf{CD}^{-1}\mathbf{C}^t$$

From Equations (6) and (7), the resulting error E is the eigenvalue λ_{min}.

Notice that the matrices $\mathbf{B}$, $\mathbf{C}$, and $\mathbf{D}$ can be easily updated when new points are added to the region or when two regions are merged, but the errors cannot be iteratively updated since they are derived from an eigenvalue calculation. We are investigating efficient iterative methods that could be applied to the surface fitting problem.

Results of the region-growing algorithm on several objects are presented in Figures 7-9.

3. Recognition and Positioning

The surface description we use along with the corresponding extraction algorithms have been presented in the previous sections. We now describe the way of using this description in the 3-D object recognition algorithm. We first investigate the possible strategies for the 3-D object recognition. Then the mathematical aspects of the algorithm are presented. Finally, the high-level control process is described.

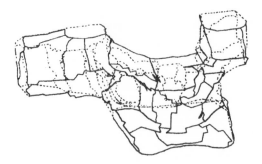

Figure 7:   Results of the segmentation into planes obtained from the triangulation of Figure 6 into 60 planar patches

3.1. Possible Strategies for the Search Problem

Now that we have worked out a representation for 3-D shapes that satisfies the requirements for solving the problem of recognizing and positioning objects, we are ready to deal with the corresponding search problem.

Our goal is to produce a list of matched model and scene primitives $((M_1,S_1), \ldots ,(M_P,S_P))$ where M_1 is a model primitive and S_j is a scene primitive. Some S_i's may be the special primitive *nil*, meaning that the corresponding model primitive M_i is not present in the scene. Producing such a list is the recognition task. We also want to position the identified model in the work space, that is, compute the rigid displacement that transforms the model coordinate system into the scene reference frame.

Any rigid displacement **T** can be decomposed in number of ways as a product of a translation and a rotation. In order to make this decomposition unique, we assume that the axis of the rotation goes through the origin of coordinates and that the rotation is applied first.

The combinatorial complexity of the correspondence problem can be very high (it is an exponential function of the number of primitives in the scene and the models). Great care should be taken in the choice of the search strategy in order to allow us to use the constraint of rigidity to reduce

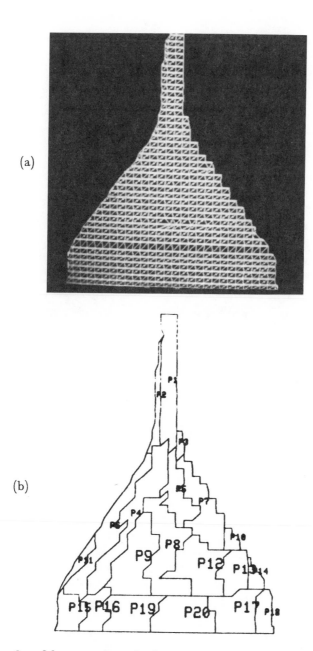

Figure 8: More results of the segmentation into planes. (a) Triangulation of a funnel; (b) Segmentation of the funnel into 20 planar patches;

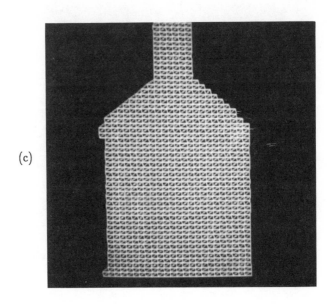

(c)

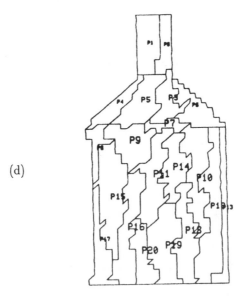

(d)

Figure 8:   (c) Triangulation of an oil bottle; (d) Segmentation into 20 planar patches

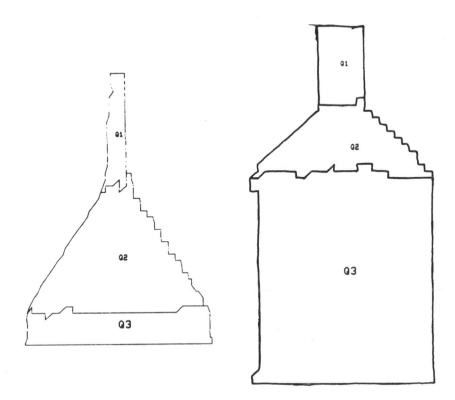

Figure 9: Results of the segmentation into quadric patches

as much as possible the size of the search space. Several techniques can be used and we review some of them.

3.1.1. Relational Matching

Relaxation techniques have been developed to solve a number of matching tasks. The matching is seen as a labeling problem where a model primitive M_i is labeled as a scene primitive S_i. If we use only region based measurements (the numerical or topological features mentioned above), there are many possible matches for a given model primitive, each one being given a quality measure $p(M_i, S_i)$.

The goal of the relaxation is then, starting from these initial measurements, to iteratively reduce the ambiguity and the incoherence of the initial matches. This is done by bringing in another piece of information,

a numerical measure of the coherence of an n-tuple of matches
$c(M_1, S_1, \ldots, M_n, S_n)$.

A number of different techniques have been proposed to achieve this.
They fall in two broad categories. The so-called discrete relaxation
techniques assume that the functions p and c can only be equal to 0 or 1 [26].
They can be used for problems like subgraph isomorphisms when the
connectivity of the graphs is not too high, but are not appropriate to our
problem where richer metric information is available. In continuous
relaxation techniques, the functions p and c take real values and $p(M_i, S_i)$
can be considered as the likelihood that S_i corresponds to M_i. The general
approach then consists of an iterative modification of the likelihoods by
combining those computed at the previous step with the coherence measures
c. Convergence toward a best match cannot in general be guaranteed [9],
[18] except in some very special cases.

Generally speaking, relaxation techniques have two main drawbacks.
First, the result is very sensitive to the quality of the initial matches (the
initial likelihoods $p(M_i, S_i)$) since convergence is toward a local maximum of
some global average measure of coherence. Second, they take into account
only local coherence (otherwise the functions c become untractable) whereas
the basic constraint of our problem, rigidity, is global. For these reasons we
have decided not to employ relaxation techniques.

3.1.2. Hough transform and clustering

Hough transform techniques have become popular for recognizing planar
shapes [4]. The basic idea of the Hough transform is to quantize the space of
the relevant transformations to be used in matching models and scenes
(scalings, translations, rotations, etc.) and use that space as an accumulator.
Each match between a model and a scene primitive corresponds to a family
of permitted transformations, and the corresponding cells in the
accumulator are increased. Best matches are then identified by searching
for maxima in the accumulator space. A verification step must generally still

be performed to break ties and improve accuracy.

These ideas could be applied to our recognition problem but with the following pitfalls. First, the transformation space is 6-dimensional, therefore implying a large size accumulator or a poor precision. Second, matching two primitives (points, lines or planes) does not completely define the transformation, as it only constrains it in ways that make the updating of the accumulator a bit complicated (Figure 10). For these reasons, and

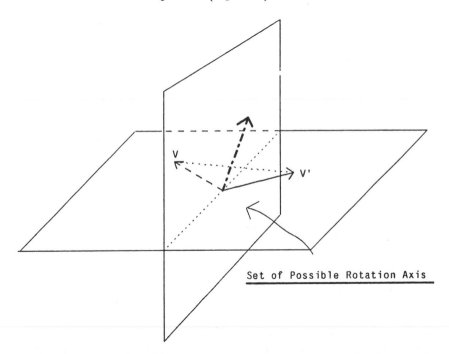

Set of Possible Rotation Axis

Figure 10: Constraints on the rotation axis resulting from the match of two vectors

because there is no easy way to exploit the rigidity constraint efficiently in the Hough transform approach, we have not implemented it.

3.1.3. Tree search

A tree-search technique explores the space of solutions; i.e., the set of lists of pairs $((M_1,S_1), \ldots ,(M_p,S_p))$, by traversing the tree of Figure 11. The

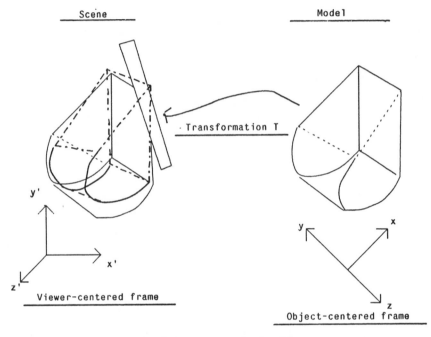

Figure 11: Principle of the tree-search algorithm

key issue is to avoid traversing the whole tree, that is, generating all solutions (even the wrong ones) while guaranteeing that the best ones will be found. The answer to this comes from exploiting the basic constraint of our problem, rigidity, and applying the paradigm of recognizing while positioning (Figure 12). For every path in the tree corresponding to a partial recognition $((M_1,S_i), \ldots ,(M_k,S_l))$, we compute the best rigid displacement T_k from model to scene for those primitives. We then apply T_k to the next unmatched model primitive M_{k+1} and consider as possible candidates for M_{k+1} only those unmatched scene primitives which are sufficiently close to $T_k(M_{k+1})$.

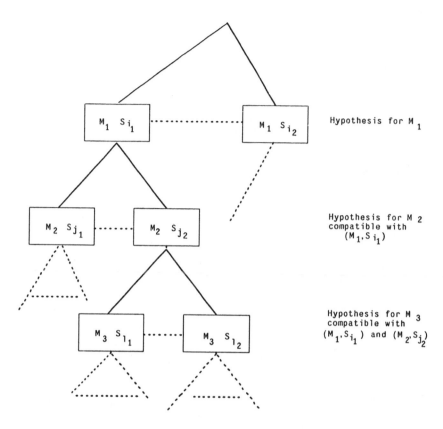

Figure 12: The rigidity constraint

This allows us to drastically reduce the breadth of the search tree, and demonstrates the power of the recognition/positioning paradigm [10]. Several important issues such as how to order the model primitives, and how to reduce the depth of the search tree, are explored later.

3.2. Mathematical Representation of the Rigidity Constraint

In this section, we describe the way of evaluating the transformation corresponding to a matching problem. The transformation estimation is first cast as a minimization problem. Then a suitable representation of the transformation is proposed. Finally, a closed-form solution and a non-exact iterative algorithm are described.

3.2.1. Estimation of the Position as a Minimization Problem

The main part of the application of the rigidity constraint is the estimation of the transformation given a partial match. Precisely, given a set of pairings (M_i, S_i) where the M_i's and S_i's are primitives of the model and the scene respectively, the problem is to compute the "best" transformation T which transforms the object model into the scene reference frame. The word "best" means that the sum of some distance D between $\mathbf{T}(M_i)$ and S_i is minimum. The rigidity constraint propagation is stated as a minimization problem

$$Min \sum_{i=1}^{N} D(\mathbf{T}(M_i), S_i) \, ,$$

where $\mathbf{T}$ is a rotation, $\mathbf{R}$, with axis through the origin, followed by a translation, $\mathbf{t}$. The distance D and the corresponding minimization problem depend the on three type of primitives: points, line segments, and planes.

1. points. The distance between points is simply the usual distance between the scene point and the transformed model point. The square Euclidean distance is preferred in order to apply a least-squares method. Therefore, the minimization problem is

$$Min \sum_{i=1}^{N} \|\mathbf{R}\mathbf{x}_i + \mathbf{t} - \mathbf{x}_i'\|^2 \, . \tag{9}$$

2. line segments. The distance between line segments contains two terms corresponding to the two components of the line representation, the direction and the distance vector. When a line segment $(\mathbf{v}, \mathbf{d})$ is transformed by $\mathbf{T}$, the transformed line $(\mathbf{v}', \mathbf{d}') = \mathbf{T}(\mathbf{v}, \mathbf{d})$ is

$$\begin{cases} \mathbf{v}' = \mathbf{R}\mathbf{v} \\ \mathbf{d}' = \mathbf{R}\mathbf{d} + \mathbf{t} - (\mathbf{t} \cdot \mathbf{v}')\mathbf{v}' \end{cases} . \tag{10}$$

The corresponding criterion is

$$Min \sum_{i=1}^{N} K\|v'_i - Rv_i\|^2 + (1-K)\|d'_i - Rd_i - t + (t \cdot v'_i)v'_i\|^2$$

where K is a weighting constant between 0 and 1.

3. **planes**. The transformed plane (v',d') of (v,d) is given by

$$\begin{cases} v' = Rv \\ t \cdot v' = d' - d \end{cases} \tag{11}$$

and the corresponding criteria are

$$Min \sum_{i=1}^{N} \|v'_i - Rv_i\|^2 \tag{12}$$

and

$$Min \sum_{i=1}^{N} (d'_i - d_i - t \cdot v'_i)^2. \tag{13}$$

Notice that an underdetermination problem may occur because a minimum number of pairings is required in order to solve the minimization problem numerically.

In the following sections, we first describe the representation of the rotations used for solving the optimization problem, then an exact solution is presented in the cases of planes and points. Finally an iterative solution is proposed for the cases which include line segments.

3.2.2. Quaternions: A Representation of Rotation

The three minimization problems are constrained by the fact that $\mathbf{R}$ is a rotation. Several representations of rotation are possible:

- An orthonormal matrix $\mathbf{R}$: $\mathbf{RR}^t = \mathbf{I}$.

- An axis k and an angle θ.

- A quaternion [13]: the product $\mathbf{Rv}$ is a product of quaternions.

The first representation leads to a high-dimensional space of constraints, while the second one leads to a non-polynomial criterion. In both cases, we cannot find an exact solution to the minimization problems, and the computation of an iterative solution is expensive. The third representation provides the simplest way of solving the problem and is described in the remaining part of this section.

A quaternion q can be defined as a pair $(\mathbf{w}, s)$ where $\mathbf{w}$ is a vector of $\mathcal{R}^3$ and s is a number. A multiplication is defined over the set of quaternions as

$$(\mathbf{w}, s) \cdot (\mathbf{w}', s') = (\mathbf{w} \wedge \mathbf{w}' + s\mathbf{w}' + s'\mathbf{w}, \quad ss' - \mathbf{w} \cdot \mathbf{w}') \qquad (14)$$

where $\wedge$ is the cross product.

The definitions of the conjugate $\bar{\mathbf{q}}$ and the module $\| \mathbf{q} \|$ of a quaternion are similar to the ones for the complex numbers:

$$\bar{\mathbf{q}} = (-\mathbf{w}, s)$$

and

$$|\mathbf{q}|^2 = \bar{\mathbf{q}} \cdot \mathbf{q} = \|\mathbf{w}\|^2 + \mathbf{s}^2 .$$

Notice that $\mathcal{R}^3$ is a subspace of $\mathbf{Q}$ thanks to the identification $\mathbf{v} \approx (\mathbf{v}, 0)$. Similarly, the module is the extension of the Euclidean norm and is multiplicative:

$$|q \cdot q'|^2 = |q|^2 \cdot |q'|^2 .$$

A rotation $\mathbf{R}$ of axis $\mathbf{a}$ and angle θ can be represented by two quaternions $\mathbf{q} = (\mathbf{w,s})$ and $\bar{\mathbf{q}}$, the application of the rotation being translated into a quaternion product by the relation

$$\mathbf{Ru} = \mathbf{q} \cdot \mathbf{u} \cdot \bar{\mathbf{q}} \tag{15}$$

where the mapping between the rotations and the quaternions is defined by

$$\begin{cases} \mathbf{w} = sin(\theta/2)\mathbf{a} \\ s = cos(\theta/2) \end{cases} \tag{16}$$

Similarly, for any quaternion of module 1, there exists a rotation satisfying the relation in Equation (15). The relations defining the three minimization problems can be translated into minimizations in the space of the quaternions, the new constraint being $\| \mathbf{q} \| = 1$.

3.2.3. Exact Noniterative Solution of the Minimization Problem

1. <u>planes</u>. The minimization problem stated in (13) can be restated in quaternion notation as

$$Min \sum_{i=1}^{N} |\mathbf{q} \cdot \mathbf{v}_i \cdot \bar{\mathbf{q}} - \mathbf{v}_i'|^2 \tag{17}$$

subject to the constraint $|\mathbf{q}| = 1$.

Since the module is multiplicative and $|\mathbf{q}| = 1$, Equation (17) can be rewritten as

$$Min \sum_{i=1}^{N} |\mathbf{q} \cdot \mathbf{v}_i - \mathbf{v}'_i \cdot \mathbf{q}|^2 \qquad\qquad (18)$$

The definition Equation (14) of the quaternion product shows that the expression $\mathbf{q} \cdot \mathbf{v}_i - \mathbf{v}'_i \cdot \mathbf{q}$ is a linear function of $\mathbf{q}$. Therefore, there exist symmetric matrices $\mathbf{A}_i$ such that

$$\sum_{i=1}^{N} |\mathbf{q} \cdot \mathbf{v}_i - \mathbf{v}'_i \cdot \mathbf{q}|^2 = \sum_{i=1}^{N} \mathbf{q}^t \mathbf{A}_i \mathbf{q}$$

where $\mathbf{q}$ is being considered as a column vector.

If $\mathbf{B} = \sum_{i=1}^{N} \mathbf{A}_i$, the minimization problem for the rotation part in the case of planes can be restated as

$$Min \sum_{i=1}^{N} \mathbf{q}^t \mathbf{B} \mathbf{q} \qquad\qquad (19)$$

subject to $|\mathbf{q}| = 1$.

Since $\mathbf{B}$ is a symmetric matrix, the solution to this problem is the 4-vector $\mathbf{q}_{min}$ corresponding to the the smallest eigenvalue λ_{min}. The matrix $\mathbf{B}$ can be incrementally computed from the pairings $(\mathbf{v}_i, \mathbf{v}'_i)$, for more details on the computation of $\mathbf{B}$ (see [14]).

2. **lines**. We do not have a noniterative solution for this case, but an iterative one is presented in the next section.

3. **points**. The transformation estimation problem in the case of points can also be restated in quaternion rotation by rewriting Equation (9) as

$$Min \sum_{i=1}^{N} |\mathbf{q} \cdot \mathbf{x}_i \cdot \bar{\mathbf{q}} + \mathbf{t} - \mathbf{x}'_i|^2. \qquad\qquad (20)$$

By the same trick used for deriving Equation (18) from (17), Equation (20) becomes

$$Min \sum_{i=1}^{N} |\mathbf{q} \cdot \mathbf{x}_i - \mathbf{x}'_i \cdot \mathbf{q} + \mathbf{t} \cdot \mathbf{q}|^2 \ . \tag{21}$$

As previously, there exist matrices $\mathbf{A}_i$ such that

$$\mathbf{q} \cdot \mathbf{x}_i - \mathbf{x}'_i \cdot \mathbf{q} = \mathbf{A}_i \mathbf{q} \tag{22}$$

Let us define $\mathbf{t} \cdot \mathbf{q}$ as a new quaternion $\mathbf{t}'$ and the 8-vector $\mathbf{V} = (\mathbf{q}, \mathbf{t}')^t$, the problem in Equation (20) becomes

$$Min \ \mathbf{V}^t \mathbf{B} \mathbf{V} \tag{23}$$

where $\mathbf{B}$ is a symmetric matrix of the form

$$\mathbf{B} = \begin{bmatrix} \mathbf{A} & \mathbf{C} \\ \mathbf{C}^t & \mathbf{N} \times \mathbf{I} \end{bmatrix}$$

and

$$\mathbf{A} = \sum_{i=1}^{N} \mathbf{A}_i^t \mathbf{A}_i \ , \qquad \mathbf{C} = \sum_{i=1}^{N} \mathbf{A}_i \ . \tag{24}$$

and N is the number of pairings. After some algebra, the solution is

$$\mathbf{t}' = \mathbf{C} \mathbf{q}_{min} / N \tag{25}$$

and

$$\mathbf{t} = \mathbf{t}' \cdot \bar{\mathbf{q}}$$

where q_{min} is the 4-vector of unit norm corresponding to the smallest eigenvalue λ_{min} of the symmetric matrix $A - C^t C / N$.

As in the case of planes, the matrices A_i depend only of the two terms of the pairing M_i and S_i.

3.2.4. Iterative Solution

As we discussed above, the estimation of the transformation is used to guide a search process and has to be repeated a large number of times. In particular, minimization Equations (9), (11), (13) and (14) have to be solved repeatedly on data sets that differ only in one term. Techniques such as recursive least-squares are good candidates for dealing with this problem and indeed turn out to be quite effective.

The basic idea is as follows. Given a set of matrices H_i and a set of measured vectors z_i, we want to find the minimum with respect to x of the following criterion

$$E_N(x) = \sum_{i=1}^{N} \|z_i - H_i x\|^2 \tag{26}$$

If x_{k-1} gives the minimum of $E_{k-1}(x)$, then x_k can be computed as

$$x_k = x_{k-1} + K_k(z_k - H_k x_{k-1})$$

where matrix K_k is given by

$$K_k = P_k H_k^t$$

$$P_k = [P_{k-1}^{-1} + H_k^t H_k]^{-1}$$

The criteria for (14), (19), and (21) can be put in the form of (27) with

$$x = t, \; z_i = d'_i - d_i, \; H_i = v'^t_i \qquad \text{for (13)},$$

$$x = q, \; z_i = 0, \; H_i q = v'_i \cdot q - q \cdot v_i \qquad \text{for (18)},$$

and

$$x = (q,t'), z_i = 0, \; H_i(q,t')^t = q \cdot x_i - x'_i \cdot q + t' \qquad \text{for Equation (20)}.$$

The criterion for (10) can be rewritten as follows

$$Min \sum_{i=1}^{N} K|v'_i \cdot q - q \cdot v_i|^2 + (1-K)|d'_i - q \cdot d_i \cdot \bar{q} - t + (t \cdot v'_i)v'_i|^2$$

The first part of the criterion does not pose any specific problem; the second does because the term containing t is a function of the index i and therefore the trick used for the points cannot be applied. What we do then is linearize the quadratic term $h_i(q) = q \cdot d_i \cdot \bar{q}$ in the vicinity of the solution q_{i-1} and apply the recursive least-squares to the modified criterion. This approach is similar to the one used in the extended Kalman filtering [19]. Convergence theorems do not exist for this particular application of the Kalman filtering theorem; we cannot, therefore, guarantee that the criterion of Equation (12) and the iterative solution is exact. This is also true for the criteria of (18) and (21) since the recursive least-squares does not take into account the constraint $|q|^2 = 1$.

Nonetheless, both on synthetic and real examples the results have been excellent as indicated in Tables 1 and 2 (in both figures errors are measured in degrees for the rotation angle and the orientation of the axis of rotation). When a solution exists with a small error, then the iterative technique will find it, whereas if no acceptable transformation exists, the quaternion norm will differ widely from 1. Therefore, the current value of the quaternion is

Iteration	Initial errors : angle 45° - axis 90°									
	1		2		3		4		5	
	angle error	axis error	angle error	axis error	angle error	axis error	angle error	axis error	angle error	axis error
3-9-1	1.98	21.94	1.17	10.55	.54	6.55	.24	4.66	.07	3.58
3-9-2	53.36	48.38	11.92	10.16	3.72	3.51	1.94	.89	1.27	1.25
3-9-3	13.11	13.58	5.53	6.06	3.30	3.83	2.31	2.79	1.76	2.19

Iteration	Initial errors : angle 20° - axis 45°									
	1		2		3		4		5	
	angle error	axis error	angle error	axis error	angle error	axis error	angle error	axis error	angle error	axis error
3-9-1	2.43	6.52	1.01	3.46	.52	2.32	.27	1.75	.13	1.40
3-9-2	1.59	2.75	.71	1.28	.45	.82	.33	.60	.26	.47
3-9-3	2.31	5.11	1.04	2.71	.65	1.84	.47	1.40	.36	1.12

Iteration	Initial errors : angle 20° - axis 20°									
	1		2		3		4		5	
	angle error	axis error	angle error	axis error	angle error	axis error	angle error	axis error	angle error	axis error
3-9-1	.52	2.35	.45	1.26	.43	.65	.42	.53	.42	.44
3-9-2	1.07	1.13	.54	.56	.36	.36	.26	.27	.21	.21
3-9-3	1.47	1.49	.73	.78	.48	.53	.36	.41	.29	.33

Table 1: Iterative estimation of the rotation for the example of Figures 17-19. An iteration uses all the primitives effectively matched (5, 6, and 8). Errors are in degrees.

an important clue about whether or not to pursue the exploration of a particular partial solution to the matching process.

3.3. Control Strategy

We are now equipped with reliable primitive extraction algorithms as well as transformation estimation algorithms. We now present the higher level control process of the recognition programs. The control process is divided into three phases: hypothesis, prediction, and verification. The results of the recognition program are presented in Figures 15-20.

Iteration	Initial errors : angle 45° - axis 90°									
	1		2		3		4		5	
	angle error	axis error	angle error	axis error	angle error	axis error	angle error	axis error	angle error	axis error
3-10-1	3.89	2.14	1.60	.95	.98	.61	.70	.45	.54	.35
3-10-2	23.64	13.15	11.12	6.52	7.05	4.18	5.11	3.04	3.99	2.37
3-10-3	4.34	2.61	2.21	1.25	1.47	.82	1.09	.60	.87	.48

Iteration	Initial errors : angle 20° - axis 45°									
	1		2		3		4		5	
	angle error	axis error	angle error	axis error	angle error	axis error	angle error	axis error	angle error	axis error
3-10-1	1.21	3.79	.78	2.01	.56	1.36	.43	1.03	.35	.83
3-10-2	7.73	6.18	4.05	3.07	2.72	2.00	2.04	1.47	1.63	1.16
3-10-3	2.74	2.68	1.39	1.33	.92	.88	.69	.65	.54	.52

Iteration	Initial errors : angle 20° - axis 20°									
	1		2		3		4		5	
	angle error	axis error	angle error	axis error	angle error	axis error	angle error	axis error	angle error	axis error
3-10-1	.71	1.27	.33	.67	.22	.46	.16	.35	.13	.29
3-10-2	2.84	2.02	1.47	1.01	.98	.66	.73	.48	.58	.37
3-10-3	1.94	1.41	1.00	.72	.66	.48	.50	.36	.39	.29

Table 2: Iterative estimation of the rotation for the example of Figures 20-22. An iteration uses all the primitives effectively matched (5, 6, and 8). Errors are in degrees.

3.3.1. Hypothesis Formation

The first step of the recognition process is the hypothesis formation. This consists of the search for sets of consistent pairings that provide enough information for computing the transformation and applying the prediction/verification phase. The hypothesis phase is crucial because it determines the number of branches of the tree that have to be explored. The actual number of pairings required depends on the type of primitives used (as shown below) ; however, two pairings can always be used in order to estimate the rotation.

	Translation	Rotation
Points	3	3
Lines	2	2
Planes	3	2

Table 3: Minimum Number of Primitives to Estimate the Rigid Transformation

The hypothesis formation proceeds in three steps.

1. selection of a first pairing. For each primitive of the model, the compatible primitives of the scene are listed. The choice of these primitives cannot make use of the rigidity constraint. Only the position-invariant features such as the length of the segments or the area of surface patches can be used. Most of these features are highly sensitive to occlusion and noise, therefore they should be used carefully and with large tolerances.

2. selection of a second pairing. Given a first pairing (M_1,S_1) and a second model primitive M_2, the candidates for the second scene primitive S_2 must satisfy the rigidity constraint. It turns out that this choice is quite simple. In the case of points the constraint on S_2 is that $D(S_1,S_2) = D(M_1,M_2)$ where D is the usual Euclidean distance. In the case of planes, the constraint is on the angle between the normals. S_2 must be chosen such that $(v_2',v_1') = (v_2,v_1)$ since for any such scene primitives the rotation is fixed and we can always find a translation (in fact, the coordinates of the translation along v_1' and v_2' are fixed). Indeed, let us write using relation (12):

$$t = \alpha v_1' + \beta v_2' + \gamma v_1' \wedge v_2' \qquad (27)$$

Assuming that v_1' and v_2' are not parallel, we obtain by multiplying (27) by v_1' and v_2':

$$d'_1 - d_1 = \alpha + \beta s \tag{28}$$

and

$$d'_2 - d_2 = \alpha s + \beta$$

where

$$s = \mathbf{v}'_1 \cdot \mathbf{v}'_2 \; .$$

Relations in Equation (28) always yield a unique solution in α and β.

In the case of lines, the situation $(\mathbf{v}'_2, \mathbf{v}'_1) = (\mathbf{v}_2, \mathbf{v}_1)$ is slightly different. Again, we have a constraint on the angles and S_2 must be chosen such that $(\mathbf{v}'_2, \mathbf{v}'_1) = (\mathbf{v}_2, \mathbf{v}_1)$. But we also have a distance constraint, for example that the shortest distance between M_1 and M_2 (notedby $d(M_1, M_2)$) be equal to $d(S_1, S_2)$.

This can be seen as follows. First we recall the result that if $L = (\mathbf{v}, d)$ and $M = (\mathbf{w}, e)$ are two non-parallel straight lines, then the algebraic shortest distance between them is given by

$$d(L, M) = [\mathbf{d\text{-}e}, \mathbf{v}, \mathbf{w}] / sin(\mathbf{v}, \mathbf{w}) \tag{29}$$

where $[\mathbf{x}, \mathbf{y}, \mathbf{z}]$ is the determinant of the three vectors $\mathbf{x}, \mathbf{y}, \mathbf{z}$. To convince ourselves of this, let us write the equation of a line intersecting L and parallel to $\mathbf{v} \wedge \mathbf{w}$, the direction of the shortest distance:

$$\mathbf{d} + \alpha \mathbf{v} + \gamma \mathbf{v} \wedge \mathbf{w} \; .$$

The equation of the line corresponding to the shortest distance is obtained by solving the following vector equation with respect to α, β and γ

$$d + \alpha v + \gamma v \wedge w = e + \beta w .\tag{30}$$

From Equation (29) we derive the algebraic distance γ, which is the projection of $d-e$ on the line of direction $v \wedge w$;

$$\gamma = (d-e) \cdot (v \wedge w)/\|v \wedge w\| ,$$

which yields to Equation (28).

Let us now go back to our constraint. Given M_1 and M_2, S_1 and S_2 satisfying $(v'_2, v'_1) = (v_2, v_1)$ we know that the rotation is determined. Let us characterize the translation. We write the translation vector t as

$$t = \alpha v'_1 + \beta v'_2 + \gamma v'_1 \wedge v'_2$$

and we use relations from Equation (10)

$$d'_1 = Rd_1 + t - (t \cdot v'_1)v'_1 ,$$

$$d'_2 = Rd_2 + t - (t \cdot v'_2)v'_2 .$$

Since R is known, we define $u_1 = d'_1 - Rd_1$ and $u_2 = d'_2 - Rd_2$. Multiplying the first equation by v'_2 and the second by v'_1 and letting $C = v'_1 \cdot v'_2$, we obtain:

$$\alpha = (u_2 \cdot v'_1)/(1 - C^2)$$

$$\beta = (u_1 \cdot v'_2)/(1 - C^2)$$

The value $1 - C^2$ is nonzero since M_1 and M_2 are not parallel. To obtain the coordinate of t along $v'_1 \wedge v'_2$, we multiply both equations by $v'_1 \wedge v'_2$; letting $S = sin(v'_1, v'_2)$, we obtain:

$$[u_1, v'_1, v'_2] = \gamma S^2 ,$$

$$[u_2, v'_1, v'_2] = \gamma S^2 .$$

In order to be consistent, we need

$$[u_1, v'_1, v'_2] = [u_2, v'_1, v'_2] .$$

After some algebra, we find this to be equivalent to

$$[d'_1 - d'_2, v'_1, v'_2] = [R(d_1 - d_2), v'_1, v'_2]$$

Since R^{-1} does not change determinants by using the relations $v_1 = R^{-1}v'_1$ and $v_2 = R^{-1}v'_2$, we obtain

$$d(S_1, S_2) = d(M_1, M_2)$$

3. <u>estimation of the transformation</u>. The transformation is estimated (or partially estimated in the case of planar patches) using the techniques described in the previous sections. As previously mentioned, some primitives (e.g., lines) do not have a canonic orientation. Therefore, the transformation estimated from an initial hypothesis is not unique and several equivalent transformations are generated. The number of possible transformations depends on the type of primitive; the most important cases are a pair of lines which produces two solutions, and a pair of quadrics which produces eight transformations when the three elongations $\lambda_{1, \ldots, 3}$ are not zero.

One important part of this step is the order in which the primitives of the model are considered for matching. Obviously non-interesting branches might be explored if the order is not carefully determined. Consider for example the case in which the two first primitives are parallel planes; then the estimated rotation about the axis perpendicular to the planes is arbitrary. Eventually a complete branch of the tree might be explored based on a wrong estimation of the transformation. Three basic rules must be applied in the ordering of the primitives.

- The small primitives (in terms of area or length) should be avoided.

- The first two or three primitives must be linearly independent in order to avoid indetermination. Actually, the best estimation is produced when the primitives are nearly orthogonal.

- If local symmetries exist in the object, the primitives that could best discriminate between almost equivalent positions of the object should the among first ones considered for matching. Notice that in some cases this might contradict the first two rules.

3.3.2. Prediction and Verification

Given an initial hypothesis and the associated transformation $T = (R, t)$, we want to predict the set of candidate primitives of the scene that can be matched with each primitive of the model in order to verify the validity of the initial hypothesis.

The way of doing that is to apply the transformation T to every model primitive M_i and find the primitives of the scene that are close enough to $T(M_i)$ (Figure 13). However, a sequential exploration of the scene description for each model primitive would increase drastically the combinatorics of the algorithm. Moreover, since we have computed a first estimate of the transformation, the rigidity constraint would reduce the number of candidates.

Therefore, we need a representation of the space of parameters which permits a direct access to the S_k such that $D(S_k, T(M_i)) < \epsilon$. Generally speaking, such a structure could be implemented as a discretized version of the space of parameters, but each "cell" of the space containing the list of

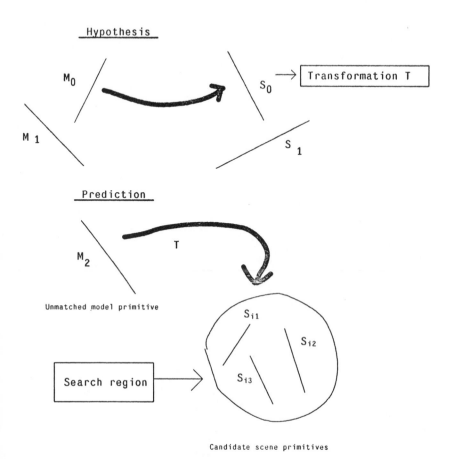

Figure 13: The prediction step

the primitives with the corresponding parameters. This structure is built only once. Then, the list of candidates is determined by reporting the cell c_i to which $T(M_i)$ belongs. This operation is made in constant time and does not depend on the initial number of scene primitives in order to minimize the number of explored branches.

Obviously, it is impossible to completely implement the previous scheme; the dimension of the parameter space is six which leads to an array of intractable size. But it is possible to discretize only part of the space. One of the easiest and most effective solutions is to discretize the spherical coordinates of the normals, direction, and principal directions in the cases of planes, lines and quadrics. The resulting data structure is the discretized unit sphere containing pointers to lists of primitives (Figure 14). The dimension of this subspace is only two, and it is efficient because the rotation usually provides a strong enough constraint to remove most of the incompatible pairings.

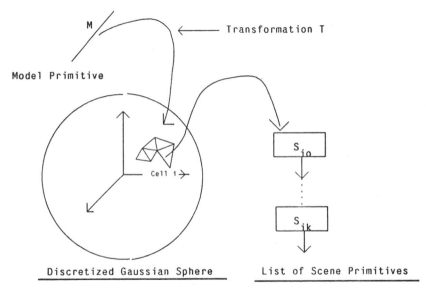

Figure 14:   Using the discretized unit sphere

Another possibility is to sort the values of the parameters of the scene primitives; then the candidates can be retrieved by a binary search technique. This second method is less efficient since we lose the direct access to the lists of candidates. On the other hand, it produces shorter lists because a wider set of parameters are taken into account.

3.3.3. Controlling the Depth of the Search

We have decided that, in order to be recognized, an object must have some fixed percentage of its surface visible (50% for example). If at some level of the tree of Figure 11 the number of *nil* assignments is such that even if all the remaining model primitives are matched, the required area percentage cannot be reached, then it is not necessary to explore further down. This allows the search program to prune entire subtrees and improve efficiency at the cost of missing a few correct interpretations of the data.

4. Conclusion

In this paper, we have presented a number of ideas and results related to the problem of recognizing and positioning 3-D rigid objects from range measurements.

We have discussed the need for representing surface information, specifically curves and surface patches. We have described a number of simple algorithms for extracting such information from range data and argued for a representation in terms of linear primitives constructed from curves and surface patches.

We have also discussed the representation of the constraint of rigidity and proposed to exploit it to guide the recognition process. The resulting paradigm consists of recognizing while positioning, and has been implemented as a hypothesis formation and verification process which has proved extremely efficient in practice.

Further work is needed in order to explore other 3-D object representations both for the application described in this paper and the more general problems of dealing with articulated objects or with classes of objects rather than specific instances. In such cases the rigidity constraint cannot be exploited as fully as we have done, and more powerful matching mechanisms and other constraints must be brought in.

Acknowledgments

We are thankful to Nicholas Ayache, Bernard Faverjon, Francis Lustman, and Fabrice Clara for many fruitful discussions. We have also benefited from discussions with Michael Brady, Eric Grimson, Masaki Oshima, Tomás Lozano-Pérez, and Yoshiaki Shirai.

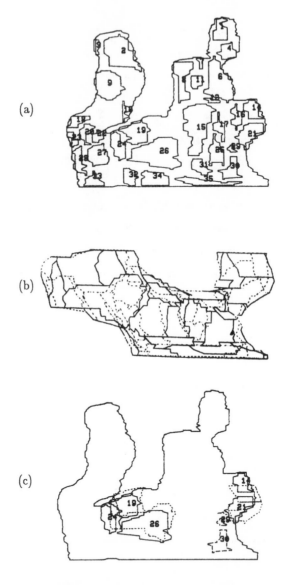

Figure 15:   Results of the recognition algorithm on a first scene (the model used is that of Figure 7(a)). (a) Scene segmentation; (b) First identified model after rotation with the estimated R; (c) Superposition of identified scene and model primitives

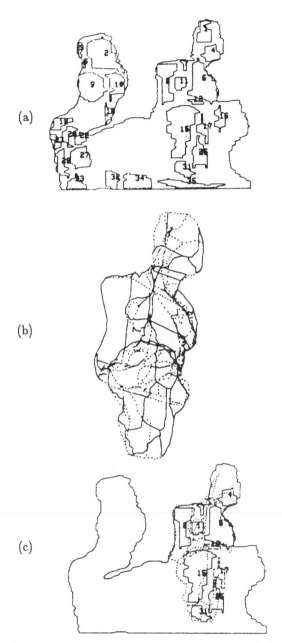

(a)

(b)

(c)

Figure 16: Results of the recognition algorithm on a first scene (continued)

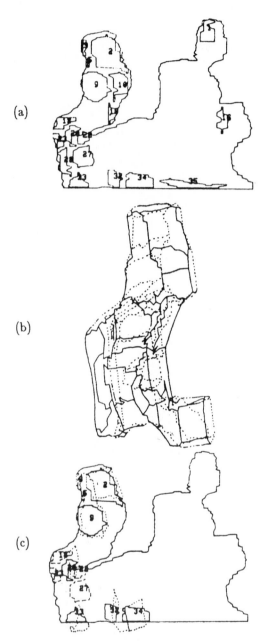

Figure 17: Results of the recognition algorithm on a first scene (continued)

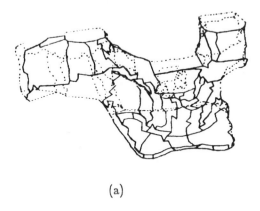

(a)

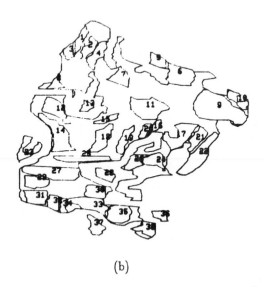

(b)

Figure 18: Results of the recognition program on a second scene. (a) Model of the object to recognize; (b) Scene segmentation;

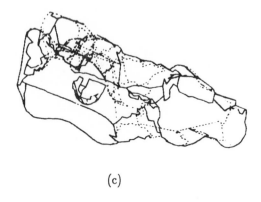

(c)

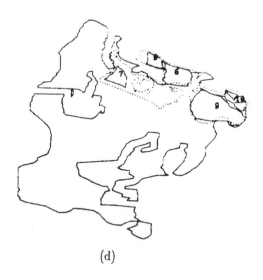

(d)

Figure 18: (c) First identified model after rotation with the estimated R; (d) Superposition of identified scene and model primitives

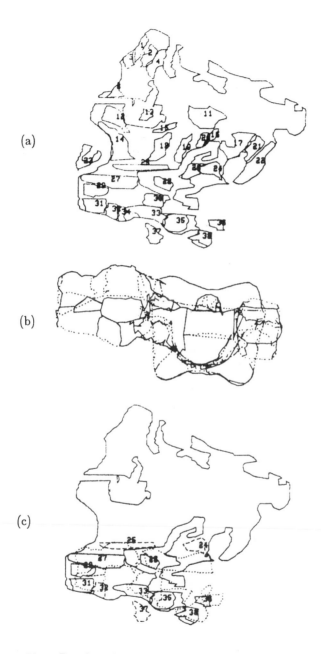

Figure 19: Results of the recognition program on a second scene (continued)

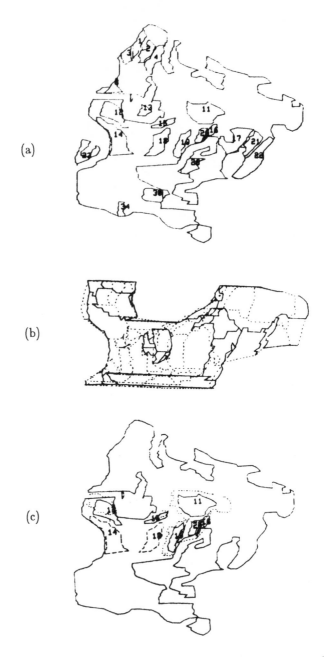

(a)

(b)

(c)

Figure 20:   Results of the recognition program on a second scene (continued)

References

[1] Agin, G.J.
Representation and description of curved objects.
Technical Report AIM-173, Stanford University, 1972.

[2] Ayache, N. and Faugeras, O.D.
A new method for the recognition and positioning of 2-D objects.
Proc. Seventh Int. Conf. on Pattern Recognition :1274-1280,
August, 1984.

[3] Baker, H.H. and Binford, T.O.
A system for automated stereo mapping.
In *Proc. Image Understanding Workshop*, pages 215-222. Science
Applications, Inc., 1982.

[4] Ballard, D.H.
Generalizing the Hough transform to arbitrary shapes.
Pattern Recognition 13(2):111-122, 1981.

[5] Boissonnat, J.D.
Representing 2-D and 3-D shapes with the Delaunay triangulation.
In *Proc. Seventh Int. Conf. on Pattern Recognition*, pages 745-748.
Montreal, Canada, August, 1984.

[6] Brady, M. and Asada, H.
Smoothed local symmetries and their implementation.
Int. J. Robotics Research 3(3), 1984.

[7] Brady, M., Ponce, J., Yuille, A. and Asada, H.
Describing surfaces.
In *Proc. Seond Int. Symp. on Robotics Research*, pages 434-445.
Kyoto, Japan, 1984.

[8] Faugeras, O.D., et. al.
Towards a flexible vision system.
In A. Pugh (editor), *Robot Vision*. IPS, UK, 1982.

[9] Faugeras, O.D., Berthot, M.
Improving consistency and reducing ambiguity in stochastic labeling :
an optimization approach.
IEEE Trans. on Pattern Analysis and Machine Intelligence
PAMI-3(4):412-424, 1980.

[10]   Faugeras O.D., Hebert, M.
A 3-D recognition and positioning algorithm using geometrical
      matching between primitive surfaces.
In *Proc. Eighth Int. Joint Conf. On Artificial Intelligence*, pages
      996-1002. Los Altos: William Kaufmann, August, 1983.

[11]   Grimson, W.E.L.
*From images to surfaces : a computational study of the human
      early visual system.*
MIT Press, Cambridge, Mass., 1981.

[12]   Grimson, W.E.L. and Lozano-Perez, T.
Model-based recognition and localization from sparse range or tactile
      data.
International Journal of Robotics Research 3(3):3-35, 1984.

[13]   Hamilton, W.R.
Elements of quaternions.
Chelsea, New York, 1969.

[14]   Hebert, M.
Reconnaissance de formes tridimensionnelles.
PhD thesis, University of Paris South, September, 1983.
Available as INRIA Tech. Rep. ISBN 2-7261-0379-0.

[15]   Hebert, M. and Kanade, T.
The 3-D profile method for object recognition.
In *Proc. CVPR'85.* San Franciso, CA, June, 1985.

[16]   Horaud, P. and Bolles. R.C.
3-DPO's strategy for matching three-dimensional data.
In *Proc. of the Int. Conf. on Robotics*, pages 78-85. Atlanta,
      Georgia, 1984.

[17]   Horn, B.K.P.
Obtaining shape from shading information.
In P.H. Winston (editor), *The Psychology of Computer Vision*,
      pages 115-156. Mc Graw-Hill, New York, 1975.

[18]   Hummel, R. and Zucker, S.
On the foundations of relaxation labeling processes.
IEEE Trans. on Pattern Analysis and Machine Intelligence
      PAMI-5:267-287, 1983.

[19]   Jazwinski, A.H.
       Stochastic processing and filtering theory.
       Academic Press, 1970.

[20]   Marr, D. and Poggio, T.
       A computational theory of human stereo vision.
       Proc. R. Soc. Lond. (B204):301-328, 1979.

[21]   Nishihara, H.K.
       PRISM: a practical realtime imaging stereo system.
       In B. Rooks (editor), *Proc. Third Int. Conf. on Robot Vision and
            Sensory Control*, pages 121-130. IFS publications and North-
            Holland, 1983.

[22]   Ohta, Y. and Kanade, T.
       *Stereo by intra- and inter-scanline search using dynamic
            progamming.*
       Tech Rep. CMU-CS-83-162, Carnegie Mellon University, 1983.

[23]   Oshima, M. and Shirai, Y.
       Object recognition using three-dimensional information.
       IEEE Transactions on Pattern Analysis and Machine Intelligence
            PAMI-5(4):353-361, July, 1983.

[24]   Ponce, J.
       Representation et manipulation d'objets tridimensionnels.
       PhD thesis, University of Paris South, 1983.
       Also available as INRIA Tech. Rep. ISBN 2-7261-0378-2.

[25]   Ponce, J. and Brady, M.
       Toward a surface primal sketch.
       1985.
       submitted to IJCAI 1985.

[26]   Rosenfeld, A., Hummel, R., and Zucker, S.
       Scene labeling by relaxation operations.
       IEEE Trans. on SMC (6):420-433, 1979.

[27]   Witkin, A.P.
       Recovering surface shape and orientation from texture.
       In M. Brady (editor), *Computer Vision*, pages 17-47. North-Holland,
            Amsterdam, 1981.

An Object Recognition System Using Three-Dimensional Information[1]

Masaki Oshima
Yoshiaki Shirai

Electrotechnical Lab
1-1-4 Umezono
Sakura-mura, Niihari-gun
Ibaraki 305, Japan

Abstract

This paper describes an approach to the recognition of stacked objects with planar and curved surfaces. The system works in two phases. In the learning phase, a scene containing a single object is shown one at a time. The range data of a scene are obtained by a range finder. The data points are grouped into many surface elements consisting of several points. The surface elements are merged together into regions. The regions are classified into three classes: plane, curved and undefined. The program extends the curved regions by merging adjacent curved and undefined regions. Thus the scene is represented by plane regions and smoothly curved regions. The description of each scene is built in terms of properties of regions and relations between them. This description is stored as an object model. In the recognition phase, an unknown scene is described in the same way as in the learning phase. Then the description is matched to the object models so that the stacked objects are recognized sequentially. Efficient matching is achieved by a combination of data-driven and model-driven search processes. Experimental results for blocks and machine parts are shown.

---

[1]This paper is based on three previous papers by the authors [18], [19], [20], and is reproduced here with their permission. Copyrights go to Pergamon Press, Institute of Electronics and Communication Engineers of Japan, and IEEE, respectively.

1. Introduction

Computer recognition of objects aims at the realization of automatic assembly or inspection. One of the most important problems in object recognition is to establish an efficient and flexible recognition method applicable to various scenes. The method must deal with scenes which have the following characteristics:

1. Objects are placed at any 3-D position with any orientation.
2. Objects may be stacked in a scene.
3. Objects have planar and/or smoothly curved surfaces.

Object recognition can be based upon two kinds of input data: an intensity picture of a scene and three-dimensional (range) data of objects in a scene. A recognition system using an intensity picture usually assumes that light intensity changes abruptly at the edge between surfaces. This assumption is not always justified. Moreover, there is no guarantee that changes in light intensity correspond to geometrical features. These facts make it difficult to develop a flexible recognition system solely by using intensity pictures. On the other hand, if three-dimensional data of objects are available [24], [17], more flexible recognition can be achieved because 3-D shapes of objects are directly obtained. Let us call this the 3-D approach.

The 3-D approach to recognizing objects has been studied since the early 1970's. Shirai [24] employed a range finder consisting of a vertical slit projector and a TV camera. Each reflected slit was segmented into sections corresponding to a surface of an object. Several adjacent sections from different slits were grouped so that planar surfaces were found. Popplestone, et al. [22] extracted cylindrical surfaces as well as planar surfaces using a similar process. Agin [1] and Nevatia and Binford [16] described a scene including curved objects with generalized cylinders, which are advantageous in describing elongated bodies. In these earlier works, the recognition programs first approximated the images of slits by straight lines or curves. The process relied only on the data points along a one-dimensional slit image, and tended to be sensitive to noise. Also, the sequential nature of the processing made it difficult to recover from the mistakes in the earlier stages of processing. To overcome this difficulty, Kyura and Shirai [13] proposed to use a surface element, which consists of a set of data points in a small region, as a unit for representing the surface. They showed that this approach was effective for scenes with complicated polyhedra such as a dodecahedron or an icosahedron. Milgram and Bjorklund [14] applied a similar method to extract surfaces of buildings seen from an airplane.

In this paper, we extend the Kyura and Shirai method [13]. A scene represented by surface elements is segmented into regions, each of which is approximated by a planar or smoothly curved surface. This method is applicable to general scenes with real objects. Other methods for feature extraction and description using 3-D data can be seen in [12], [25], [6], [8], [11], [3], [15], [5], [7].

In order to recognize a scene with multiple objects, an image ought to be segmented into parts, each of which corresponds to an object. Some systems use the knowledge of vertex type and the like [9], [10], [27], [23], [25] for segmentation. In contrast, our system segments an image by finding known objects.

Recognition of objects usually requires a pattern matching process: matching a part of a scene description to a part of an object model. The control strategy for matching is important particularly in processing a complex scene, since it is very time consuming to match each part of a scene against every possible model. To reduce the number of trials, the present system first tries to extract reliable and useful features of a scene and then find models including these features so that the earlier result may guide further processing.

Shape discrimination classifies objects in a scene into predetermined classes based on their geometrical features. Few studies on shape discrimination in the 3-D approach have been reported. In shape discrimination, by using intensity images, the placement of objects is often restricted. For instance, objects are neither allowed to rotate in a 3-D space nor to occlude one another [2], [28]. Some systems use three dimensional models of objects [26], [4]. Tsuji and Nakamura [26] allowed the rotation of an object in a 3-D space with a restriction that the ellipse in an image is in correspondence with a circle in the scene.

Using 3-D data of objects, Shirai [24] classified single prismatic body in a scene into one of a few classes according to the plane shapes of the body. Nevatia and Binford [16] proposed to recognize objects with a collection of elongated parts, by matching properties of the parts and relations between them with those in models. Although these studies showed the usefulness of the 3-D approach, the objects were considerably restricted. In our system, shape discrimination is performed by matching 3-D descriptions consisting of surface (region) properties and their relations to 3-D descriptions of models. We can use the shapes of surfaces and their placement in 3-D. This scheme is useful for recognizing a wide variety of objects.

The system to be described in this paper has been developed along this line. We will describe the outline of the system in Section 2, the description method in Section 3, and the matching method in Section 4. The experimental results and discussions will be presented in Section 5.

2. Outline of the System

The system works in two phases: learning and recognition as shown in Figure 1. In the learning phase, known objects are shown to the system one at a time. The system produces a description of the object in terms of properties of regions (surfaces) and relations between them. The description is a model of the object. If one view is not enough to describe the object, several typical views are shown and multiple models are produced. In the recognition phase, the system produces a description of an unknown scene in the same way as in the learning phase. The description is matched to the

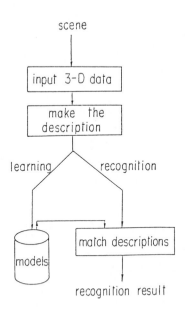

Figure 1: Main Control of the System

models and the objects which have been previously learned are found sequentially. The following sections present details of description and matching.

3. Description of Scenes

The range finder [24], [21] in the present system employs a vertical slit projector and a TV camera to pick up reflected light (see Figure 2). By rotating a mirror from left to right, many points in a field of view are observed. Starting with the three-dimensional coordinates of the points (Figure 3(a)), the processing proceeds as follows:

1. Group the points into small surface elements and, assuming each element to be a plane, get the equations of the surface elements: Figure 3(b).
2. Merge the surface elements together into elementary regions: Figure 3(c).
3. Classify the elementary regions into planar and curved regions: Figure 3(d).
4. Extend the curved regions by merging adjacent curved regions to produce larger regions (global regions) and fit quadratic surfaces

to them: Figure 3(e).

5. Describe the scene in terms of properties of regions and relations between them: Figure 3(f).

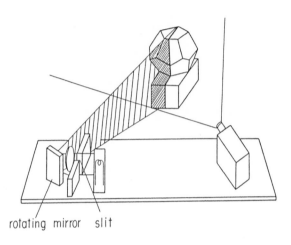

rotating mirror   slit

Figure 2:   Construction of the Range Finder

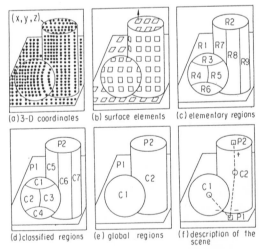

Figure 3:   Conceptual Scheme of the Scene Description Process

3.1. Construction of Surface Elements

3.1.1. Three-dimensional Coordinates

A vertical slit light is projected on the object through a rotating mirror. The signal from the TV camera is digitized by a real time A/D converter and is sent to a special piece of hardware (a SLITTER) [21]. The SLITTER extracts the center point of a slit image for every scanning line (Figure 4)

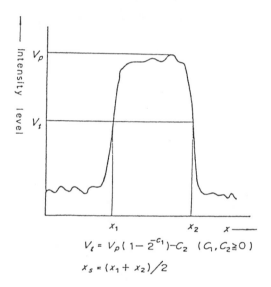

$$V_t = V_p(1 - 2^{-C_1}) - C_2 \quad (C_1, C_2 \geq 0)$$

$$x_s = (x_1 + x_2)/2$$

Figure 4: Video Signal Corresponding to a Slit Light

in real time and sends it to the computer: the time required to get a set of data for a slit is 16ms. A stepping motor rotates the mirror to change the direction of the slit beam. The raw data are represented by a 400×240 array of the slit positions. The (i,j) element represents the horizontal position of the point on the j-th scanning line in the TV image of the i-th slit. An example of a scene and the data are shown in Figure 5: the scene consists of a dodecahedron, an icosahedron, and a cylinder. From the slit positions, the three-dimensional coordinates are calculated by triangulation and are stored in the same way as the slit image.

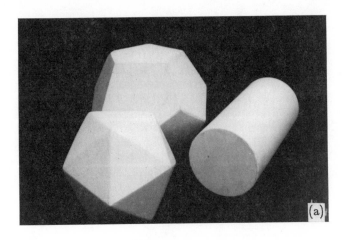

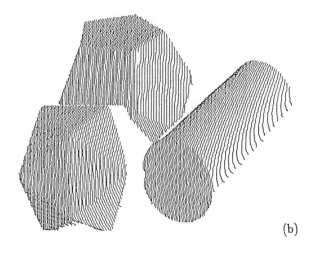

(b)

Figure 5: Example Scene. (a) Original Scene; (b) Slit Image (This example is used throughout the paper)

3.1.2. Surface Element

The adjoining 8×8 points on the object surface are grouped into surface elements. As shown in Figure 6, the 8×8 subarrays are overlapped. Assuming each element to be a plane, the equation of the plane is found by the least square method from its member points. We denote a surface element by the equation:

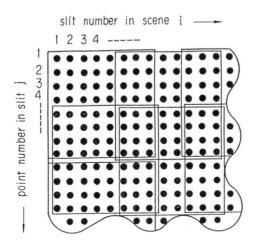

Figure 6:   Placement of 8 × 8 Arrays for Surface Elements

$$f(x,y,z) = \lambda x + \mu y + \nu z - p = 0$$
$$\lambda^2 + \mu^2 + \nu^2 = 1 \tag{1}$$

where x, y, and z are coordinates of the point on the plane. The calculation for plane fitting is shown in Appendix A and B. The scene is now represented by 80×60 surface elements. Some of the surface elements, for example those at an edge, are meaningless. Therefore, the elements with too large a standard deviation s in the fitting are ignored in further processing. Figure 7 shows surface elements obtained for the scene of Figure 5.

3.2. Merging Surface Elements

The next step is to merge the surface elements into larger but approximate plane regions (elementary regions). The merging process is divided into two stages: (1) search for a kernel of an elementary region in the scene; (2) merge surface elements into an elementary region around the kernel. These stages are then repeated.

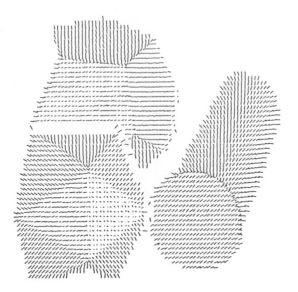

Figure 7:   Surface Elements

3.2.1. Searching for a Kernel

The candidates of a kernel are those elements which do not belong to any region already found. In finding a plane surface, it is desirable that a kernel is a part of a smooth surface. For this purpose, the following evaluation function m is calculated for every candidate:

$$m = w/n + \bar{s}, \tag{2}$$

where n is the number of the other candidates within its eight neighbors, $\bar{s}$ is the average of their standard deviations, and w is a constant. If m is small it means that the element has many candidates in its neighbor and constitutes a smooth surface. The element with the minimum m is chosen as the kernel.

3.2.2. Making a Region

The program extends a region around the kernel by merging its neighboring elements. The candidate element to be merged must currently be adjacent to the region. If the plane equation of the candidate is similar to that of the region, it is merged into the region. The conditions are:

$$d^2{}_k = (\lambda_k - \lambda_r)^2 + (\mu_k - \mu_r)^2 + (\nu_k - \nu_r)^2 < d^2{}_t, \tag{3}$$

$$|p_k - p_r| < p_t, \tag{4}$$

where λ, μ, ν, and p are parameters of Equation (1), k denotes the element, r denotes the region, and t denotes thresholds. In Figure 8, the dotted elements are already connected to the kernel and a set of marginal elements (hatched in Figure 8) are under examination. Each time a set of marginal elements is fixed, plane equation of the region is updated. The program repeats this process until no extension is possible.

The result of merging regions is illustrated in Figure 9 where each character specifies the group to which the surface element belongs. This result is called a region map.

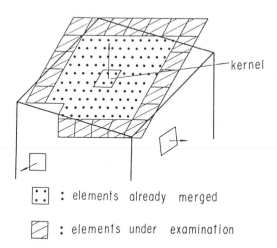

$\boxed{\cdot\,\cdot}$: elements already merged

$\boxed{\diagdown}$: elements under examination

Figure 8: Region Growing Process

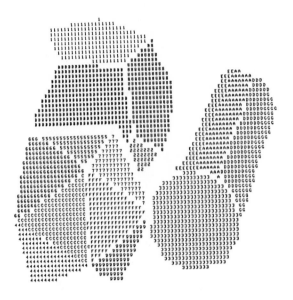

Figure 9:   Elementary Regions

3.3. Classification of Elementary Regions

Each elementary region can correspond to a plane surface or a part of a curved surface. In this stage the elementary regions are discriminated based on their parameters. It is difficult to draw a strict distinction between plane and curved regions on the basis of local criteria. A smooth plane, for example, does not always yield a region with uniformly distributed points due to noise and digitization error.

At this stage, therefore, regions are classified into plane, curved, and undefined ones. The undefined regions are considered later with the assistance of global information.

We should first consider the reliability of each region. In Section 3.1, surface elements with too large a standard deviation s are rejected, mainly to discard meaningless elements around the edges of a plane. Some of the elements which are left and have been merged may not be reliable, especially where the slit light is broad and weak. The elementary region which consists of these kinds of elements should be marked before classification. Therefore, by using the average of standard deviation $\bar{s}$ of the member elements, a region with large $\bar{s}$ is classified as an undefined region. The rest of the elementary regions are classified based on two parameters. The first parameter $\overline{d}^2$ is the mean variance of the angle of the region which is given by:

$$\overline{d}^2 = \frac{\sum\limits_{k=1}^{n} d^2{}_k}{n} \qquad (5)$$

where $d^2{}_k$ is defined by Equation (3) and n is the number of elements in the region. The second parameter, ϕ_e, is the effective diameter of a region, which is defined as follows (refer to Figure 10):

$$\phi_e = min(max(\phi_{11}, \dots, \phi_{1n_1}), \dots, max(\phi_{41}, \dots, \phi_{4n_4})). \qquad (6)$$

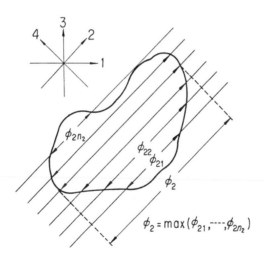

Figure 10: Effective Diameter

A curved surface with the larger curvature has the smaller ϕ_e. A very small value of ϕ_e means small or slender regions. Regions with small ϕ_e are classified as undefined. The rest are classified by the following criteria:

$$g = \alpha \cdot \overline{d}^2 - \phi_e, \qquad\qquad (7)$$

$$g \begin{cases} < -g_t & \text{plane region,} \\ > g_t & \text{curved region,} \\ \text{otherwise} & \text{undefined region,} \end{cases} \qquad (8)$$

where α is a constant, and $g_t(> 0)$ is a threshold. Results of classifying regions from several scenes by these criteria are summarized in Figure 11.

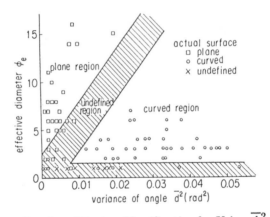

Figure 11: Results of Region Classification by Using $\overline{d}^2$ and ϕ_e

Figure 12 shows the classification result for Figure 9. The marks □ and ∘ mean that the regions are planar and curved, respectively.

In our experiments for various scenes including planar and simple smoothly-curved objects, the results have been satisfactory. Most of the undefined regions consist of a small number of elements. Misclassification of important regions rarely happens.

3.4. Curved Regions

The program attempts to merge curved and undefined regions into global, curved regions. The process is divided into two stages: (1) search for a kernel region of a global region in the scene; (2) merge curved or undefined regions around it into the global, curved region. These stages are repeated. After the global curved regions are found, the program fits quadratic

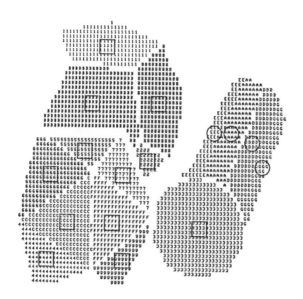

Figure 12: Classified Elementary Regions

surfaces to them and classifies the surfaces.

3.4.1. Searching for the Kernel Region

The candidates of a kernel region are curved elementary regions which do not belong to any global region already found. We calculate the following evaluation function l for every candidate region:

$$l = w_\phi \cdot \phi_e + n \tag{9}$$

where n is the number of elements in the region and w_ϕ is a constant. A region with large l means that it has a large effective diameter and has many surface elements. The candidate with the maximum l is chosen as the kernel region.

3.4.2. Region Merging

The program extends a global curved region around the kernel region by merging its neighboring elementary regions. The candidates to be merged are the curved and undefined elementary regions which are currently adjacent to the global region.

The program proceeds step by step. At first candidates are only those that are just adjacent to the kernel region. Figure 13 illustrates a global region growing. The program checks whether or not each candidate region can be merged to the global region using the elementary regions called "touch stones". The touch stones are those elementary regions which are included in the global region and adjacent to the candidate. For region 11 in

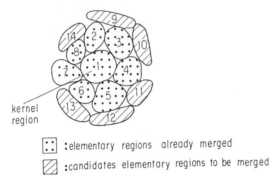

:elementary regions already merged

:candidates elementary regions to be merged

Figure 13:   Extension of Global Curved Region

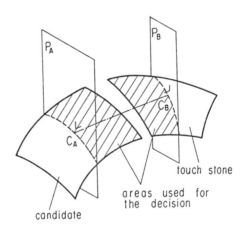

Figure 14:   Test for Region Merging

Figure 13, for example, regions 4 and 5 are the touch stones. If the candidate is determined to be smoothly connected with one of the touch stones, it is merged. The test is as follows: In Figure 14, suppose that C_A and C_B are the three-dimensional points corresponding to the centroid of the candidate and that of the touch stone, respectively. The planes P_A and P_B are placed perpendicular to the line segment $C_A C_B$. Two hatched areas in Figure 14 are approximated individually by planes. The two regions are determined to be smoothly connected if the angle θ between the normals of the planes satisfies the following condition:

$$\theta < w_q \cdot d_t, \tag{10}$$

where d_t is the same as in Equation (3) and w_θ is a constant. The smoothly curved body (for example, a sphere) may have yielded different elementary regions by Equation (3), but the expected angle between two neighboring regions is proportional to d_t. The threshold in Equation (10) is related to d_t so that the same body yields the same global region for various d_t's.

After global, curved regions are found, the unassigned surface elements are reconsidered. If the angle between an element and its neighboring elementary region is small, the element is merged into the same global region (this time the threshold is greater than that used in Equation (3)). Figure 15 shows a merged global, curved region.

Figure 15: Merged Global Curved Region

3.4.3. Fitting Quadratic Surfaces to the Curved Region

We have imposed no constraints on curved surfaces in the preceding processings. Now we try to approximate curved surfaces by quadratic ones. The quadratic surface is expressed in Equation (11), and the evaluation function d^2 to be minimized is given by Equation (12).

$$h(x,y,z) = a_{11}x^2 + a_{22}y^2 + a_{33}z^2$$
$$+2a_{12}xy + 2a_{23}yz + 2a_{31}zx$$
$$+2a_{14}x + 2a_{24}y + 2a_{34}z + a_{44} = 0, \tag{11}$$

$$d^2 = \sum_{i=1}^{n} h^2(x_i, y_i, z_i). \tag{12}$$

In Equation (12), (x_i, y_i, z_i) are the coordinates of the point in the curved global region. If the value d^2/n is greater than a threshold, the surface is not approximated by the quadratic surface. The calculation for obtaining the coefficients is shown in Appendix C.

The quadratic surface is classified using coefficients in Equation (11). The classes are a parabolic cylinder, an elliptic cylinder, a hyperbolic cylinder, a paraboloid, a cone, an ellipsoid, a hyperboloid, and one undefined. The classification is done in the conventional mathematical way with the exception of using some thresholds. Though exhaustive experiment for every class has not been done until now, there is no difficulty for typical objects. Figure 16 shows the fitting quadratic surface of Figure 15, where the quadratic surface is represented by the contour of equal z values (i.e., depth) and it is classified to an elliptic cylinder.

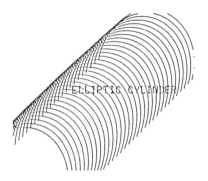

Figure 16:   Fitted Quadratic Surface of the Curved Region

3.5. Building Descriptions

Having found the map of the global regions, the program describes the scene in terms of properties of regions and relations between them.

3.5.1. Edges of Regions

Two kinds of processing are applied to find the following classes of edges of the global regions; they are applied to the whole scene sequentially:

1. The edges between neighboring regions of plane surfaces or quadratic surfaces.
2. The others.

For those in class 1, the intersecting edge is first determined by the equations of the regions. This method is less sensitive to noise than using local edge operators, because all the points in the regions concerned contribute to the determination. If more than two plane regions adjoin one another, the corresponding line segments should meet at one vertex. In this case, the coordinates (x_v, y_v, z_v) of the vertex are determined to minimize the following d^2, the total sum of the square of the distance to each plane.

$$d^2 = \sum_{i=1}^{n} f^2_i (x_v, y_v, z_v) \tag{13}$$

where f_i is the same in Equation (1) and n is the number of the related planes (see Appendix D).

The edge in class 2 is described by a series of the edge points which are extracted from region map by using a 3×3 local operator.

3.5.2. Properties of Regions and Relations Between Them

For each region, its type, curved or plane, has been given in the preceding process. For each pair of the regions, the adjacency between them has been checked. For a pair of adjacent regions the convexity or concavity of the bordering edge between them is examined as follows. An edge found in the previous subsection is at first approximated by piecewise linear segments. For each segment, let us define the border points as those which are in the vicinity of the border. Each group of the border points corresponding to each region is approximated by a plane, and the centroid of all the border points (i.e., union of the two groups) is also calculated. If the centroid is not in the origin side (the origin is at the center of the rotating mirror) of both planes, the common edge of the regions is regarded as convex. If all segments are convex (concave), the type of the intersection is convex (concave); otherwise, the type is mixed.

A scene can now be expressed as a graph whose nodes are the regions and whose arcs between nodes correspond to the relations between them. An example of the description is shown in Figure 17: the marks □ and o mean that the regions are planar and curved, respectively; the dotted line with the mark + means that the two regions are adjacent with a convex common edge. Figure 18 displays edges from another viewpoint, and demonstrates that the edges in the system are represented in three dimensions.

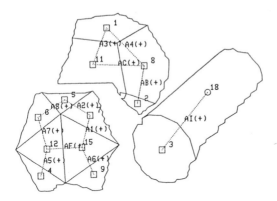

Figure 17: Scene Description

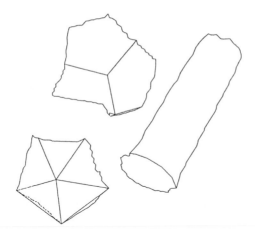

Figure 18: Edges in Figure 17 Seen From Another Position

The properties of each region S_i consist of the following:

1. $p_1(S_i)$: type of the surface fitted to the region S_i. The type is classified into planar, curved, and undefined. The curved surface is further divided into seven types of quadratic curves and other surfaces.

2. $p_2(S_i)$: the number of regions adjacent to S_i. Two regions are adjacent if any part of one region is adjacent to another region.

3. $p_3(S_i)$: area. A plane is fitted to the points in S_i. Points are projected onto the plane. These points constitute a two dimensional region (called a "silhouette"). The area $p_3(S_i)$ is the

area of the silhouette. The following p_4 through p_9 are defined
for the silhouette.

4. $p_4(S_i)$: perimeter.
5. $p_5(S_i)$: compactness ($4\pi \cdot$ area/perimeter^2). Although this is not
   independent of p_3 and p_4, it is useful for deciding if the shape is
   circular.
6. $p_6(S_i)$: mean radius. This is the mean distance from the centroid
   of the silhouette to each point on its border.
7. $p_7(S_i)$: standard deviation of radii.
8. $p_8(S_i)$: minimum radius.
9. $p_9(S_i)$: maximum radius.
10. $p_{10}(S_i)$: occlusion. The vicinity of the border points with 3-D
    discontinuity is analyzed. If there is another region in front of
    the region along a slit, the region is decided to be occluded. For
    the occluded region, the value of $p_{10}(S_i)$ is 1. Otherwise, the
    value is 0.

The relations between regions S_i and S_j consist of the following:

1. $q_1(S_i,S_j)$: type of intersection. The type can be: convex,
   concave, mixed, or without intersection.
2. $q_2(S_i,S_j)$: the angle between regions. This is the angle between
   the two normals of S_i and S_j. If a region is curved, the angle is
   calculated by using the plane which is fitted to the points in the
   region.
3. $q_3(S_i,S_j)$: relative positions of the centroids.

4. Matching

The system searches for the correspondence between the model regions
and the scene regions. It is essential to restrict the search space because the
blind search is very time-consuming. The system first selects from
unmatched scene regions those which seem to be the most reliable and useful
for recognition. A part of a scene consisting of these regions is called a
kernel (see Figure 19). Then the system selects candidate models which
include regions corresponding to the kernel. If there is an object model
which includes the kernel in the scene, then the model is selected. The
selected models can be only a fraction of the whole. This model selection is
data-driven, but the rest of the process is model driven. The system searches

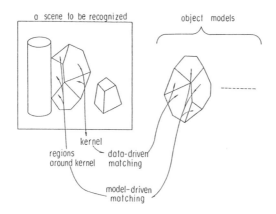

Figure 19:   Matching Process

the scene around the kernel for the regions which can correspond to the regions in the model. The recognized portion of the scene increases accordingly as the matching proceeds. Since the search is model-driven, the number of trials is expected to be considerably smaller. After an object is found, the rest of the scene regions are processed similarly. Further details are described in the following sections.

4.1. Selecting a Kernel

A kernel usually consists of two regions called the principal part and the subordinate part. Among unmatched regions, the system selects the principal part firt, and then the subordinate part, if possible.

4.1.1. Selecting the Principal Part of the Kernel

The candidates for the principal part are those regions which are unknown (that is, the regions are not decided for the correspondence between model regions) and are not occluded by other regions. The principal part of a kernel is selected on the basis of the type of a region, its area, and the number of adjacent regions. The relative angles of planar surfaces are more reliable than those of curved surfaces because they are less sensitive to rotation or partial occlusion. Thus, planar regions are given priority over curved ones. Regions with larger areas are also more useful because the properties of larger regions are less sensitive to noise. Regions which have many neighbor regions are also more useful because many relations can be used in matching. The following function is calculated for each candidate S_j. The region S_k which maximizes the function is selected as the principal part of a kernel:

$$f_1(S_j) = \lambda_1 h(p_1(S_j)) + \lambda_2 p_2(S_j) + \lambda_3 p_3(S_j) \tag{14}$$

where

$$h(p_1(S_j)) = \begin{cases} 2 & \text{if } S_j \text{ is planar.} \\ 1 & \text{if } S_j \text{ is curved} \end{cases}$$

Here p_1, p_2, p_3 are the type of the region, the number of adjacent regions, and the area of the region, respectively. In our experiments, the weights $\lambda_1, \lambda_2, \lambda_3$ are set so that the influence of p_1, p_2, p_3 on f_1 is smaller in the order p_1, p_3, p_2. If there is no candidate for the principal part, the matching procedure terminates.

4.1.2. Selecting the Subordinate Part of the Kernel

The candidates for the subordinate part are those regions which are adjacent to the principal part already selected and which can satisfy the condition of a candidate for the principal part. The candidate which maximizes f_1 is selected as the subordinate part. A kernel consists of the principal part and the subordinate part, if there is a candidate for the subordinate part. Otherwise, a kernel consists of the principal part alone.

4.2. Assuming Probable Models

The system selects probable models which include regions corresponding to the kernel. There are two cases depending on the kernel.

First, if the kernel consists of only the principal part, those models are selected which have a region corresponding to the principal part. The selection is based on the selection function which evaluates the dissimilarity between the kernel and the model region in the properties. Let M_{ij} denote the j-th region of the i-th model. The function compares the properties of S_k and M_{ij}. Let p_l denote the l-th property of the region, and let D_l denote an operator which evaluates the difference in the l-th property values. For $l = 1$

$$D_l(p_l(S_k), p_l(M_{ij})) = \begin{cases} 0 & S_k \text{ and } M_{ij} \text{ are the same type;} \\ 1 & S_k \text{ and } M_{ij} \text{ are not the same type} \end{cases} \qquad (15)$$

for $1 = 5$ and 7

$$D_l(p_l(S_k), p_l(M_{ij})) = |p_l(S_k) - p_l(M_{ij})|; \qquad (16)$$

for the other l

$$D_l(p_l(S_k), p_l(M_{ij})) = \frac{|p_l(S_k) - p_l(M_{ij})|}{p_l(M_{ij})}. \qquad (17)$$

The selection function f_2 is defined as follows:

$$f_2(S_k, M_{ij}) = \sum_{i=1}^{9} \mu_l D_l(p_l(S_k), p_l(M_{ij})). \qquad (18)$$

The weights μ_l were set identically in our experiment.

Here f_2 has a small value (ideally zero) if the kernel matches well with the region in the model. The system selects models whose f_2 values are small, and assumes the correspondences. If there is no model whose f_2 value is small, the system decides that the principal part does not coincide with any model region and returns to select another kernel.

The second case is that the kernel consists of the principal part S_k and the subordinate part $S_{k'}$. In this case, the models which have a pair of regions corresponding to the two parts are selected. The selection is based on another selection function which evaluates the dissimilarity between the kernel and the model regions for both the principal and subordinate parts and their relations. Let M_{ij} and $M_{ij'}$ denote the candidate regions in the model which may correspond to the principal part and subordinate part, respectively (see Figure 20).

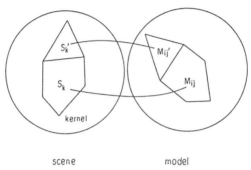

scene   model

Figure 20:   Comparison Between Kernel and Model Regions

The function compares S_k with M_{ij}, and $S_{k'}$ with $M_{ij'}$. Moreover, it also compares the relations between S_k and $S_{k'}$ with relations between M_{ij} and $M_{ij'}$. The selection function is defined as follows:

$$f'_2(S_k,S_{k'},M_{ij},M_{ij'}) = f_2(S_k,M_{ij})+f_2(S_{k'},M_{ij'}) \tag{19}$$

$$+ \sum_{m=1}^{3} \nu_m D_m(q_m(S_k,S_{k'}),q_m(M_{ij},M_{ij'}))$$

where q_1, q_2, and q_3 denote the type of intersection, the angle between regions, and the relative positions of the centroids of the regions, respectively. The weights ν_m are set so that the influence of q_1, q_2, and q_3 on f'_2 is smaller in the order q_1, q_2, and q_3. Note that f'_2 is not sensitive to the translation or rotation of objects. The system selects the models whose f'_2 values are small, and assumes the correspondences.

The value of f'_2 is not small if the subordinate part does not correspond to the entire model region, or does not correspond to the model (for example, the principal part corresponds to a certain model, but the subordinate part does not). Thus, if there is no model whose f'_2 value is small, the system decides that the subordinate part is inadequate and returns to select another subordinate part again.

If a single model is selected, the kernel is called the "kernel with a single model." The system moves on to the next matching process (matching between regions), which is described in the next subsection. If multiple models are selected, the system suspends further matching processes for that kernel until the matching processes for all the other kernels with a single model finish.

4.3. Matching Between Regions

In this matching process, the system searches the scene around the matched kernel for the regions corresponding to those in the model. Figure 21 represents the search tree for it: each node represents the correspondence between a scene region and a model region. A depth-first search process is used. Starting from the model regions corresponding to the kernel, the matching process proceeds region by region. At each step the system selects a new region, called the "front-end region," from those which are adjacent to the known regions in the model (that is, the regions which have been selected and matched in earlier steps). If there is no candidate for a front-end region, the matching process terminates.

Whenever a front-end region is selected from the model regions, the system searches the scene for the corresponding region (called "matched region"). The candidates for the matched regions are those regions adjacent to known regions in the scene. The search is based on the dissimilarity function. This evaluates the dissimilarity between the front-end region and the candidate region using the relations as well as the properties.

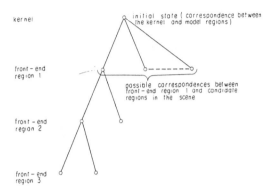

Figure 21:   Search Tree Produced by the Matching Process

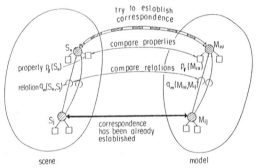

Figure 22:   Schema of the Comparison

In Figure 22, suppose M_{ij} and S_j represent a known pair of regions and that M_{iu} and S_n are the front-end region and a candidate of the matched region, respectively. The function compares the properties of the front-end region $p_l(M_{iu})$ to those of the candidate in the scene $p_l(S_n)$. The function also compares the relations $q_m(M_{iu}, M_{ij})$ between the front-end region and the known regions M_{ij} in the model to those $q_m(S_n, S_j)$ of the corresponding regions in the scene. Let K denote a set of already known regions. The dissimilarity function between M_{iu} and S_n is defined as:

$$f_3(S_n, M_{iu}) = \sum_{l=1}^{9} \rho_l D_l(p_l(S_n), p_l(M_{iu}))$$ (20)

$$+ \sum_{m=1}^{3} \frac{\tau_m \left(\sum_{j \in K} p_3(M_{ij}) D_m(q_m(S_n, S_j), q_m(M_{iu}, M_{ij})) \right)}{\sum_{j \in K} p_3(M_{ij})}$$

where weights ρ_l and τ_m are set in the same manner as in Equation (18) and Equation (19). The areas p_3 are used as weights so that more reliable regions contribute more to f_3. Note that f_3 is not sensitive to the translation or rotation of objects.

If the value of f_3 is small (ideally zero) and all terms in Equation (20) are less than the thresholds, the match between the front-end and candidate regions is decided as acceptable. In this case, the system tries to find another front-end region again. In a search tree, if there are multiple matched regions for a front-end region (for example, objects with very similar faces), one of them is selected at first and the matching process proceeds. When the subsequent matching process on this search tree terminates, other matched regions on this node are initiated and the process proceeds in the same manner.

The value f_3 is not small if a surface of an object is not seen entirely, or the region does not correspond to the assumed model. However, if a plane is partially occluded by other surfaces, the type and the relative direction of the region are not affected by the occlusion. Thus, if no matched region is found and the type of the front-end region is planar, the system tries to find a planar region among the candidates which satisfies the following conditions:

1. the region is occluded
2. the dissimilarity of the relative angle between the region and the front-end region is small
3. the 3-D position correspondence is adequate
4. the area of the region is less than that of the front-end region.

If there is such a region, it is decided that it corresponds to a portion of the front-end region.

If a model M_j is selected for a kernel S_l, the set of matched regions is called the candidate body of the model M_j with respect to the kernel S_l, and is denoted by $B(S_l, M_j)$. The set of the correspondences between the candidate body and the model regions is called the interpretation.

When the process for a kernel terminates, the candidate body is determined as acceptable if there is only one candidate body (that is, if there is only one candidate model). If there are multiple candidate bodies, the system does not accept them until the matching procedure finishes for the whole scene.

4.4. Interpreting a Scene

When the matching procedure terminates, a set of candidate bodies has been produced. The candidate body $B(S_l, M_j)$ with respect to a kernel S_l and $B(S_m, M_{j'})$ with respect to another kernel S_m are called inconsistent if they have a common element. Otherwise they are consistent. For each kernel S_l, the candidate body $B(S_l, M_j)$ is rejected if it is included in another candidate body $B(S_l, M_{j'})$ for the same kernel which is consistent with other candidate bodies. Among the rest of the candidate bodies, all consistent combinations are adopted. Each combination gives an interpretation of the scene. The positions and orientations of the objects are now easily calculated from the correspondences of the regions.

5. Experimental Results and Discussions

Most of the ideas proposed have been implemented and two kinds of experiments were performed. In the first experiment, ten kinds of objects with planar and quadratic curved surfaces (shown in Figure 23) were used. About twenty models were built. Figure 24 illustrates an example of a model description. The mark + means that the intersection between two regions is convex. The recognition results for Figure 17 are shown in Figure

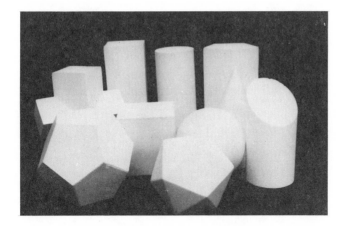

Figure 23: Objects Which Were Built into Models

Figure 24: Example of a Model Description

25. The first two letters in a region indicate the model, and the letter in the parentheses indicates the corresponding region in the model. The results for several similar scenes were also satisfactory.

In the second experiment, machine parts (a pulley, liner, piston, and connecting rod of a car, as shown in Figure 26(a)) were built into models. An example of the input image (corresponding to Figure 26(a)) is shown in Figure 26(b). The surface elements, elementary regions, and the final result are shown in Figure 26(c), (d), and (e), respectively. The processing time of

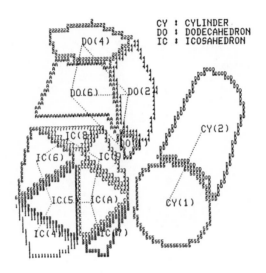

Figure 25: The Result of Recognition

the program (written in Fortran) for a typical scene such as that shown in Figure 26 was about 3 minutes for creating a description and 0.5 minutes for matching by a Tosbac 5600 computer.

Inconsistent interpretations, mentioned in Section 4, were not found in our experiments. The choice of thresholds and parameters is important for obtaining good results. When the values of thresholds for f_2, f'_2, and f_3 are set at the higher level, different surfaces tend to be recognized as identical. On the other hand, identical surfaces tend to be recognized as different when the values are set at the lower level because of noises and digitization errors. In our experiments thus far, the values of D (the operator which evaluates the difference between property values) for the parameters p_3, p_4, and p_6 have been typically below 0.2 (20 percent) when two surfaces were actually in correspondence. The choice of the parameters does not change the results drastically thus far in our experiments.

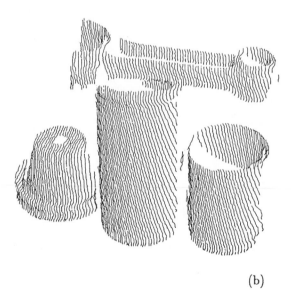

Figure 26: Processing for a Scene Containing Machine Parts. (a) Scene to be Recognized; (b) Slit Image

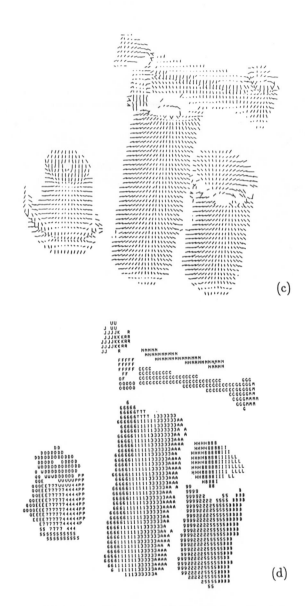

(c)

(d)

Figure 26:   (c) Surface Elements; (d) Elementary Regions

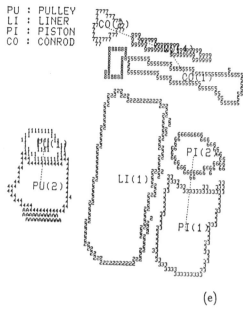

PU : PULLEY
LI : LINER
PI : PISTON
CO : CONROD

(e)

Figure 26: (e) Result of Recognition

6. Conclusion

The authors have proposed a system to recognize stacked objects using range data. The system describes a scene in terms of planes and smoothly curved surfaces. Models of objects are built in the system by showning them one at a time. Objects in an unknown scene are recognized by matching the description of the scene to those of the models. The matching program picks out regions which are most reliable and useful for recognition, and matches them to the regions of the models. Once candidate models are selected, the rest of the scene regions are searched for under the guidance of the models. Thus, the system has achieved flexible and efficient recognition. The results of our experiments show that this scheme is promising.

Future directions are as follows:

1. In the learning phase, objects are shown to the system from multiple directions and multiple models are built for an object. They are treated as if they were independent of one another. The search effort in matching tends to increase as the number of views increases. If they could be merged together into a single

model, more efficient recognition could be achieved.

2. The present system can determine correspondences for occluded planar surfaces from the relations between regions. It is desirable that these kinds of decisions also be made for occluded curved surfaces.

Acknowledgments

The authors would like to thank the members of the Computer Vision Section at the Electrotechnical Laboratory for their helpful discussions. Acknowledgment is also due to Professor S. Tsuji for his valuable comments.

Appendix A

Fitting a Plane to a Set of Points (I)

For a set of points $\{(x_i, y_i, z_i)|i=1, \ldots, n\}$ we wish to fit a plane,

$$\lambda x' + \mu' y + \upsilon' z - 1 = 0 . \tag{A1}$$

Let us define the estimate of the fit as

$$d^2 = \sum_{i=1}^{n} (\lambda' x_i + \mu' y_i + \nu' z_i - 1)^2 \tag{A2}$$

to minimize it. Conditions $\dfrac{\partial d^2}{\partial \lambda'} = 0$, $\dfrac{\partial d^2}{\partial \mu'} = 0$, $\dfrac{\partial d^2}{\partial \nu'} = 0$ give

$$
\begin{bmatrix}
\Sigma x_i^2 & \Sigma x_i y_i & \Sigma x_i z_i \\
\Sigma y_i x_i & \Sigma y_i^2 & \Sigma y_i z_i \\
\Sigma z_i x_i & \Sigma z_i y_i & \Sigma z_i^2
\end{bmatrix}
\begin{bmatrix}
\lambda' \\
\mu' \\
\upsilon'
\end{bmatrix}
=
\begin{bmatrix}
\Sigma x_i \\
\Sigma y_i \\
\Sigma y_i
\end{bmatrix}
$$

Solving this, we obtain λ', μ', and ν'. The direction cosines λ, μ, and ν and distance p from the origin in $\lambda x + \mu y + \nu z = p$ are found by comparing this with (A1). Then we have

$$\lambda = \frac{\lambda'}{\sqrt{\lambda'^2 + \mu'^2 + \nu'^2}} \quad \mu = \frac{\mu'}{\sqrt{\lambda'^2 + \mu'^2 + \nu'^2}} \quad \nu = \frac{\nu'}{\sqrt{\lambda'^2 + \mu'^2 + \nu'^2}}$$

$$p = \frac{1}{\sqrt{\lambda'^2 + \mu'^2 + \nu'^2}} \quad .$$

Appendix B

Fitting a Plane to a Set of Points (II)

For a set of points $\{x_i = (x_i, y_i, z_i)^t \mid i = 1, \ldots, n\}$, we wish to fit a plane with normal direction vector $P_o = (\lambda, \mu, \nu)^t$ ($\|P_o\| = 1$) and distance p from the origin. Here the transpose of a vector V is denoted by V^t. We define the estimate of the fit as

$$d^2 = \sum_i (x_i^t P_o - p)^2 \ . \tag{B1}$$

Note that we estimate the distance from each point to the plane.

$$\frac{\partial d^2}{\partial p} = 0$$

gives

$$\left(\sum_i x_i^t \right) P_o = \sum_i p = np$$

so

$$p = \bar{x}^t P_o \tag{B2}$$

where $\bar{x}$ denotes the average of x_i's.

Substituting (B2) with (B1), we have

$$d^2 = \sum_i \{ (x_i - \bar{x})^t P_o \}^2$$
$$= P_o^t [\sum_i (x_i - \bar{x})][\sum_i (x_i - \bar{x})]^t P_o = P_o^t A P_o , \tag{B3}$$

where $A = [\sum_i (x_i - \bar{x})][\sum_i (x_i - \bar{x})]^t$. Under the constraint $\|P_o\| = P_o^t P_o = 1$, we minimize d^2. By the Lagrange's method of indeterminate coefficients, we obtain

$$\frac{\partial d^2}{\partial P_o} - m \frac{\partial}{\partial P_o} (P_o^t P_o - 1) = 0 ,$$

where m is a constant.

Then we have

$$A \, P_o = m \, P_o \, .$$

Note that P_o and m are the eigenvector and the eigenvalue of A. Substituting this to (B3), we have

$$d^2 = m \, .$$

Thus d^2 is minimized if P_o is the eigenvector of A corresponding to the minimum eigenvalue of A.

Appendix C

Fitting a Quadratic Surface to a Set of Points

Let

$$h(x,y,z) = a_{11}x^2 + a_{22}y^2 + a_{33}z^2$$

$$+ \, 2a_{12}xy + 2a_{23}yz + 2a_{31}zx$$

$$+ \, 2a_{14}x + 2a_{24}y + 2a_{34}z - a_{44} = 0 \qquad \text{(C1)}$$

denote a surface to be fitted to a set of points $\{(x_i,y_i,z_i)|(i=1,\ldots,n)\}$. We define the estimate of the fit as

$$d^2 = \sum_{i}^{n} h^2(x_i,y_i,z_i) \, . \qquad \text{(C2)}$$

Setting $a_{44} = 1$ and we minimize d^2.

$$\frac{\partial d^2}{\partial a_{11}} = 0, \ldots, \quad \frac{\partial d^2}{\partial a_{34}} = 0$$

gives

$$
\begin{bmatrix}
\Sigma x^4 & \Sigma x^2 y^2 & \Sigma x^2 z^2 & 2\Sigma x^3 y & 2\Sigma x^2 yz & 2\Sigma x^3 z & 2\Sigma x^3 & 2\Sigma x^2 y & 2\Sigma x^2 z \\
\Sigma y^2 x^2 & \Sigma y^4 & \Sigma y^2 z^2 & 2\Sigma y^3 x & 2\Sigma y^3 z & 2\Sigma y^2 zx & 2\Sigma y^2 x & 2\Sigma y^3 & 2\Sigma y^2 z \\
\Sigma z^2 x^2 & \Sigma z^2 y^2 & \Sigma z^4 & 2\Sigma z^2 xy & 2\Sigma z^3 y & 2\Sigma z^3 x & 2\Sigma z^2 x & 2\Sigma z^2 y & 2\Sigma z^3 \\
2\Sigma x^3 y & 2\Sigma xy^3 & 2\Sigma xyz^2 & 4\Sigma x^2 y^2 & 4\Sigma xy^2 z & 4\Sigma x^2 yz & 4\Sigma x^2 y & 4\Sigma xy^2 & 4\Sigma xyz \\
2\Sigma yzx^2 & 2\Sigma y^3 z & 2\Sigma yz^3 & 4\Sigma y^2 zx & 4\Sigma y^2 z^2 & 4\Sigma yz^2 x & 4\Sigma yzx & 4\Sigma y^2 z & 4\Sigma yz^2 \\
2\Sigma zx^3 & 2\Sigma zxy^2 & 2\Sigma z^3 x & 4\Sigma zx^2 y & 4\Sigma z^2 xy & 4\Sigma z^2 x^2 & 4\Sigma zx^2 & 4\Sigma zxy & 4\Sigma z^2 x \\
2\Sigma x^3 & 2\Sigma xy^2 & 2\Sigma xz^2 & 4\Sigma x^2 y & 4\Sigma xyz & 4\Sigma x^2 z & 4\Sigma x^2 & 4\Sigma xy & 2\Sigma xz \\
2\Sigma yx^2 & 2\Sigma y^3 & 2\Sigma yz^2 & 4\Sigma y^2 x & 4\Sigma y^2 z & 4\Sigma yzx & 4\Sigma yx & 4\Sigma y^2 & 4\Sigma yz \\
2\Sigma zx^2 & 2\Sigma zy^2 & 2\Sigma z^3 & 4\Sigma zxy & 4\Sigma z^2 y & 4\Sigma z^2 x & 4\Sigma zx & 4\Sigma zy & 4\Sigma z^2
\end{bmatrix}
$$

$$
\times
\begin{bmatrix}
a_{11} \\ a_{22} \\ a_{33} \\ a_{12} \\ a_{23} \\ a_{31} \\ a_{14} \\ a_{24} \\ a_{34}
\end{bmatrix}
=
\begin{bmatrix}
-\Sigma x^2 \\ -\Sigma y^2 \\ -\Sigma z^2 \\ -2\Sigma xy \\ -2\Sigma yz \\ -2\Sigma zx \\ -2\Sigma x \\ -2\Sigma y \\ -2\Sigma z
\end{bmatrix}
\tag{C3}
$$

The coefficients a_{11} to a_{34} are found by solving these simultaneous equations.

Appendix D

Determining a Vertex From a Set of Planes

Let

$$
f_i(x,y,z) = \lambda_i x + \mu_i y + \nu_i z - p_i = 0 \qquad (i=1,\ldots,n,\ n \geq 4)
\tag{D1}
$$

denote a set of planes.

We determine the vertex (x_v, y_v, z_v), where these planes (approximately) meet, by minimizing $d^2 = \sum_{i=1}^{n} f^2_i(x_v, y_v, z_v)$.

$$\frac{\partial d^2}{\partial x_v} = 0, \quad \frac{\partial d^2}{\partial y_v} = 0, \quad \frac{\partial d^2}{\partial z_v} = 0 \tag{D2}$$

gives

$$\begin{bmatrix} \Sigma \lambda_i^2 & \Sigma \lambda_i \mu_i & \Sigma \lambda_i \upsilon_i \\ \Sigma \mu_i \lambda_i & \Sigma \mu_i^2 & \Sigma \mu_i \upsilon_i \\ \Sigma \upsilon_i \lambda_i & \Sigma \upsilon_i \mu_i & \Sigma \upsilon_i^2 \end{bmatrix} \begin{bmatrix} x_v \\ y_v \\ z_v \end{bmatrix} = \begin{bmatrix} \Sigma \lambda_i p_i \\ \Sigma \mu_i p_i \\ \Sigma \upsilon_i p_i \end{bmatrix}. \tag{D3}$$

Solving this, we find (x_v, y_v, z_v). This idea is also useful for two-dimensional cases in which we want to find a point where multiple lines (approximately) meet.

References

[1]   Agin, G.J.
        Representation and description of curved objects.
        Technical Report AIM-173, Stanford University, 1972.

[2]   Barrow, H. and Popplestone, R.
        Relational descriptions in picture processing.
        Machine Intelligence 6.
        Edinburgh University Press, Edinburgh, Scotland, 1971, pages
            377-396.

[3]   Bolles, R. C. and Fischler, M. A.
        A RANSAC-based approach to model fitting and its application to
            finding cylinders in range data.
        In *Proc. 7th Int. Joint Conf. Artificial Intell.*, pages 637-643.
            Vancouver, B.C., Canada, Aug., 1981.

[4]   Brooks, R.
        Model-based three dimensional interpretations of two dimensional
            images.
        In *Proc. 7th Int. Joint Conf. Artificial Intell.*, pages 619-624.
            Vancouver, B.C., Canada, Aug., 1981.

[5]   Dane, C.
        An object-centered three-dimensional model builder.
        PhD thesis, Univ. of Penna., Philadelphia, 1982.

[6] Duda, R. O., Nitzan, D., and Barrett, P.
Use of range and reflectance data to find planar surface regions.
IEEE Trans. Pattern Anal. Machine Intell. PAMI-1:259-271, July,
1979.

[7] Faugeras, O. D.
New steps toward a flexible 3-D vision system for robotics.
*Preprints of the Second International Symposium of Robotics
Research. Note: also see In Proc. 7th Int. Conf. Pattern
Recogn., pages 796-805; Montreal, Canada, July* :222-230, 1984.

[8] Gennery, D. B.
Object detection and measurement using stereo vision.
In *Proc. 6th Int. Joint Conf. Artificial Intell.*, pages 320-327.
Tokyo, Japan, Aug, 1979.

[9] Guzman, A.
*Computer recognition of three-dimensional objects in a visual
scene.*
Technical Report MAC-TR-59, Ph.D. thesis, Project MAC, MIT,
Mass. , 1968.

[10] Huffman, D. A.
Impossible objects as nonsense sentences.
Machine Intelligence 6.
Edinburgh University Press, Edinburgh, 1971, pages 295-323.

[11] Inokuchi, S. and Nevatia, R.
Boundary detection in range pictures.
In *Proc. 5th Int. Joint Conf. Pattern Recogn.*, pages 1301-1303.
Miami Beach, FL, Dec., 1980.

[12] Ishii, M. and Nagata, T.
Feature extraction of three-dimensional objects and visual processing
in a hand-eye system using laser tracker.
Pattern Recognition 8:229-237, Oct., 1976.

[13] Kyura, N. and Shirai, Y.
Recognition of objects using three-dimensional region method (in
Japanese).
Bull. Electrotech. Lab. 37:996-1012, Nov., 1973.

[14] Milgram, D. L. and Bjorklund, C. M.
Range image processing: planar surface extraction.
In *Proc. 5th Int. Joint Conf. Pattern Recogn.*, pages 912-919.
Miami Beach, FL, Dec., 1980.

[15]   Mitiche, A. and Aggarwal, J. K.
        Detection of edges using range information.
        In *Proc. IEEE Int. Conf. Acoust., Speech, Signal Processing*, pages
            1906-1911. Paris, France, May, 1982.
        Also appeared in IEEE Trans. on PAMI, Vol. 5, No. 2, March, 1983,
            pp. 174-178.

[16]   Nevatia, R.
        Structured descriptions of complex objects.
        In *Proc. 3rd Int. Joint Conf. AI*, pages 641-647. Stanford, CA,
            Aug., 1973.
        Also Ph.D. thesis, Stanford Univ.- AI Lab, 1974 (AIM 250).

[17]   Nitzan, D., Brain, A. E., and Duda, R. O.
        The measurement and use of registered reflectance and range data in
            scene analysis.
        Proc. IEEE 65:206-220, Feb., 1977.

[18]   Oshima, M. and Shirai, Y.
        A scene description method using three-dimensional information.
        Pattern Recognition 11:9-17, 1979.

[19]   Oshima, M. and Shirai, Y.
        Object recognition using three-dimensional information (in
            Japanese).
        *Trans. Institute of Electronics and Communication Engineers of
            Japan* J65-D(5):629-636, May, 1982.

[20]   Oshima, M. and Shirai, Y.
        Object recognition using three-dimensional information.
        IEEE Transactions on Pattern Analysis and Machine Intelligence
            PAMI-5(4):353-361, July, 1983.

[21]   Oshima, M. and Takano, Y.
        Special hardware for the recognition system of three-dimensional
            objects (in Japanese).
        Bulletin of the Electrotechnical Laboratory 37:493-501, 1973.

[22]   Popplestone, R. J., Brown, C. M., Ambler, A. P., and Crawford, G. F.
        Forming models of plane-and-cylinder faceted bodies from light
            stripes.
        In *Proc. 4th Int. Joint Conf. AI*, pages 664-668. Sept., 1975.

[23]   Shapira, R. and Freeman, H.
       Computer description of bodies bounded by quadratic surfaces from a
           set of imperfect projections.
       IEEE Trans. Comput. C-27:841-854, 1978.

[24]   Shirai, Y.
       Recognition of polyhedrons with a range finder.
       Pattern Recogn. 4:243-250, 1972.

[25]   Sugihara, K.
       Range-data analysis guided by a junction dictionary.
       Artificial Intelligence 12:41-69, May, 1979.

[26]   Tsuji, S. and Nakamura, A.
       Recognition of an object in a stack of industrial parts.
       In *Proc. 4th Int. Joint Conf. Artificial Intell.*, pages 811-818.
           Tbilisi, USSR, Sept., 1975.

[27]   Waltz, D.
       Understanding line drawings of scenes with shadows.
       The Psychology of Computer Vision.
       McGraw-Hill, New York, 1975, pages 19-91.

[28]   Yachida, M. and Tsuji, S.
       A versatile machine vision system for complex industrial parts.
       IEEE Trans. Comput. C-26:882-894, Sept., 1977.

3DPO: A Three-Dimensional Part Orientation System

Robert C. Bolles

Artificial Intelligence Center
SRI International
333 Ravenswood Ave.
Menlo Park, CA 94025

Patrice Horaud

Laboratoire d'Automatique de Grenoble
BP 46, 38402
Saint-Martin d'Heres
France

Abstract

A system that recognizes objects in a jumble, verifies them, and then determines some essential configurational information, such as which ones are on top, is presented. The approach is to use three-dimensional models of the objects to find them in range data. The matching strategy starts with a distinctive edge feature, such as the edge at the end of a cylindrical part, and then "grows" a match by adding compatible features, one at a time. (The order of features to be considered is predetermined by an interactive, off-line feature-selection process.) Once a sufficient number of compatible features has been detected to allow a hypothesis to be formed, the verification procedure evaluates it by comparing the measured range data with data predicted according to the hypothesis. When all the objects in the scene have been hypothesized and verified in this manner, a configuration-understanding procedure determines which objects are on top of others by analyzing the patterns of range data predicted from all the hypotheses. Experimental results of the system's performance in recognizing and locating castings in a bin are presented.

1. Introduction

How does a person find a pancake turner in a drawer of kitchen utensils? If it is partially occluded by other objects, as it probably is, he recognizes one of its "local" features, such the handle or the flat blade, and then uses it to hypothesize the position and orientation of the whole object. The more distinctive the visible features are, the faster a person can find it.

We are interested in tasks like these not only because we want to learn more about how people perform them, but also because we want to develop techniques for performing similar industrial tasks. In this paper we are

Figure 1:   Bin of castings

particularly concerned with the problem of recognizing and locating identical objects jumbled together in a bin (see Figure 1). By working on this class of difficult tasks we expect to develop general-purpose techniques for recognizing and locating partially visible objects. Our approach is to use three-dimensional models of the objects to find them in range data. Our rationale is that, first of all, range data simplify the locational analysis because the geometric information is encoded directly in the data. Second, range sensors will soon be economical for use in industrial tasks. Finally, familiarity with the model of a part will add enough new constraints to make it practical to locate relatively complex parts jumbled together in a bin.

In the realm of industrial tasks, one of the most important merits of a technique is its speed. One way to achieve speed is to locate a minimum of object features and extract as much information as possible. For example, after a feature has been found, its identity can be used to suggest the next feature to be located and its position can constrain the region to be searched. A typical strategy of this type is as follows:

- Locate a distinctive feature of the object to be found.
- Use that feature's position to suggest where to look for a second feature to verify the first.
- Use the two features to predict a third feature that completely constrains the position and orientation of the object.

Since some of the predicted features may not be visible or the feature detectors may miss some visible ones, alternatives have to be provided. Therefore, a complete strategy is an ordered tree of features and the recognition process is a tree search.

As with all tree searches, it is important to order the alternatives according to their expected utility. This may be done in advance to minimize execution time. Thus, the system is naturally partitioned into two

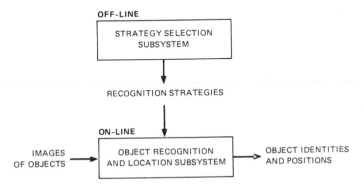

Figure 2: Top-level block diagram of the 3DPO system

stages (see Figure 2). The first stage selects the recognition strategies and the features to be used in them, while the second stage applies the strategies to locate objects in range data. Since the selection process is done only once for each task, it can be performed off-line and pass its results to the on-line system to be used repeatedly.

One of our goals for the 3DPO (Three-Dimensional Part Orientation) system is to explore ways to select strategies and features automatically. In this paper, however, we concentrate on the development of an aggressive recognition system that tries to use as much of the available information as possible at each step of the process. For example, in addition to the constraints derived from previously located features, the system uses (1) the geometry of the sensor to classify range discontinuities, (2) the intrinsic properties of the object features to reduce the list of candidate matches, and (3) the fact that the objects are opaque to predict range images, which in turn are used to verify hypothesized objects.

In this paper we describe a system that recognizes objects in a jumble, verifies them, and determines some essential configuration information, such as which ones are on top. In the next section we present the approach and show how it relates to others. In the third section we describe the object representation scheme, which is designed to support both the off-line and on-line portions of the system. In the fourth section we discuss the rationale behind selection of the strategies and features to recognize a particular object. In sections five through nine we describe the steps of the on-line recognition procedure. In each section we include a discussion of the strengths and weaknesses of the techniques employed. We conclude with a brief outline of work projected for the future.

Throughout the paper we illustrate the techniques being described by applying them to the task of recognizing the casting in Figure 1. We chose this object because it is a moderately complex part that is typical of a large class of industrial parts.

2. Approach

We have partitioned the on-line recognition process into five steps:
1. Data acquisition
2. Feature detection
3. Hypothesis generation
4. Hypothesis verification
5. Configuration understanding.

First, we gather a "dense" range image of the scene (see Figure 3). In Figure 3(a), image points closer to the sensor are lighter. Second, we locate features in the data, such as an arc at the end of a cylinder or a straight edge formed by the intersection of two planes. Third, we "grow" a match around a distinct feature by adding compatible features, one at a time. The order of features to be considered is predetermined by the off-line feature selection process. As each new addition is made to the set of mutually consistent features, the system refines its estimate of the object's pose (i.e., its position and orientation). When a sufficient number of features has been added to determine the pose, the matching system forms a hypothesis and passes it to the verification procedure for assessment. The verification procedure uses the hypothesis to predict the range data that should have been measured if the object were in fact at the indicated pose, and compares these predictions to the measured data. If the predictions match the data, the verifier accepts the hypothesis; if not, it rejects it. When all the objects have been thus hypothesized and verified, the configuration-understanding procedure determines which objects are on top of other objects by analyzing the patterns of range data predicted from the hypotheses.

Industrial vision researchers are converging to a matching strategy that is a tree search (e.g., [13], [8], and [9]). These groups have concentrated on different types of low-level features and applied somewhat different compatibility checks, but the matching methods are essentially the same. Our approach is of this type. It requires more preliminary analysis of the objects than some of the former approaches because it emphasizes key features, but the gain in run-time efficiency is often worth the effort.

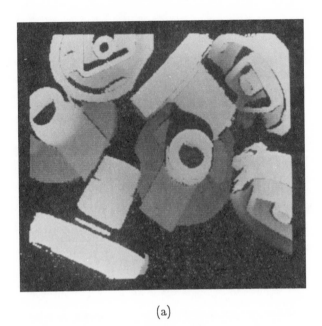

(a)

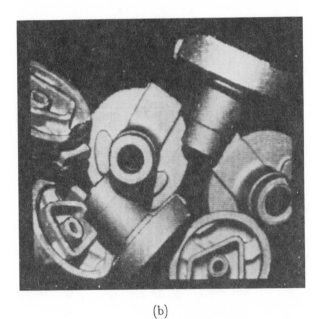

(b)

Figure 3:   Registered range and intensity images of a jumble.   (a) Height image of a jumble; (b) Intensity image of jumble

Range analysis programs have generally concentrated on one type of feature. Some have worked with surface patches (e.g., [6], [13], and [8]), some with edges (e.g., [11]), and some with other simple shapes (e.g., [15], [17], and [12]). The 3DPO system starts with edges, but quickly expands its analysis to include the adjoining surfaces. This approach is particularly well suited to industrial parts that have distinct edges. Dave Smith at CMU has recently developed a similar system that builds descriptions of objects, such as pans and shovels, from a combination of surface and edge features (described in an unpublished manuscript).

Most object recognition research has concentrated on making hypotheses, not verifying them. In this paper we include techniques for verification, as well as for analyzing sets of hypotheses to ascertain which parts are on top of others.

3. Object Modeling

Computer-aided design (CAD) systems and their representations are intended for constructing and displaying objects, not recognizing them. Even though a CAD model may be complete in the sense that it contains all the three-dimensional information about an object, there are more convenient representations for a recognition system. For example, a CAD model might state the size and position of a hole, but a recognition system needs a list of all the holes of that size. In general, a recognition system must be able to provide answers to such questions as:

- How many features are there of this type and size?
- Which surfaces intersect to form this edge?
- What other features lie in this plane?
- What neighboring feature could be used to distinguish this feature from others like it?

To answer these questions efficiently, each object feature should be listed under several different classifications. For example, a dihedral edge should be in the list of edges bounding the two surfaces that meet to form the edge; it should be in the list of all dihedral edges (classified according to their included angles); and it may also be in other lists, such as the list of features in a specific plane. This redundancy is the key to efficient processing.

In the 3DPO system the model of a part consists of four components: an extended CAD model, a set of feature classification networks, a planar patch model, and a wire-frame model (see Figure 4). The extended CAD model is the primary model of the part; the other models are derived from it. Each is designed for a specific function. The feature networks support the off-line feature selection procedure; the planar patch model is used to predict range data for the verification and configuration-understanding procedures; the wire-frame model is used to display hypotheses.

The extended CAD model consists of a standard volume-surface-edge-vertex description and pointers linking topologically connected features. For example, the representation of an edge includes pointers to the two surfaces that intersect to form it, while the representation of a surface contains an ordered list of its boundary edges and a list of its holes. We have implemented a version of this modeling system that uses a pointer structure similar to Baumgart's winged-edge representation [1]. The primitives are cylindrical and planar surfaces surrounded by circular arcs and straight lines. We selected this limited set of primitives because they are common components of machined and cast objects and can be easily modeled mathematically.

Figure 5 shows two views of the casting model built in terms of these primitives. It contains 7 full cylinders, 8 partial cylinders, and 25 planar patches, all of which are bounded by 32 circular arcs and 28 straight lines. These numbers are large enough to exclude the straightforward matching strategy that compares each observed feature (i.e., line, arc, plane, or cylinder) with each model feature and tries to find the largest set of consistent matches. The combinatorial explosion inherent in this search, however, can be reduced dramatically by carefully selecting the order of the features to be considered, measuring certain properties of the observed

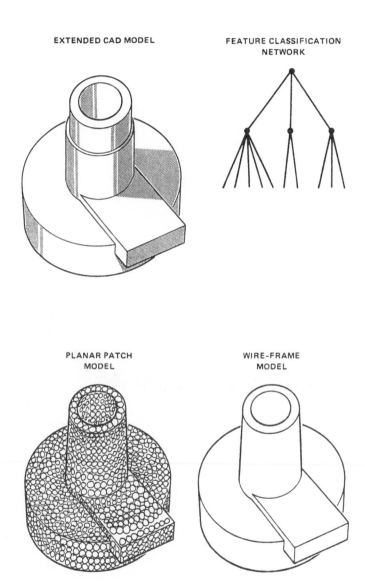

EXTENDED CAD MODEL

FEATURE CLASSIFICATION
NETWORK

PLANAR PATCH
MODEL

WIRE-FRAME
MODEL

Figure 4: Four-part object model

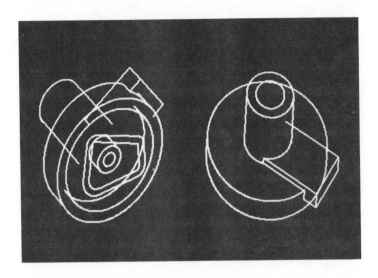

Figure 5:   Model of casting (bottom and top)

features, and then restricting the matches to those between features with similar properties.

The second component of an object model in the 3DPO system is a set of feature networks. The networks are primarily designed to support the off-line feature selection portion of the system. In the current implementation they classify the features according to their types and sizes. There are several different lists, including a list of circular arcs, sorted by radius, and a list of straight edges, sorted by length. Given these ordered lists as a data base, we have implemented a set of routines that can answer some of the questions mentioned earlier. These routines extract entries from these lists, and then analyze the topology of the object in the vicinity of each extracted feature. For example, these routines can quickly produce a list of all circular edges that are concave dihedral angles and have radii between .8 and 1.0 inch.

We plan to include other feature groupings, such as lists of surface elements that have a common normal or lists of cylinders that have a common axis. Each representation will be designed for a set of special-purpose procedures that analyze the data in terms of a single property; yet all the representations will contain pointers back to the extended CAD model, which serves as the core representation.

The third component of a 3DPO object model is a planar patch description of the object. It is composed of lists of small planar patches on the object's surface. Figure 6 shows two views of the model of the casting,

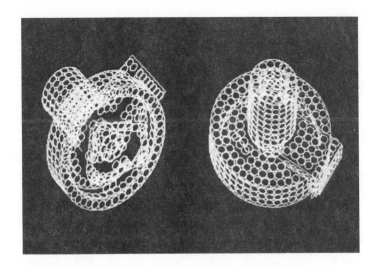

Figure 6: Planar patch model of casting (bottom and top)

which consists of approximately 500 patches. The patches are grouped into sublists of those lying on individual object features, such as a cylindrical or planar surface. In the current system each planar patch is the same size, and is represented by a three-dimensional location for its center and a surface normal. The surface normals are used by the range prediction routines to eliminate quickly those surfaces that face away from the sensor. This simple surface model was used instead of a more complete CAD model because it provided an easy way to predict range values.

The fourth and final component of an object model is the wire-frame model. It is a list of object features, such as cylinders and planes, to be displayed for hypothesized objects. Figure 5 is actually a view, not of the

CAD model, but of the wire-frame model of the casting. The CAD model includes significantly more features than the ones shown. Given a hypothesis to display, the current display routine applies a few simple rules to determine which lines to draw for each feature. A more complete graphics system would produce a more accurate rendering of the object. With our relatively crude graphics system, we have found it helpful to simplify the display by showing only a few key features.

In the 3DPO modeling system we differentiate between view-independent relationships, such as topological connectivity, and those that are view-dependent, such as image proximity (in pixels). The view-independent relationships are functions of the inherent geometric characteristics of the part, such as its size and topology, and can easily be enumerated by an analysis of the part model. The view-dependent relationships, on the other hand, are functions of the sensor, its type, and its location; they are significantly more difficult to list. Even a simple part can exhibit fifty or more structurally different views [5]. (Also see [10] for a discussion of the changes in appearance of an object as it is rotated in front of a viewer.) We plan to investigate techniques similar to those used in ACRONYM [4] for representing classes of similar views.

4. Feature Selection

The recognition strategy, as outlined above, is to locate a key feature, and then add one feature at a time until the object can be reliably and precisely located. We refer to the first feature to be located as the "focus feature", which is consistent with the terminology used in the two-dimensional local-feature-focus method [3]. The selection of the focus feature is a function of several factors; among these are the uniqueness of the feature, its expected contribution, the cost of detecting it, and the likelihood of detection. Some features are inherently easier to find. If two features provide information that is essentially of equal value, but one is easier to find, it should be ranked ahead of the other one. If a feature is often missed by the feature detectors, its rank should be lower than for one that is consistently found.

Selection of the second, third, or fourth features is even more complex than the first because it is a function of the set of features already detected as well as its own characteristics. For example, if one or more features have been found, some may be eliminated from the list to be considered because they are on the side of the object away from the sensor. Faugeras, et al. [7] have used this type of reasoning to reduce their tree searches.

In the 3DPO system we concentrate on edges because they contain more information than surface patches and are relatively easy to detect in range data. We have worked with three types of edges, straight dihedrals, circular

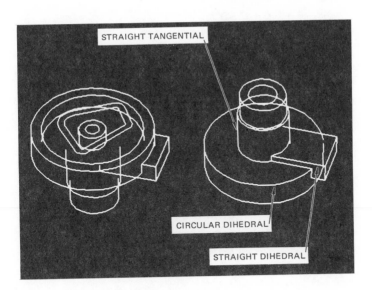

Figure 7: Three types of edges

dihedrals, and straight tangentials (see Figure 7). Each type has a set of intrinsic properties that can be used to identify matching model edges. For example, a straight dihedral has its length, the size of the included angle (if both surfaces are detected), and the properties of the adjacent surfaces, such

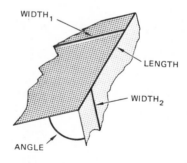

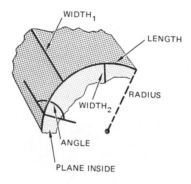

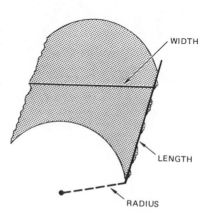

Figure 8:   Properties of the three edge types

as their widths and areas (see Figure 8(a)). Circular dihedrals have the same properties as straight dihedrals plus two additional ones: the radius of the circle and the binary property of whether or not the planar surface is inside the circle, as shown in Figure 8(b), or outside. Note that it is possible to establish these last two properties even if only one surface is detected. Straight tangential edges (see Figure 8(c)) have fewer intrinsic properties than circular dihedrals, besides which they are often difficult to locate in range data because they are noisiest on surfaces curving away from the sensor. Therefore, as a general rule, circular dihedrals are more distinct and easier to find than the two types of straight edges.

In the grow-a-match approach that we are pursuing, there are four contributions a new feature can make:
- Reduce the number of interpretations for a cluster of located features.
- Verify an interpretation.
- Determine some unknown degrees of freedom associated with an interpretation.
- Increase the precision with which the degrees of freedom can be computed.

A single feature can make more than one contribution. For example, a new feature could verify an interpretation and increase the precision.

An unconstrained object in a jumble has six degrees of freedom associated with its position and orientation, three displacements and three rotations. Different types of object features constrain different degrees of freedom. For example, a circular dihedral determines all but one of the object's six degrees of freedom. The only unknown is the rotation about the axis of the cylinder. Locating a straight dihedral on an object also determines five degrees of freedom, except that there is a binary choice of the orientation along the edge. The one undetermined degree of freedom is the position of the edge along the matching line. A straight tangential determines only four degrees of freedom because the object can rotate about the cylinder and slide up and down the matching edge. Thus, circular dihedrals, in addition to being more distinct than the straight edges, also provide more constraints on the object's position and orientation.

Given the task of recognizing and locating an object or set of objects, the 3DPO system partitions the set of object features into subsets with similar intrinsic properties, and applies a different matching strategy for each subset. Thus, if a two-inch circular dihedral is found as the focus feature, one strategy is adopted; if a one-inch circular dihedral is found, a different strategy is used. The strategies in the current system are determined interactively and are implemented as procedures. As mentioned earlier, we plan to develop an automatic selection procedure that builds trees of features for subsequent interpretation by a general-purpose procedure.

The circular dihedrals are the best focus features for the casting in Figure 1 because their intrinsic properties generally reduce the number of possible interpretations to one or two. They are the best even though the casting has more circular edges than straight edges. Figure 9 lists 15 of the

NAME	RADIUS	PLANE	DIHEDRAL ANGLE	ARC ANGLE	LENGTH	PLANE WIDTH	CYLINDER WIDTH
Base-bottom	2.30	inside	90	360	14.4	0.4	1.0
Base-top	2.30	inside	90	317	12.7	1.3	1.0
Shelf-bottom	2.30	outside	-90	43	1.7	0.4	1.2
Inside-base	1.90	outside	90	360	11.9	0.4	0.7
Pipe-shoulder	1.00	inside	90	360	6.3	0.1	1.8
Pipe-top	0.90	inside	90	360	5.7	0.3	0.9
Pipe-base	1.00	outside	-90	243	4.2	1.3	1.8
Pipe-on-shelf	1.00	outside	-90	116	2.0	1.6	1.5
Pipe-joint	0.90	outside	-90	360	5.7	0.1	0.9
corner-1	0.68	inside	90	100	1.2	0.2	0.7
Small-cylinder	0.50	inside	90	360	3.1	0.2	0.7
corner-2	0.38	inside	90	89	0.6	0.2	0.7
corner-3	0.33	inside	90	103	0.6	0.2	0.7
Inside-pipe	0.55	outside	90	360	3.5	0.3	2.6
corner-4	0.43	outside	90	100	0.7	0.2	0.7

Figure 9:   Table of circular arcs

circular arcs on the casting and their intrinsic properties. (The other 17 arcs are not included because they are either too small to detect or are not visible.) The horizontal lines in the table delimit subsets of features that have similar values for the first two properties listed across the top of the table. Note, for example, that there are four circles with radii of approximately 2 inches, but they are partitioned into two subsets because two of them have a planar surface inside the circle and two of them have it on the outside. The first two properties are used because they can be computed even if only one of the surfaces adjacent to the edge is visible. Figure 9 illustrates the importance of knowing the sizes of model features and the ability of the feature detection procedures to classify features according to their sizes.

The 3DPO system employs a different strategy for each of the six subsets of features in Figure 9. Each strategy depends on the number of features in the subset and the structure of the object in the vicinity of the feature. The strategy for the first subset initially tries to ascertain which one of the two features has been found. If the surfaces adjacent to the edge have been located, their respective properties suffice to differentiate the two. If not, the system tries to find either the PIPE-BASE arc or the INSIDE-BASE arc (shown in Figure 10),

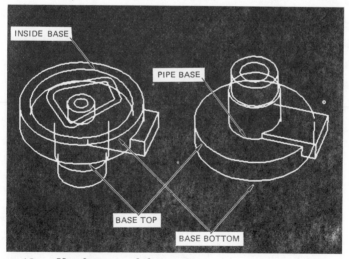

Figure 10: Key features of the casting

either of which would identify the focus feature. Once this has been done, the next subgoal of the strategy is to compute the rotation about the axis of the cylinder. The strategy provides for this by locating some of the straight dihedral edges shown in Figure 10. This is a typical strategy. It starts with a focus feature that is relatively distinct; it next locates a second feature to identify the focus feature; then it finds a feature or two to determine the final degree of freedom.

These edge-based strategies are made possible by the fact that we can reliably locate edges and measure properties of the adjacent surfaces. However, since our data are gathered by a triangulation system, they rarely contains more than two surfaces that are connected directly. Usually a "missing data" area intervenes between surfaces. Therefore, any attempt to grow matches topologically past a couple of surfaces is likely to fail.

In the past, most vision researchers have avoided objects with several features because they wanted to keep the images as simple as possible. We, on the other hand, have found that an abundance of features is generally helpful because it means that there is almost always a nearby feature to help identify one for which there are many interpretations. If all the objects are parallelepipeds, all the features look the same. The nearby features are not much help either, since they all look the same too. However, if an object has 30 different features, finding one is often enough to form a hypothesis, while finding two generally suffices to both form a hypothesis and verify it. The fact that the average depth of a strategy tree is about three or four indicates that the features on the casting provide significant information.

5. Data Acquisition

We use the White Scanner, built by Technical Arts, Inc. of Seattle, Washington, to gather our range data. It is a triangulation device that projects a plane of light onto the scene, as shown in Figure 11. It measures the x, y, z, and intensity of points along the intersection of the light plane and the objects in the scene. To gather a "dense" range image the plane is

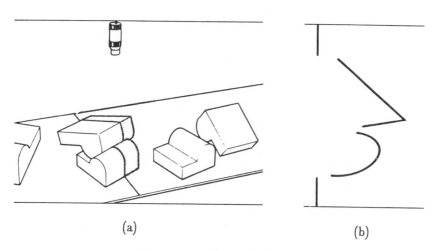

(a)   (b)

Figure 11: Diagram of a triangulation-based range sensor. (a) Range sensor based on a camera and a plane of light; (b) Image of the intersection of the light plane and the objects

scanned across the scene and the image built up one line at a time. Figure 12 is an overview of the work station showing the laser and the camera used to gather raw data for processing by the White Scanner's computer.

Figure 13 shows the data obtained by this sensor. Figure 13(a) contains the intensity data, Figure 13(b) is the z component of the range data, Figure 13(c) is the x component (the y component is not shown), and Figure 13(d) is the code image. The coordinate system of the data has its origin on the table top and the z axis points up toward the sensor. Thus, z values correspond to heights above the table. The data are encoded in these images so that larger values are lighter, which means that the z component is encoded so that higher points are lighter. The black regions in the images are areas of "no data" that are usually due to occlusion; the camera cannot see the intersection of the light plane and the objects in the scene. The code image specifies the sensor's behavior at each point. It indicates such things as "no visible stripe", "multiple visible stripes", and "blooming".

It takes approximately five minutes to gather a 240 x 240 dense range image, such as the one shown in Figure 13. The field of view of the sensor is approximately 12 inches by 12 inches. The relative precision of the height

Figure 12:   Overview of the work station

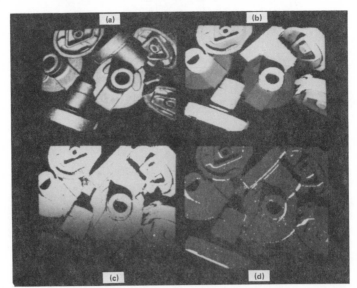

Figure 13:   Raw data produced by the White Scanner. (a) Intensity image; (b) z image; (c) x image; (d) Code image

measurements is about .020 of an inch. At the time we took these images we had a problem with the laser light plane drifting relative to the camera, which had the effect of reducing the absolute precision somewhat.

Since the surfaces of the castings are dull (i.e., they have a large Lambertian component in their reflectivity function), the sensor is able to measure data for surface patches whose normals point as much as 45 or 50 degrees away from the line of sight. A shinier surface would reflect less energy back to the camera, which would shrink the range of orientations for measurable surfaces.

Shiny surfaces cause two additional problems, multiple reflections and blooming. Multiple reflections occur when the light plane intersects a shiny surface, bounces off it and intersects another surface. If the camera sees light reflected from both surfaces it cannot distinguish the primary reflection from the secondary one; consequently, the sensor is unable to compute a unique range value.

The second problem, blooming, arises when the camera's dynamic range is insufficient to handle the amount of light reflected from the surface. This is more of a problem for shiny surfaces because they have a large specular component in their reflectivity function, which means they reflect a larger portion of the incoming light at the specular angle. If the camera happens to view a shiny surface patch from the specular direction, it receives significantly more light than at other angles, which may cause it to bloom. The effect of blooming on the range data depends on how the camera blooms. Some cameras, such as the Fairchild CCDs, fill a whole column of the image with white if they bloom, which causes "multiple reflections" on all lines. Other cameras, such as the GE CIDs, just spill white into adjacent pixels. Unfortunately, they don't always do it symmetrically, which would be relatively easy to correct.

6. Feature Detection

We locate local features, such as the arc at the end of a cylinder, by performing the following sequence of operations:

- Detect discontinuities in the range data.
- Classify them as jump discontinuities, convex discontinuities, shadows, etc.
- Discard artifacts, such as shadows.
- Link the discontinuities into edge chains.
- Partition the chains into subchains lying in planes.
- Segment the planar subchains into arcs and lines.
- Analyze the surfaces adjacent to the arcs and lines.
- Refine the locations of the arcs and lines by intersecting the adjacent surfaces.

We locate discontinuities by analyzing the data one row at a time and then one column at a time. The row analysis detects vertical edges, while the column analysis detects horizontal edges. In each direction we mark jump and slope discontinuities. We locate slope discontinuities by recursively partitioning a sequence of contiguous points into line segments and then evaluating each corner that is introduced. To evaluate a corner, we fit portions of the data on both sides of it with lines and mark it as a slope discontinuity if the angle between the two lines is less than some threshold, such as 150 degrees. In effect, we use the recursive line-fitting procedure to suggest possible discontinuities, which we then evaluate further.

Figure 14 illustrates this edge detection process. Figure 14(a) shows a typical slice of range data in the coordinate system of the light plane. The ordered list of points is first partitioned into six sublists of contiguous points, and then line segments are recursively fitted to each sublist; this fitting process introduces the circled corners in Figure 14(b). After these are evaluated, the two corners on the arc are discarded because they are not sharp enough.

The next step is to classify each discontinuity. Figure 15 shows the types of classifications used by the 3DPO system. Figure 16 shows the classifications of the discontinuities in Figure 14. A tangential discontinuity

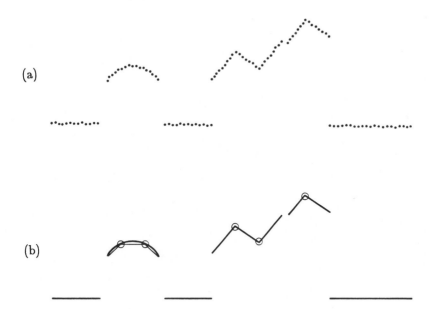

Figure 14: (a) Slice of range data; (b) Corners inserted by the recursive line-fitting routine

is formed by a surface curving away from the line of sight (e.g., along the side of a cylinder). A shadow discontinuity occurs when one surface occludes another. In Figure 16 the two rightmost shadows are easily detected because they lie directly under a jump discontinuity. The other two are more difficult to detect. To do so, the program uses a heuristic for tangential discontinuities that, in effect, allows for a wider gap between light rays than at jump discontinuities.

Once the discontinuities have been classified, the artifacts, including shadows and picture edges, are removed from further consideration. This elimination of artifacts typically reduces the number of edge points by about thirty percent. It is, therefore, an important step in the process of locating valid object features.

After the individual edge points have been detected, classified, and filtered, they are linked together to form edge chains ("edges"). The linking procedure starts with an unattached edge point and tries to form a string of

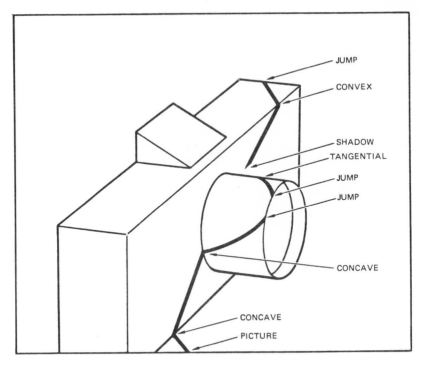

Figure 15:   Discontinuity classifications along a slice

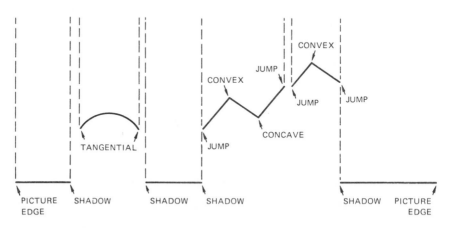

Figure 16:   Discontinuity classifications along a slice

connected edge points. It adds one point at a time to its current list, giving preference to points of the same type, points in line with the current edge, and points close to the preceding one. Figure 17 shows the edges located in

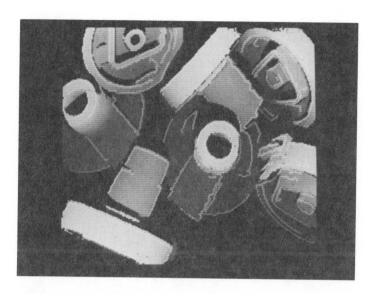

Figure 17:   Nonshadow range edges

the data in Figure 13. Notice that the top edge of the shadow cast by the vertical pipe near the center of the picture has been eliminated by the shadow filter. A small portion at the right of that shadow edge was missed by the filter.

As indicated previously, the 3DPO system concentrates on three types of object features: straight dihedrals, circular dihedrals, and straight tangentials. Each of them lies in a plane. A straight dihedral lies in both the planes that intersect to form the edge; a circular dihedral lies in the plane that intersects the cylinder at right angles; a tangential edge of a cylinder lies in a plane tangent to the cylinder. Therefore, to locate such features, the program partitions edges into planar subedges. It does this by recursively fitting planes to smaller and smaller portions of the edges until the points along each subedge lie in a plane.

The next step is to partition the planar subedges into circular arcs and straight lines. To do this, the program maps the three-dimensional points along a subedge into the two-dimensional coordinates of the plane passing through them, and then partitions the two-dimensional curve into circular arcs and straight lines. It applies a technique similar to Pavlidis [14] to accomplish this final segmentation. Figure 18 shows the straight lines and the points along the circular arcs found for the edges in Figure 17.

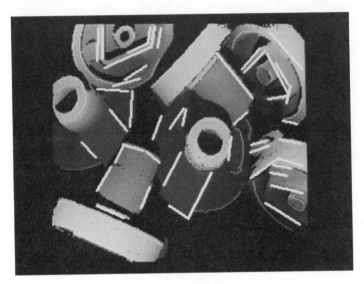

Figure 18:   Circular arcs and straight lines

The last two steps of the feature detection process are to analyze the surfaces on either side of an edge, if there are data on them, and intersect them to improve the estimates of the edge's location. The analysis of the surfaces classifies each one according to its type and measures a few simple properties, such as the maximum excursion of the surface from the edge. A circular edge is expected to have one planar surface and one cylindrical surface adjacent to it. Straight segments may be the tangential edge of a cylinder, the intersection of two planes, or the intersection of a plane and a cylinder. After the surfaces have been classified, they are refitted with as much data as possible and the updated surface equations are used to improve the estimates of the parameters of the lines and circles. This completes the feature detection process.

The most time-consuming part of the feature detection process is the analysis of the surfaces adjacent to the edges. While it is important to determine surface type, it is less essential that exact fits be produced. Rough estimates of the surface parameters may be sufficient. Therefore, it appears that it would be more efficient to develop some techniques that characterize surface types quickly than to spend the time required to perform costly iterative fitting.

Extra features slow the matching process a little because they force the program to explore hypotheses that it cannot verify. However, we have not found this to be a significant problem. In fact, we usually set the thresholds for accepting features relatively low and rely on the high-level system to filter out the extra ones. Low thresholds are beneficial because they let in features that otherwise would have been missed, whereas the false features are quickly rejected because they do not form clusters of features that are consistent with the objects.

7. Hypothesis Generation

The top-level strategy for hypothesizing object locations tries to grow a match from the most distinctive feature found by the feature detection procedure, then from the second most distinctive feature, and so on. When a hypothesized object is verified, its features are removed from the global list of features, which, in effect, simplifies the scene to be analyzed. This process of hypothesizing and verifying objects continues until everything in the scene is understood or no more matches can be formed.

The most distinctive features for the casting are the large to medium-sized circular arcs listed in Figure 9, followed by the longer tangential edges. Therefore, the program starts its analysis of a scene, such as the one shown in Figure 18, by looking for the larger circular arcs. When it finds one, it applies the strategy that was established for that type of focus feature by the off-line strategy selection process. Figure 19 shows one of these large arcs.

Figure 19:   A circle fitted to an arc

It has a radius of approximately 2 inches and a planar surface on the outside, which means it must be either part of the SHELF-BOTTOM arc or the INSIDE-BASE arc. Since its length is significantly longer than 1.7 inches, the program eliminates the SHELF-BOTTOM interpretation.

To verify the INSIDE-BASE interpretation, the strategy associated with this focus feature applies a routine to locate one of the concentric circles on the bottom of the casting. This routine is essentially a verification routine that tries to find a circle with a specific radius at a specific three-dimensional position and orientation. To do this it first looks through the list of features already detected and then, if the circle being sought is not in that list, the routine analyzes the range data for evidence of it. Figure 20 shows two concentric circles located in this manner. The small one had already been detected by the initial feature detection process. The larger one was detected by reanalysis of the data. The system uses the data along these circles to refit the plane passing through them and to improve its estimate of their common center's location. These parameters determine five of the

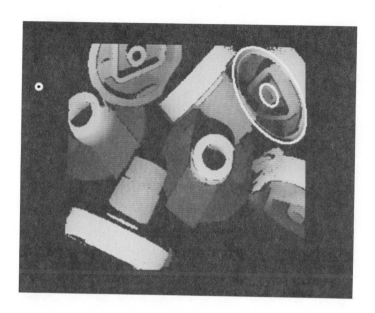

Figure 20:   Concentric circles

casting's six degrees of freedom.

To determine the sixth degree of freedom, the program has to locate a feature that is not symmetric about the axis of the cylinder. The best features for this purpose are the line segments that are part of the design on the bottom of the casting. Unfortunately, they are all quite similar. Therefore, instead of selecting one feature for the next branch of the tree search, the strategy includes a multifeature step that involves locating several lines and determining the largest subset of them that is consistent with the model. The maximal-clique algorithm used to find the largest subset of consistent matches is itself a tree search [2]. However, the system views the location of this subset as only one step in the strategy, not several, so that it can evaluate the success of the maximal-clique algorithm separately.

Figure 21 shows the line segments that are in the same plane as the circle and may be part of the pattern. There are five lines, each with two or three interpretations; these imply a graph containing 12 nodes to be analyzed by

Figure 21:   Candidate lines for determining the rotation

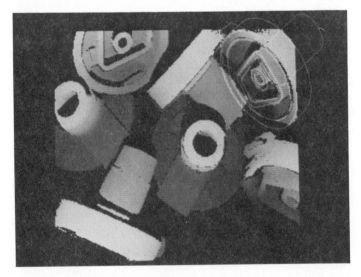

Figure 22:   Hypothesized casting

the maximal-clique algorithm. Figure 22 shows the final hypothesis, which
is based on the initial circle and four of the line segments.

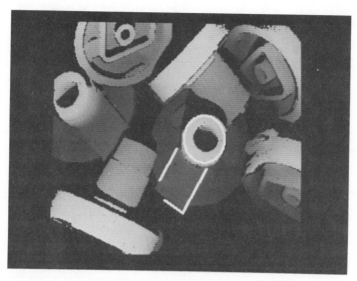

Figure 23: Circle and candidate lines

In Figure 23 another circular arc is shown with its set of line segments for computing the rotation. In this case, the maximal-clique algorithm finds three line segments that are mutually compatible with the circle. Figure 24 depicts the final hypothesis.

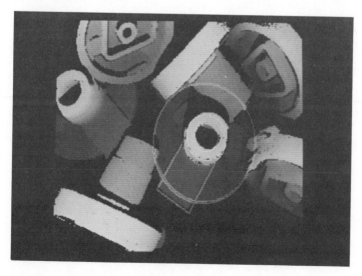

Figure 24: Hypothesized casting

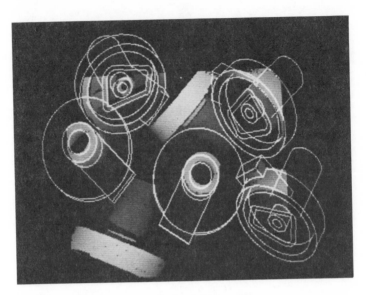

Figure 25:   Five hypotheses derived from arcs

The program arrived at five hypotheses by starting with circular arcs (see Figure 25). It then started considering tangential edges. The strategy for dealing with them is to locate other cylinders and circles that have axes almost collinear with the initial cylinder. Locating one additional feature is generally enough to compute five degrees of freedom, whereupon the strategy is the same as for circles: locate a feature to determine the sixth degree of freedom. Figure 26 shows a cylinder and a compatible circle. Figure 27 shows a hypothesis based on the five degrees of freedom computed from the cylinder and circle. The system missed the seventh casting in the scene because it only found one of its features.

The tests we use for checking the compatibility of one feature with another are extensions of the two-dimensional point-to-point tests used in the local-feature-focus system [3] and the three-dimensional point-to-plane tests used by [9]. Since the observed features are segments of lines, circles, and cylinders, the tests are segment-to-segment matches in the sense that they can use the lengths of the features to constrain the extent to which one feature can slide along a matching feature. Long features constrain the sliding more than short ones. So far we have not tried to develop a

Figure 26:   Cylinder and compatible circle

Figure 27:   Hypothesized casting

minimum set of tests. We have simply implemented a set of inexpensive tests to eliminate obvious mismatches.

Industrial systems, to be practical, have to be robust. For the 3DPO system, this means that the high-level system has to work even when the low-level system misses features and finds extra ones. It recovers from most missing features by focusing on other features of the object. If several features on an object are missed or are obscured by other objects, the system may take a while to recognize the object because it has to cycle through its list of features. Of course, if all the major features are missed, the system inevitably will miss the object altogether.

8. Hypothesis Verification

After the system hypothesizes an object's pose, there are three things it can do to increase an arm's chances of acquiring the object correctly:
- Verify the hypothesis
- Refine the pose estimate
- Determine the object's configuration

In this section we describe a verification technique. In the next section we show how it can be extended to determine which objects are on top of the jumble. Although the pose refinement step is an essential component of a complete system, it will not be discussed here. We simply observe that an approach to pose refinement described by [16] looks promising.

There is only one way to check a hypothesis: compare predictions based on it with data gathered from the scene. Predictions may differ in type, but the process is nonetheless identical. If too many predictions disagree with the data, the hypothesis is rejected.

Predictions can be object features, such as holes, corners, or surface patches, or they can be sensor data, such as the expected intensity of a point on the surface of an object. Most matching strategies have feature-level verification built into the matching process. They use the first few features to narrow the number of possible matches down to one, equivalent to making a hypothesis, and then match additional features to increase confidence in that hypothesis. These systems generally report the hypotheses that contain the most matching features to be the best matches.

Data-level comparison is another type of verification that can be done. In this kind of comparison, the program employs a hypothesis to predict the data that would been measured by the sensor if the object had been in the hypothesized pose, and then compares these predictions with the data actually measured by the sensor. In this paper we describe data-level techniques that complement traditional feature-level techniques. We concentrate on range data because they encode the geometry of an object directly and are relatively easy to predict.

The 3DPO system forms one hypothesis at a time (such as the one shown in Figure 22) and then tries to verify it. To check a hypothesis, the program predicts the range data, compares it with the actual data, and then makes decisions based on the correlation between the predicted and actual data. The predictions are an estimate of what the sensor would have seen if the objects had been in the hypothesized poses. To make the predictions, the program uses the planar-patch model shown in Figure 6. Given a hypothesis or set of hypotheses, the program builds an image by painting in regions corresponding to the surface patches in the scene that are closest to the sensor. This is essentially the same as the z buffer technique used by computer graphics systems. Figure 28(b) shows a predicted range image that corresponds to the measured data in Figure 28(c). It was produced from the seven hypotheses shown in Figure 28(a). (The seventh casting which was missed by the hypothesis generation procedure, was added interactively to the scene description to illustrate this range prediction procedure and the configuration understanding procedure discussed in the next section.)

When a measured range value is compared with a predicted value, three situations can occur:

1. The measured data are approximately equal to the predicted data.
2. The measured data are significantly farther from the sensor than

Figure 28: (a) Seven hypothesized castings; (b) Predicted height image

(c)

Figure 28: (c) Measured height image

the predicted data.

3. The measured data are significantly closer to the sensor than the predicted data.

These situations are presented in Figure 29. In the first case, the measured data agree with the prediction, and the system increases its confidence in the hypothesis that led to that prediction. In the second case, the sensor appears to have seen through the predicted object because the measured data are farther from the sensor than the predicted data. This is strong negative evidence, since the objects are assumed to be opaque. In the third case, there appears to be an object between the sensor and the hypothesized object. By itself, this situation is inconclusive. It neither supports nor refutes the hypothesis. However, given a set of hypothesized objects for a scene, it is possible to determine whether or not the measured data belong to any one of them. If so, the program marks the data as explained and treats them as neutral evidence. If not, it marks the data as unexplained and treats them as weak negative evidence. The three types of predictions are referred to as positive, negative, and neutral evidence, respectively.

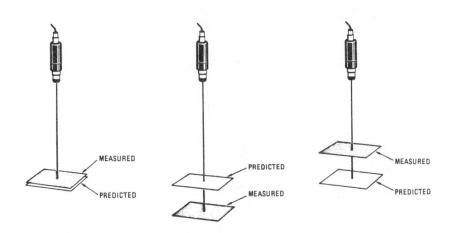

Figure 29:   Possible relationships between predicted and measured range data. (a) Measured data are approximately equal to predicted data; (b) Measured data are significantly farther from the sensor than predicted; (c) Measured data are significantly closer to the sensor than predicted

To illustrate the classification of predicted surface patches, consider the range image in Figure 30(a). In the middle of the image there is a casting that is different from the model. It has a pipelike portion that is about the same diameter as the one on the expected casting, but is significantly longer. Let us assume that the system finds the end of that pipe and hypothesizes a pose, such as the one shown in Figure 30(b). Since the hypothesis is based on the data near the end of the pipe, the predictions in that region agree with the measured data. In some of the other regions, however, the predictions disagree. Figure 30(c) shows the negative evidence (i.e., the predicted data that are farther from the sensor than the measured data). Figure 30(d) displays the neutral evidence (i.e., the predicted data that are closer than the measured data). This hypothesis would be rejected because of the large region of negative evidence.

The 3DPO system makes hypotheses one at a time, checking each individually as it is formed. Figure 31 depicts a good hypothesis. Figure

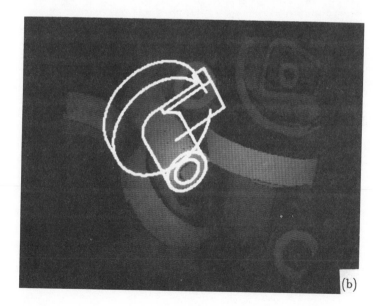

Figure 30: A bad hypothesis. (a) Measured range data; (b) Hypothesized casting;

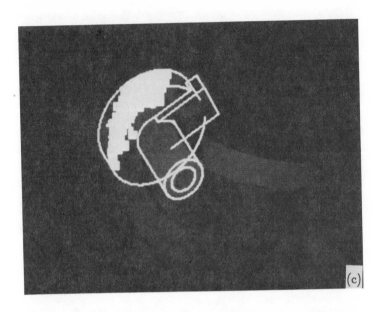

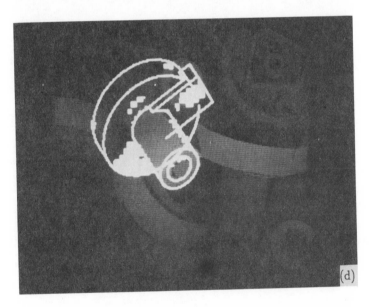

Figure 30:   (c) Negative evidence; (d) Neutral evidence

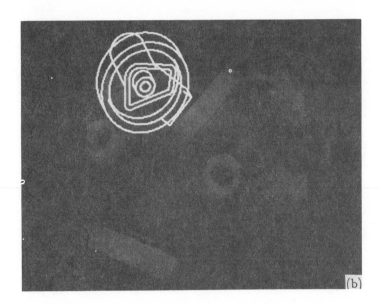

Figure 31: A good hypothesis. (a) Measured range data; (b) Hypothesized casting;

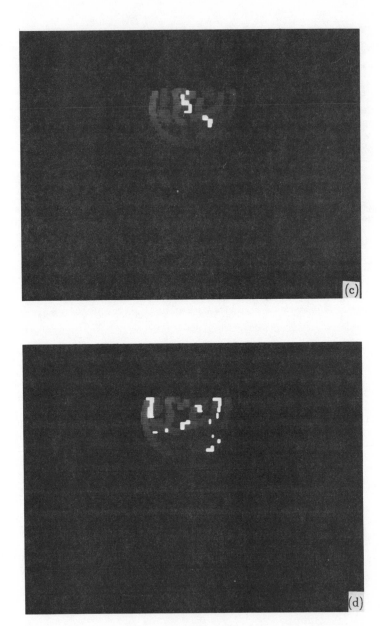

Figure 31:   (c) Negative evidence; (d) Neutral evidence

31(a) shows the range data, Figure 31(b) the hypothesis, Figure 31(c) the negative evidence, and Figure 31(d) the neutral evidence. There are several small discrepancies along the edges of the object that are due to a slight misalignment of the hypothesis. Because they are all small, however, the system still accepts this hypothesis. Nevertheless, the large number of edge effects emphasizes the need for a technique to refine pose estimates. A better pose estimate would eliminate most of the discrepancies.

The feature-based approach to hypothesis verification is convenient because the features have already been detected as part of the matching process; essentially, all one has to do is match a few additional features. However, there are at least two disadvantages. First, this method requires that an object possess several distinct features. Consequently, for example, it doesn't work very well for smooth objects such as apples. Second, it places a burden on the feature detectors in that they must locate all the features and describe them in a canonical way. If any features are missed or described incorrectly, a more complicated matching strategy will be necessary. For example, if a feature is missed, the program may have to apply a specially tuned feature detector to find the feature so as to include it in the verification process. (We have used this approach in the 3DPO system to find such things as concentric circles.) If a feature detector happens to describe the low-level data in a way that is different from what was expected, the matcher will have to be smart enough to recognize alternative descriptions. For example, if the line-and-arc fitter should segment an arc predicted from the model into a sequence of short lines, the matcher would have to recognize the pieces as parts of the arc.

One of the advantages of a data-level approach to hypothesis verification is that it is a homogeneous process that works for all types of objects. Another advantage is that it produces explanations in terms of regions, which is a convenient form for deciding how much of a scene is explained by a set of hypotheses. One disadvantage is that this approach requires a detailed model of the physics of the sensor. While this is relatively straightforward for range data, our current capabilities cannot handle intensity images. We can of course predict the locations of edges in an intensity image, but even then we lack models of the detectors employed to locate the edges. Since each detector has side effects, such as displacing edges and inserting new ones, precise predictions are still not possible.

A second disadvantage of data-level verification is that it is sensitive to misalignment. Slight offsets lead to discrepancies along the edges of an object. One way to reduce misalignment is to apply an iterative technique that assigns measured data points to surfaces of the object, uses these assignments to update the pose estimate, and then repeats the process until it minimizes the sum of the errors. Given a good estimate of the pose, the program can reject the hypothesis if the sum of the errors is too large or, alternatively, it can make the data-level comparison to perform a more structural evaluation of the match.

In the future we plan to investigate ways of combining feature-level verification with data-level techniques. This combination would utilize features to help develop a region-based explanation of a scene. For example, the location of a feature could be used for local correction of a global pose estimate so as to avoid the edge effects resulting from an unguided data-level comparison. Such a system would also provide a way of extending the technique to gray-scale analysis when it is impossible to predict absolute intensity values, yet possible to predict intensity edges and approximate intensity values.

9. Configuration Understanding

There are several reasons it is better to pick up an object from the top of a pile than one that is partially buried. First, the topmost object usually has more surfaces exposed and hence provides more ways in which it can be grasped. Second, its relatively accessible location minimizes the force required to extract it. This also tends to minimize the forces that might change the object's pose in the hand of the robot. This is important because the goal of the 3DPO system is to ascertain the pose of an object before grasping it so that the arm can select a grasping position that will be compatible with the pose required at the time the object is set down. If the pose of the object in the hand changes as the object is being pulled out of the pile, the system loses some essential positional information. A third reason for selecting the top object is that its removal generally causes minimal disruption in the rest of the pile, thus simplifying the analysis necessary for selection of the next object to be acquired.

The 3DPO system determines which object is at the top of a pile by predicting a range image from all the verified hypotheses, tagging each projected range value with the number of the hypothesized object it was derived from, and then checking to see which one is on top whenever two predictions are made at the same place in the image. As it makes predictions and compares them with the partially completed image, the program gathers statistics on the number of overlapping patches between the members of each object pair. After completing this analysis, it uses the statistics to construct a graph that represents the significant occlusions.

Figure 32 illustrates a typical occlusion. In this example, at least four range values would be predicted along the indicated ray, which corresponds to one pixel in the synthetic image. It is easy to determine that, at this particular pixel Object 2 is on top of Object 1. However, when three or more objects are present along a ray, the amount of configurational information extracted depends on the amount of data the program stores for each pixel. The current program keeps track of the range value closest to the sensor and the object it belongs to. When a new range value is predicted, the program checks the synthetic image to ascertain whether a value has already been predicted for that pixel. If not, it inserts one. If so, it compares the object identification numbers of the old and new predictions. If they are the same,

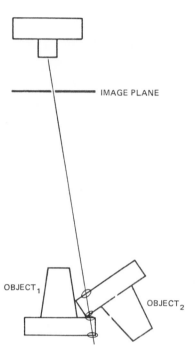

Figure 32:   One object occluding another

the program updates the predicted range value, if necessary, and continues. If the objects are different, it notes which of them is on top and updates both the range value and the object number.

This process gathers all the occlusion information if there are no more than two objects along any one ray. If there are three or more, however, the occlusion relationships obtained depend on the order in which the hypothesized objects are processed. For example, let us consider Figure 33. If Object 1 is processed by the range prediction software first, Object 2 second, and Object 3 third, the relationships that would then be computed are that Object 2 is on top of Object 1 and that Object 3 is on top of Object 2. The fact that Object 3 is on top of Object 1 is missed. If the program kept track of all range values along each ray, together with their objects, it would be possible to compute all such relationships. However, the data structures required to store this information are unwieldy. Fortunately, since the two top objects usually occlude lower objects completely, stacks of

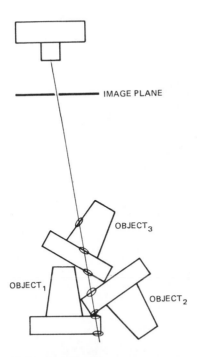

IMAGE PLANE

OBJECT$_3$

OBJECT$_1$

OBJECT$_2$

Figure 33: Stack of three objects

three or more objects are not often detected in range images. In addition, missing an occlusion relationship at a pixel is normally not critical because the relationship generally occurs at other places in the image where the program can detect it. In this case, missing one occurrence of a relationship simply reduces the estimate of the amount of overlap between the two objects; this is significantly less detrimental than missing the crucial fact that they are overlapping.

After the program has gathered statistics about occlusions in the image, it builds a graph to represent object interrelationships. Figure 34 displays the graph constructed for the seven hypotheses of Figure 28(a). Figure 35 shows the information from that graph in terms of arrows superimposed upon the intensity image returned by the range sensor. An arrow points from an occluded object to the occluding one. The arrows indicate, for example, that the object at the top center of the image is lying atop two other objects. The three objects at the butts of the arrows would not be

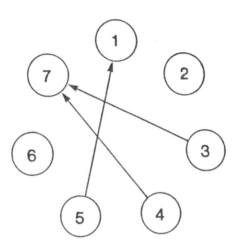

Figure 34:   Graph of the "on top of" relationships for the hypotheses
in Figure 28(a)

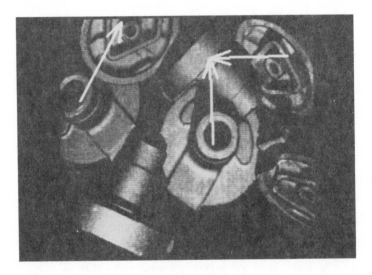

Figure 35:   Arrows point from occluded castings to occluding ones

good choices to be picked up first because other objects are known to be **on**
top of them.

It should be borne in mind that there might not be an object on top. **All** of the objects might be partially occluded by other objects. In that case, the graph would be cyclic. Figure 36 shows two configurations of objects in which none of them is on top. In Figure 36(a) two concave objects are arranged so that they overlap each other. In Figure 36(b) three convex objects are arranged so that all of them are partially occluded.

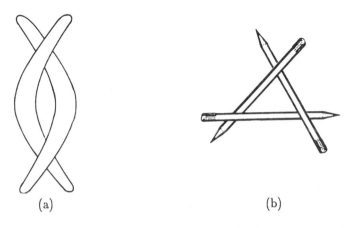

(a)   (b)

Figure 36: Cyclic occlusion relationships. (a) Two concave objects that overlap each other; (b) Three convex objects that overlap one another

The information as to what is on top of what represents the first-level understanding of an object configuration. A second level might be a specification stating exactly which objects are resting on or leaning against other objects and where they touch one another. This information would make it possible to perform a more detailed analysis of which objects would be moved in the course of extracting one of them from the pile. Unfortunately, it appears too difficult to compute these relationships from predicted range images or from three-dimensional models without a spatial-reasoning system that understands gravity and the geometric constraints associated with mutual contact between objects.

10. Discussion

It is interesting to note that the problem of locating objects with six degrees of freedom in range data is remarkably similar to the problem of using gray-scale images to locate objects constrained to lie on a plane parallel to the image plane. In both cases, one feature determines most of the degrees of freedom. For example, a circular hole in a gray-scale image determines two of the three degrees of freedom. A circular edge in range data determines five of the six degrees of freedom. Another similarity is that a partial topology of features can be determined. In range data, the edge-surface-edge connectivity provides a direct way to grow clusters of related features. In gray-scale images, the corner-line-corner connectivity along edges provides a similar capability.

In the future we plan to continue our investigations into ways of detecting features and growing clusters so that more and more of the recognition process can be shifted from search methods to cluster formation procedures. This shift will lead to an increase in efficiency because it reduces the amount of unconstrained search required to recognize an object. We also plan to explore techniques for analyzing CAD models and selecting recognition strategies and features automatically.

Acknowledgments

The work reported herein was supported by the National Science Foundation under Grant DAR-7922248. This article appeared originally in the International Journal of Robotics Research, Vol. 5, No. 2. Copyright, 1986, Massachusetts Institute of Technology. Reprinted by permission of the MIT Press.

References

[1]   Baumgart, B. G.
      Winged-edge polyhedron representation.
      Technical Report Memo # AIM-179, Stanford University, 1972.

[2]   Bolles, R. C.
        Robust features matching through maximal cliques.
        In *Proceedings of SPIE's Technical Symposium on Imaging
            Applications for Automated Industrial Inspection and
            Assembly.* Washington, DC, 1979.

[3]   Bolles, R. C. and Cain. R. A.
        Recognizing and locating partially visible objects: The local-feature-
            focus method.
        International Journal of Robotics Research 1(3):57-82, 1982.

[4]   Brooks, R.
        Model-based three dimensional interpretations of two dimensional
            images.
        In *Proc. 7th Int. Joint Conf. Artificial Intell.*, pages 619-624.
            Vancouver, B.C., Canada, Aug., 1981.

[5]   Chakravarty, I.
        The use of characteristic views as a basis for recognition of three-
            dimensional objects.
        1982.
        Ph.D. thesis, Rensselaer Polytechnic Inst.

[6]   Duda, R. O., Nitzan, D., and Barrett, P.
        Use of range and reflectance data to find planar surface regions.
        IEEE Trans. Pattern Anal. Machine Intell. PAMI-1:259-271, July,
            1979.

[7]   Faugeras, O. D., et al.
        Toward a flexible vision system.
        Robot Vision.
        IFS (Publications) Ltd., United Kingdom, 1983.

[8]   Faugeras O.D., Hebert, M.
        A 3-D recognition and positioning algorithm using geometrical
            matching between primitive surfaces.
        In *Proc. Eighth Int. Joint Conf. On Artificial Intelligence*, pages
            996-1002. Los Altos: William Kaufmann, August, 1983.

[9]   Grimson, W.E.L. and Lozano-Perez, T.
        Model-based recognition and localization from sparse range or tactile
            data.
        International Journal of Robotics Research 3(3):3-35, 1984.

[10]   Koenderink, J. J. and Van Doorn, A. J.
        The singularities of the visual mapping.
        Biological Cybernetics 24(1):51-59, 1976.

[11]   Nevatia, R. and Binford, T. O.
        Description and recognition of curved objects.
        Artificial Intelligence 8:77-98, 1977.

[12]   Nitzan, D., Brain, A. E., and Duda, R. O.
        The measurement and use of registered reflectance and range data in
            scene analysis.
        Proc. IEEE 65:206-220, Feb., 1977.

[13]   Oshima, M. and Shirai, Y.
        A scene description method using three-dimensional information.
        Pattern Recognition 11:9-17, 1978.

[14]   Pavlidis, T.
        Curve fitting as a pattern recognition problem.
        In *Proc. of the Sixth Conference on Pattern Recognition*. Munich,
            Germany, October, 1982.

[15]   Popplestone, R. J., Brown, C. M., Ambler, A. P., and Crawford, G. F.
        Forming models of plane-and-cylinder faceted bodies from light
            stripes.
        In *Proc. 4th Int. Joint Conf. AI*, pages 664-668. Sept., 1975.

[16]   Rutkowski, W. and Benton, R.
        *Determination of object pose by fitting a model to sparse range
            data.*
        Technical Report Interim Technical Report, Intelligent Task
            Automation Program, Honeywell, Inc., January, 1984.

[17]   Shirai, Y. and Suwa, M.
        Recognition of polyhedrons with a range finder.
        In *Proceedings of the Second International Joint Conference on
            Artificial Intelligence*, pages 80-87. 1971.

Recognition and Localization
of Overlapping Parts From Sparse Data

W. Eric L. Grimson
Tomás Lozano-Pérez

Massachusetts Institute of Technology
Artificial Intelligence Laboratory

Abstract

This paper discusses how sparse local measurements of positions and surface normals may be used to identify and locate overlapping objects. The objects are modeled as polyhedra (or polygons) having up to six degrees of positional freedom relative to the sensors. The approach operates by examining all hypotheses about pairings between sensed data and object surfaces and efficiently discarding inconsistent ones by using local constraints on: distances between faces, angles between face normals, and angles (relative to the surface normals) of vectors between sensed points. The method described here is an extension of a method for recognition and localization of non-overlapping parts previously developed by the authors.

1. Background

In order to interact intelligently with its environment, a robot must know *what* objects are *where*; that is, it must be able to identify and locate objects in its workspace. In this paper, we treat these two tasks under the title of the *recognition problem*. We will stress localization over identification since in most industrial robotics tasks the identity of the objects is known.

A solution to the recognition problem should satisfy the following criteria:

1. A recognition algorithm must degrade gracefully with increasing noise in the sensory measurements.

2. A recognition algorithm should be able to deal with partially occluded objects.

3. A recognition algorithm should be able to identify and locate objects from relatively sparse information. The sparseness of the data may be due to inherent sensor sparseness, occlusion, or noise.

4. A recognition algorithm should be applicable to different sensor types. While particular optimizations will be possible for specific sensors, one would like a recognition technique that serves as a common core for recognition from tactile, ranging, sonar and visual sensors.

This paper presents an approach to the recognition problem that satisfies these criteria. The approach operates by examining all hypotheses about pairings between sensed data and object surfaces and efficiently discarding inconsistent ones by using local constraints on distances between faces, angles between face normals, and angles (relative to the surface normals) of vectors between sensed points.

The method described here is an extension of a method for recognition and localization of non-overlapping parts previously described in [14] and reviewed in Section 3. The new method is described in Section 4; it handles highly overlapped parts using either two-dimensional visual data or three-dimensional range data (see Figures 1-4). We also report in Section 5 our experience with a new formulation of the geometric constraints that does not decouple position and orientation. A number of other extensions are described in Section 6, including a technique for dealing with objects of unknown size. Section 7 is a discussion of our approach to recognition and a review of related work.

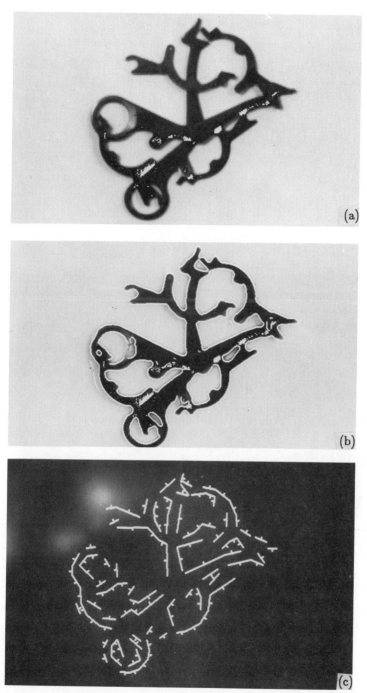

Figure 1: Two dimensional edge data. (a) grey level images; (b) zero crossings; (c) edge fragments.

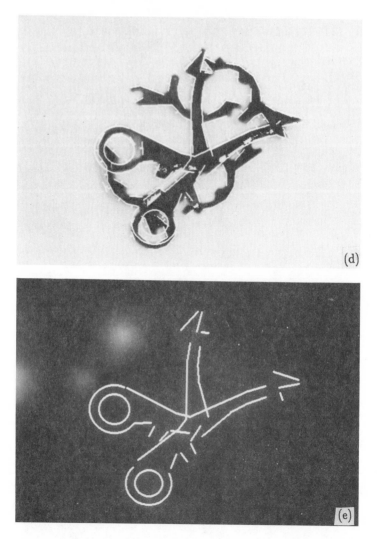

Figure 1:   (d) located object in image; (e) located object

2. Problem Definition

The specific problem considered in this paper is how to identify a known object and locate it, relative to a sensor, using relatively few measurements. Because we seek a technique that is applicable to a wide range of sensors, we make few assumptions about the sensory data. We assume only that the sensory data can be processed to obtain sparse measurements of the positions and surface orientations of small planar patches of object surfaces

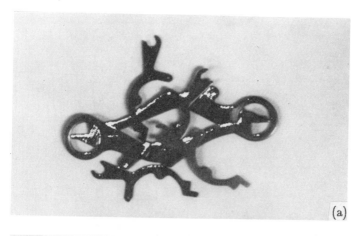

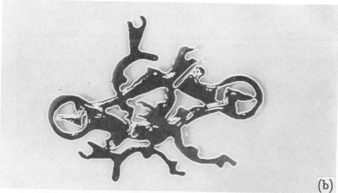

Figure 2: Two dimensional edge data. (a) grey level images; (b) zero crossings; (c) edge fragments

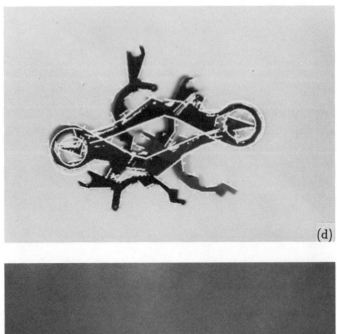

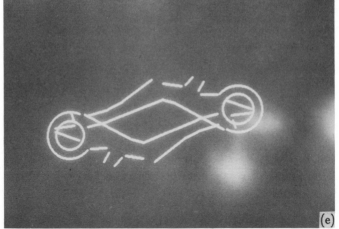

Figure 2:   (d) located object in image; (e) located object

in some coordinate frame defined relative to the sensor. The measured positions are assumed to be within a known error volume, and measured surface orientations to be within a known error cone. Furthermore, the object may be overlapped by other unknown objects, so that much of the data does not arise from the object of interest.

If the objects have only three degrees of freedom relative to the sensor (two translational and one rotational), then the positions and surface

(a)   (b)

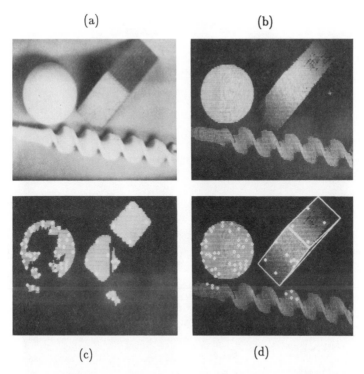

(c)   (d)

Figure 3:   Three dimensional range data. (a) Original scene; (b) range data where brightness encodes height; (c) planar patches with representative points; (d) located object superimposed on range data (filled in circles are data points accounted for)

normals need only be two-dimensional. If the objects have more than three degrees of freedom (up to three translational and three rotational), then the position and orientation data must be three-dimensional.

Since the measured data approximate small planar patches of the object's surface, we assume that an object can be modeled as sets of planar faces. Only the individual plane equations and dimensions of the model faces are used for recognition and localization. No face, edge, or vertex connectivity information is used or assumed; the model faces do not even have to be connected, and their detailed shape is not used. This is important. It is easy to build polyhedral approximations of moderately curved objects, but we cannot expect these approximations to be perfectly stable under sensor variations. The positions of vertices, orientations and

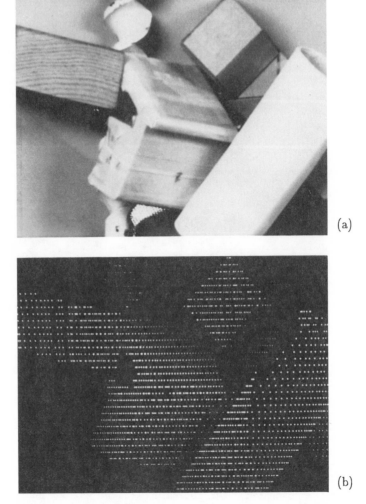

(a)

(b)

Figure 4: Three dimensional range data. (a) Original scene; (b) range data where brightness encodes height

lengths of edges, and areas of faces will all vary depending on the position of the object relative to the sensor. Our recognition method does not rely on the detailed polyhedral model; it relies instead on aggregate geometric information about the object's faces. As a result, the method can be readily applied to curved objects approximated by planar patches.

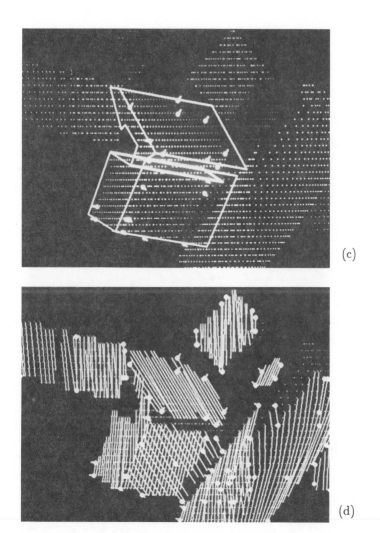

(c)

(d)

Figure 4: (c) planar patches with representative points; (d) located object superimposed on range data (filled in circles are data points accounted for)

2.1. Basic Approach

Our approach to model-based recognition is to cast it as a search for a consistent matching between the measured surface patches and the surfaces of known object models. The search proceeds in two steps:

1. *Generate Feasible Interpretations:* A set of feasible

interpretations of the sense data is constructed. Interpretations consist of pairings of each sensed patch with some object surface on one of the known objects. Interpretations in which the sensed data are inconsistent with local constraints derived from the model are discarded.

2. *Model Test:* The feasible interpretations are tested for consistency with surface equations obtained from the object models. An interpretation is legal if it is possible to solve for a rotation and translation that would place each sensed patch on an object surface. The sensed patch must lie *inside* the object face, not just on the surface defined by the face equation.

There are several possible methods of actually searching for consistent matches. For example, in Grimson and Lozano-Pérez [14] we chose to structure the search as the generation and exploration of an *interpretation tree* (see Figure 5). That is, starting at a root node, we construct a tree in a depth first fashion, assigning measured patches to model faces. At the first level of the tree, we consider assigning the first measured patch to all possible faces, at the next level, we assign the second measured patch to all possible faces, and so on. The number of possible interpretations in this tree, given s sensed patches and n surfaces, is n^s. Therefore, it is not feasible to explore the entire search space in order to apply a model test to all possible interpretations. Moreover, since the computation of coordinate frame transformations tends to be expensive, we want to apply this part of the technique only as needed.

The goal of the recognition algorithm is thus to exploit local geometric constraints to minimize the number of interpretations that need to be tested, while keeping the computational cost of each constraint small. In the case of the interpretation tree, we need constraints between the data elements and the model elements that will allow us to remove entire subtrees from consideration without explicitly having to search those subtrees. In our case, we require that the distances and angles between all pairs of data elements be consistent with the distances and angles possible between their

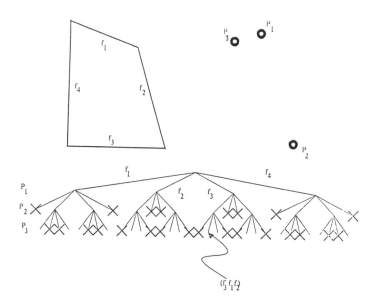

Figure 5: A simple example of constrained search. We want to find consistent matchings of the three data points to the edges of the indicated quadrilateral. If we only use distance between data points, then the table of possible ranges between the edges of the object is given by

	1	2	3	4
1	[0, 1.5]	[0, 3.25]	[2, 3.25]	[0, 2.5]
2	[0, 3.25]	[0, 2]	[0, 2.5]	[1.4, 3.25]
3	[2, 3.25]	[0, 2.5]	[0, 2]	[0, 3.25]
4	[0, 2.5]	[1.4, 3.25]	[0, 3.25]	[0, 2.5]

The tree indicates the set of possible assignments of data points to object edges, given distance as the only constraint. One can see that only 16 out of 64 possible interpretations are consistent with this very simple constraint.

assigned model elements.

In general, the constraints at the generation stage should satisfy the following criteria:

1. The constraints should be coordinate-frame independent. That is, the constraints should embody restrictions due to object shape and not to sensing geometry.

2. The constraints should be simple to compute and, at the same time, able to prune large portions of the search space.

3. The constraints should degrade gracefully in the presence of error in the sensory measurements.

4. The constraints should be independent of the specifics of the sensor from which the data came, so that they will apply equally to different types of sensors.

In this paper, we deal with two different, but related, sets of geometric constraints. In the first set position and orientation are decoupled. The decoupling leads to very efficient implementations, but reduces their pruning power. The second set retains the natural coupling between positions and orientations. This set is more powerful, but computationally more complex. Both sets are developed first for the case of a single, isolated object, and then for the case of overlapping objects.

3. Decoupled Constraints

In this section we review the method presented in [14] for recognizing non-overlapping parts. We begin by deriving a set of coordinate-frame-independent constraints that were presented there. The sensory data are sparse planar patches, each consisting of a position measurement and a unit surface normal (the extent of a patch is assumed small). The first question to ask is what types of coordinate-frame-independent constraints are possible. Clearly, a single patch provides no constraint on the model faces that could consistently be matched to it. Therefore, we consider pairs of patches. Each such pair can be characterized by the pair of unit normals, n_1

and n_2, and the separation vector between the patch centers, **d**, as shown in Figure 6.

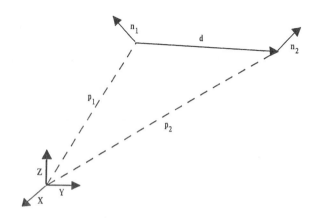

Figure 6: The constraints between pairs of measured surface patches. A given pair of sensory points P_1, P_2 can be characterized by the components of the vector **d** between them, in the direction of each of the surface normals n_1, n_2, and in the direction of their cross product, $n_1 \times n_2$, and by the angle between the two normals, $n_1 \cdot n_2$.

3.1. The Constraints

First we construct a local coordinate frame relative to the sensed data; we use both unit normals as basis vectors. In two dimensions, these define a local system, except in the degenerate case of the unit normals being (anti-)parallel. In three dimensions, the third component of the local coordinate frame can be taken as the unit vector in the direction of the cross product of the normal vectors. In this frame, one set of coordinate-frame-independent measurements is the components of the vector **d** along each of the basis directions and the angle between the two measured normals. More formally, our measurements are $n_1 \cdot n_2$, $d \cdot n_1$, $d \cdot n_2$, $d \cdot u$ where u is a unit vector in the direction of $n_1 \times n_2$.

These measurements are equivalent, but not identical to the set used in [14]. In the earlier paper, we used the magnitude of d and two of its components; this is equivalent, up to a possible sign ambiguity, to using the three components of the vector. This possible ambiguity was resolved using a triple product constraint.

To turn these measurements into constraints on the search process, we must relate them to measurements on the model elements. Since objects are modeled as sets of planar faces, the relationship is straightforward. Consider the first measurement, $n_1 \cdot n_2$. If this is to correspond to a measurement between two faces in the model, then the dot product of the model normals must agree with this measurement. If they do not agree, then no interpretation that assigns those patches to these model faces need be considered. In the interpretation tree, this corresponds to pruning the entire subtree below the node corresponding to that assignment. The test can be implemented efficiently by precomputing the dot product between all pairs of faces in the models. Of course, for the case of exact measurements, the dot product of the measured normals must be identical to that of the associated model normals. In practice, exact measurements are not possible, and we must take possible sensor errors into account. Given bounds on the amount of error in a sensory measurement, we can compute a range of possible values associated with the dot product of two sensed normals (see [14] for details). In this case, if the dot product of the associated model normals is contained in the range of possible values associated with the dot product of the sensed normals, then the corresponding assignment of model faces to sensed points is consistent.

Similar constraints can be derived for the components of the separation vector in the directions of the unit normals. Each pair of model faces defines an infinite set of possible separation vectors, each one having its head on one face and its tail in the other. We can compute bounds on the components of this set of vectors in the direction of each of the face normals. Again, for an assignment of sensed patches to model faces to be consistent, the measured value must agree with the precomputed model values. As in the case of surface normals above, we can incorporate the effects of error in the measurements by using bounds on the magnitude of the error to compute a range of possible values for the components of the sensed vectors, and this range must be consistent with the associated model range in order for an interpretation to be consistent.

It is easy to see that these constraints satisfy most of our criteria: they are independent of coordinate frames, simple, and general. It remains to establish that they are powerful and degrade gracefully with noise. Grimson [13] argues from a combinatorial analysis that these constraints are very powerful, and in the case of data all obtained from a single object, will converge quickly to a small set of interpretations. The analysis also shows that the constraints should exhibit a graceful degradation with increasing sensor noise. These predictions have been verified empirically, both by simulation and by processing real sensory data. Grimson and Lozano-Pérez [14] report on a large set of simulations run on a series of test objects, for varying types of error conditions.

3.2. Adding a Model Test

Once the interpretation tree has been pruned, there are typically only a few non-symmetric interpretations of the data remaining. It is important to realize, however, that these constraints are not guaranteed to reject all impossible interpretations. Let d_{ij} be the distance between two sensed patches, P_i and P_j. It could be the case that this measured distance is consistent with the range of distances between faces f_u and f_v, but only if the patches are inside of small regions on the candidate surfaces. Now consider what happens when adding another patch-surface pairing, (P_k, f_w), to an interpretation that already includes (P_i, f_u) and (P_j, f_v). Our constraints permit adding this pairing only if the distances d_{ik} and d_{jk} are consistent with the range of distances between f_u, f_w and f_v, f_w, respectively. In doing this, however, it uses the ranges of distances possible between any pair of points on these faces. It does not take into account the fact that only small regions of f_u and f_v are actually eligible by previous patch-surface pairings.

Because of this decoupling of the constraints, the fact that all pairs of patch-surface assignments are consistent does not imply that the global assignment is consistent. To determine global consistency, we solve for a transformation from model coordinates to sensor coordinates that maps each of the sensed patches to the interior of the appropriate face. There are many methods for actually solving for the transformation; one is described in Grimson and Lozano-Pérez [14]. This model test is applied to interpretations surviving pruning so as to guarantee that all the available geometric constraints are satisfied. As a side effect, the model test also provides a solution to the localization problem.

When a model test is applied to the feasible interpretations, two effects are noticed. First, some locally consistent interpretations are discarded as being globally inconsistent. Second, of the remaining interpretations, many are observed to be almost identical, differing, for example, only in the assignment of one or two data points to nearby faces. To get a more accurate picture of the effectiveness of the recognition technique, we can cluster the interpretations that satisfy the model test on the basis of their associated transformations. In particular, we cluster solutions whose direction of rotation is within $1.5°$ of each other. The number of distinct interpretations is greatly reduced.

4. Data from Overlapping Objects

The method described in the previous section assumes that all the data come from a single object. In this section we show how the method can be extended to handle data from multiple overlapping objects.

Assume that all of the sensed patches, except one, originate from the same object. Let P_i be the extraneous measurement. Usually, it will be impossible to find an interpretation that includes this measurement. But, not all interpretations will fail at level i in the tree; it may require adding a few more data points to the interpretation before the inconsistency is noted. It is only when all possible single-object interpretations fail that we are certain to have at least one extraneous data point.

It may still be possible to find an interpretation of all the data, including extraneous measurements, that is consistent with the pairwise constraints. It is even possible, by a fortuitous alignment of the data, for interpretations involving extraneous data to pass the model test. There is nothing within the approach described here to exclude this possibility. Of course, the larger the number of patch-surface pairings in the interpretation, the less likely this is to happen. In many cases, it may be necessary to verify the interpretation by acquiring more data. We will not pursue this point here; we will assume, instead, that the presence of extraneous points will cause all interpretations to fail either the local constraints or the model test.

4.1. Generating Interpretations for Subsets of the Data

One straightforward approach to handling extraneous data points is to apply the recognition process to all subsets of the data, possibly ordered by some heuristic. But, of course, this approach wastes much work determining the feasibility of the same partial interpretations. Can we consider all subsets of the data without wasting the work of testing partial interpretations? The simple way we have done this is by adding one more branch to each node of the interpretation tree. This branch represents the possibility of discarding the sensed patch as extraneous. Call this branch the *null face*. The remainder of the process operates as before except that, when applying the local constraints, the null face behaves as a "wild card"; assigning a patch to the null face will never cause the failure of an interpretation.

It is easy to see that if an interpretation is legal, the process described above will generate all subsets of this interpretation as leaves of the tree. This is true of partial interpretations as well as full interpretations since every combination of assignments of the null face to the sensed patches will still produce a valid interpretation.

The same condition that ensures the validity of this process guarantees its inefficiency. We do not want to generate all subsets of a valid interpretation. In general, we want to generate the interpretations that are consistent with as much as possible of the sensed data. The following simple method guarantees that we find only the most complete interpretations.

The interpretation tree is explored in a depth-first fashion, with the null face considered last when expanding a node. In addition, the model test is applied to any complete interpretations, that is, any that reach a leaf of the tree. This choice of tree traversal has the effect of considering the legal interpretations essentially by length order (where length is taken to be the number of non-null faces paired with sensed data by the interpretation).

Consider the points, P_1 to P_s in order, where s is the total number of sensed patches, and let a variable *MAX* keep track of the longest valid interpretation found so far. At any node in the tree, let M denote the number of non-null faces in the partial match associated with that node. It is only worth assigning a null face to point P_i if $s-i+M \geq MAX$; otherwise, the length of the interpretations at all the leaves below this node will be less than that of the longest interpretation already found. If we initialize *MAX* to some non-zero value, then only interpretations of this length or longer will be found. As longer interpretations are found, the value of *MAX* is incremented, thus ensuring that we find the most complete interpretations of the data. Note that if an interpretation of length s is found, then no null-face assignments will be considered after that point.

Looking for the longest consistent interpretations allows the matching algorithm to overcome the most severe combinatorial problems of the null-face scheme, but it makes the algorithm susceptible to a potentially serious problem. One of the bases of our approach to recognition has been to avoid any global notion of "quality" of match. We have simply defined generous error bounds and found all interpretations that did not violate these bounds. Once all the valid interpretations have been found, a choice between them can be made on a comparative basis rather than on some arbitrary quality measure. The modified algorithm, however, discards valid interpretations that are shorter than the longest valid interpretation. Therefore, a long interpretation on the margin of the error bounds can force us to ignore a shorter interpretation that may be the correct match.

We know of no general solutions to this problem. Quality measures such as how well the transformation maps the measured patches onto the faces [10] are useful but also susceptible to error. Our choice would be to consider all the valid interpretations whose length is within one or two of the longest interpretation and which are not subsets of a longer interpretation. This is also heuristic. We have avoided this issue in the rest of the paper and simply coped with the occasional recognition error.

4.2. Heuristics for Limiting Search

The presence of overlapping objects dramatically increases the search space for recognition. One effect is simply the need for more measurements to ensure that there are enough measurements on the object of interest. The other effect is the need to consider many alternative subsets of the measurements. Even when using local constraints and focusing on longest interpretations, complete exploration of the interpretation tree is extremely time consuming. We have achieved a significant improvement in recognition time by using two heuristic techniques, both of which reduce the size of the interpretation tree that is explored. Both of these techniques may cause the recognition process to fail but, in most applications, one prefers to reduce the recognition time even at the expense of occasional failures.

The first heuristic technique avoids considering some subsets of the measurements that are unlikely to lead to a longest interpretation. In particular, once an interpretation I is longer than some threshold, it ignores interpretations in which data points matched to faces by I are matched to the null face. This heuristic can be easily incorporated in the depth-first tree-exploration we use. Our experience is that this technique when used with a conservative threshold is quite effective in pruning the search without noticeably increasing the failure rate. We have used this technique in all our experiments with real data reported later in this section.

The other heuristic technique is based on the Hough transform [1]. It does not affect the search algorithm; instead, it drastically reduces the size of the initial interpretation tree. The method works as follows for two-dimensional data. We are given a set of measured edge fragments (see Section 4.5) and a set of model edges. For each pair of model edge and data edge, there is a rotation, η, of the model's coordinate system that aligns the model edge to the data edge. Then, there is a range of translations, x,y, that displace the model so that the chosen model edge overlaps the chosen data edge. If the pairing of model edge and data edge is part of the correct interpretation of the data, then one of the range of combinations of x,y,η obtained in this way will describe the correct transformation from model to sensor coordinates. All the model/data edge-pairings corresponding to that legal interpretation will also produce the correct x,y,η combination (modulo measurement error). We keep track of the range of x,y,η values produced by each model/data edge-pairing; this can be done with a three-dimensional array, with each dimension representing the quantized values of one of

x, y, and η. Clusters of pairings with nearly the same values define candidate interpretations of the data (see Figure 7).

This technique can serve as a recognition method all by itself [1], but in that context it has some important drawbacks. One problem is simply the growth in memory requirements as the degrees of freedom increase. A related but more important problem is the difficulty of characterizing the range of transformations that map the model into the data in three dimensions. Consider the case of mapping a planar model face into a measured planar patch. The orientation of the model coordinate system relative to the sensor coordinate system is specified by three independent parameters, but the constraint of making a model plane parallel to a data plane only provides two constraints (rotation around the plane normal is unconstrained). Therefore, each model/data pairing generates a one parameter family of rotations. Associated with each rotation, there is a range of displacements that manage to overlap the model face with the measured patch. Computing these ranges exactly is quite difficult and time consuming.

What we have done is to use the Hough transform as a coarse filter to produce an initial set of possible model/data pairings -- not to localize the objects. First, each potential model/data pairing is used to define a range of parameter values related to the position and orientation of the model relative to the sensor. These parameters, however, need not be the full set of parameters that define the coordinate transform between model and sensor. In two dimensions, for example, the parameter set may contain only the rotation angle, or both the angle and the magnitude of the displacement vector, or the full set of angle and displacement parameters. In three dimensions, we can use, for example, the magnitude of the displacement vector and the two angles of the spherical coordinate representation of some fixed vector on the model. The model/data pairings are clustered on the basis of a coarse quantization of these parameters. Each cluster associates

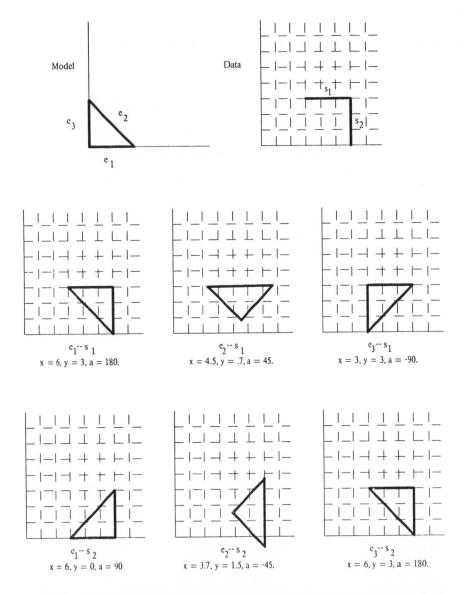

Figure 7: Hough transform preprocessing. Given a set of model edges e_i and a set of data edges s_j, we can consider all pairings of such edges. Each one describes a coordinate frame transform, as shown. By hashing over x, y, a, we can collect sets of data-model pairing that correspond to similarly predicted poses of the object.

with each data edge a set of candidate model edges; this is precisely what defines the interpretation tree for our method.

The two-stage matching method, interpretation generation followed by model test, is then applied to the largest clusters until a valid interpretation is found. The effect of the initial clustering is to drastically reduce the size of the search space at the expense of initial preprocessing. Typically, data edges in a cluster are associated with only one or two model edges out of the possible thirty or forty edges in the model. Therefore, the branching factor in the interpretation tree is cut down from thirty or forty to one or two. Predictably, this has a noticeable impact on performance. Many of the pairings, however, are still spurious due to noise and the fact that the parameters do not completely characterize position and orientation. Therefore, it is necessary to use the null-face technique described earlier.

We have tested the resulting method on simulated data as well as on actual data from three types of sensors. The results are described in the following sections.

4.3. Simulations with Two-Dimensional Data

We have done extensive testing of the algorithm with simulated two-dimensional data of the type illustrated in Figure 8. A number of polygons, representing the outlines of parts, are overlapped at random. The position and orientation of a number of data points are determined by computing the outermost intersection of randomly-chosen rays with the polygon boundaries. The position and normal information is then corrupted by random errors designed to simulate the effect of imperfect sensors.

For example, the data reported in Table 1 were obtained by the following simulation technique. Three objects, two of which were not the object of interest, were overlapped at random. Given the overlapping parts, twenty points of contact were chosen at random and then corrupted by random error whose magnitude was bounded by the limits indicated in the table. Note that the diameter of the object was 2.506 units, so that an error in

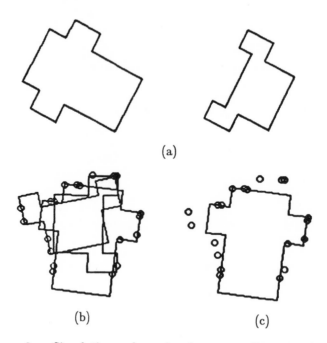

(a)

(b)   (c)

Figure 8: Simulations of overlapping two dimensional parts. A collection of copies of objects selected from the set illustrated in part (a) was overlapped at random, as illustrated in part (b) Points of contact were selected at random along the perimeter of the overlapping group, and corrupted with random error. The circles in part (b) indicate an example of sensory data. The recognition and localization algorithm then searched for interpretations of the data consistent with a specific model, as shown in part (c).

positional measurement of 0.10 is roughly 4 percent of the total diameter of the object. Once the interpretations, with their associated coordinate transformations, were found, they were clustered together into distinct interpretations. To do this, transforms whose difference in rotational angle was less than $\pi/18$ and whose translational components differed by less than 0.025 were considered to be the same. Table 1 lists the results of running this simulation one hundred times for each of the error ranges illustrated. Similar results using arbitrary numbers of overlapping parts and different numbers of contact points have also been obtained.

The algorithm performs quite well in this application, even without using the heuristic techniques described earlier. The extensive simulations

No. of Interpretations For Multiple Objects – 2D					
Object	Angle	Dist	Min	50th	95th
Housing Top	$\pi/10$	.01	1	1	3
		.05	1	1	2
		.10	1	1	5
	$\pi/5$	.01	1	1	7
		.05	1	1	5
		.10	1	2	6
	$\pi/4$	.01	1	1	4
		.05	1	2	4
		.10	1	2	14

Table 1: Illustration of statistics in the performance of the localization process for three overlapping two-dimensional objects, two of which are not the object of interest. Each row lists parameters of a histogram of the number of interpretations consistent with the local constraints, based on 100 trials with the object randomly oriented with 3 degrees of freedom. In the table, the *angle* column lists the radius of the error cone about the measured surface normal; the *dist* column lists the error range of the distance sensing. The *min* column list the minimum number of interpretations observed without a model test; the *50th* column lists the median point of the set of simulations with a model test and clustering; the *95th* column lists the 95th percentile of the set of simulations with a model test and clustering. As well, the object was overlapped with several other objects at random. In each case, 20 sensory data points were used.

reported in Table 1 indicate that the technique can reliably recognize and locate overlapping two-dimensional objects, even when only using a few points of sensory data obtained along the silhouette of the overlapping parts. As the amount of the overlapping of the parts becomes more extensive, the algorithm may begin to find more than one feasible interpretation, but the number of solutions degrades gracefully. In general, as long as enough points are sensed on the desired object, the algorithm can locate it. Furthermore, the time to do the recognition and localization is relatively low--on the order of two or three seconds on a Symbolics 3600 Lisp Machine. The time grows when the measurement error grows, in the manner illustrated by the simulations reported in [14].

4.4. Simulations with Three-Dimensional Data

We have performed similar simulations with overlapping three dimensional objects, each with six degrees of freedom, as illustrated in Figure 9. In this case, a number of polyhedra are overlapped at random, and

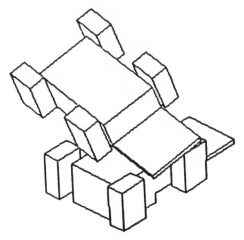

Figure 9: Simulations of overlapping three   dimensional parts. Simulations similar to those shown in Figure 8 were performed in three dimensions, with overlapping parts such as those shown above.

the position and orientation of a number of data points are determined by computing the outermost intersection of randomly-chosen rays with the polyhedral boundaries. The position and orientation information is then corrupted by random errors.

The data reported in Table 2 were obtained using the following technique. Two objects were overlapped at random, and fifteen points of contact were chosen at random and then corrupted by random error whose magnitude was bounded by the limits indicated in the table. In this case, the diameter of the object was 3.95 units, so that an error in positional measurement of 0.15 is roughly 4 percent of the total diameter of the object.

In the two dimensional case reported in Table 1, model tests were directly intermixed with the generation of the tree of interpretations. That is, whenever a leaf of the tree was reached, a model transformation was

– No. of Interpretations For Multiple Objects – 3D						
Object	Angle	Dist	Min	50th	95th	Missed
Housing	$\pi/10$	.03	1	2	9	.14
		.07	1	2	13	.07
		.15	1	5	44	.09
	$\pi/5$	.03	1	5	57	.02
		.07	1	7	47	.03

Table 2: Illustration of statistics in the performance of the localization process for overlapping three-dimensional objects. Each row lists parameters of a histogram of the number of interpretations consistent with the local constraints, based on 100 trials with the object randomly oriented with 6 degrees of freedom. As well, the object was overlapped with a second object at random. In each case, 15 sensory data points were used. The *missed* column reports the percentage of cases in which no interpretation was found, due to the separation of the model test from the application of the constraints; other columns are as in Table 1.

computed and the sensory points were checked to ensure that the transformation did correctly transform the points onto their associated faces. While this ensures that the use of the external variable MAX will not exclude any correct interpretations, the algorithm is potentially slow because of the expense of computing and applying model transformations. A faster but less reliable alternative is to generate the tree of interpretations, using the variable MAX to cutoff search as before, but in this case to simply collect all pairwise consistent interpretations, ordered by the number of non-null matches. That is, any leaf in the interpretation tree is added to a list of possible interpretations, ordered by the number of real matches in the interpretation. Once all such interpretations have been collected into sets of equal length (as defined by the number of non-null matches), we apply a model test to each interpretation of a set, starting at the set of interpretations with the largest number of real matches, and continuing until at least one interpretation is judged to be globally consistent. The interpretations from the corresponding set that pass a model test constitute the set of interpretations of the data.

While such a technique is clearly expected to be faster than applying a model test at each leaf of the tree, it is not guaranteed to find the correct interpretation. In particular, suppose that the correct interpretation has m real points of contact, but that the pairwise constraints allow an interpretation with $m+1$ non-null matches to pass through. This will cause *MAX* to be at least $m+1$, and may therefore prevent the search process from finding the correct interpretation. When the model test is applied after the generation of the interpretation tree, it will exclude the $m+1$ match, and this will cause the process to find no consistent interpretation.

Since the technique is much faster, it is of interest to know how often such a method will fail to find the correct answer. Thus, in our simulations of 3D overlapping parts, we have applied this faster, but less reliable technique, and recorded the percentage of cases in which no interpretation was found. The data reported in Table 2 illustrate the results of running 100 simulations for each of the ranges of error listed. As in the 2D case, once the interpretations together with their associated transformations, were found, they were clustered on the basis of their transformations, and the statistics of histogramming these clustered interpretations are reported. As well, the percentage of cases in which no interpretation was found are also listed.

We note that the algorithm appears to be less effective at finding unique interpretations of three-dimensional objects than of two-dimensional ones. In part this follows from the slightly weaker form of the constraints in three dimensions, especially when using only points of contact, rather than extended regions. As well, objects which exhibit partial symmetries (especially relative to the amount of error inherent in the sensory data) can frequently lead to multiple interpretations when using sparse sensor information. For example, for the case illustrated in Figure 9, if the sensory data all happen to lie on the block-like central portion of the object, and do not sample the projecting lip, the algorithm will discover several interpretations of the data, consisting of symmetric rotations of the object. Clearly, additional sampling of the object should reduce this ambiguity.

4.5. Edge Fragments from Gray-Level Images

A modified version of the algorithm described here has been applied to locating a complex object in cluttered scenes, using edge fragments from images obtained by a camera located (almost) directly overhead. The images are obtained under lighting from several overhead fluorescent lights. The camera is a standard vidicon located approximately five feet above the scene. The edge fragments are obtained by linking edge points marked as zero crossings in the Laplacian of Gaussian-smoothed images [20]. Edge points are marked only when the gradient at that point exceeds a predefined threshold; this is done to eliminate some shallow edges due to shadows. The algorithm is applied to some predefined number of the longest edge fragments.

This application requires extensions to the general method. One point to notice is that we have large edge fragments rather than small patches; therefore, we can use the length of the fragments as an additional local constraint. In our implementation, we do not assign edge fragments to model edges that are shorter than the measured fragment; we do assign small edge fragments to long model edges. More importantly, we could compute whole ranges of measurements from the edge fragments (as we do from model edges) rather than the single values from point-like patches we assume elsewhere. The constraints would then require that the measured range be contained in the model range. An easy way of approximating these stronger constraints is by treating the edge as two small patches located at endpoints of the edges, but constraining both patches to be assigned to the same model edge. Both of these approaches can be generalized to three dimensions.

The most difficult problem faced in this application is that we cannot reliably tell which side of the edge contains the object; that is, the edge normals can be determined only up to a sign ambiguity. Although region brightness can sometimes be used to separate figure from ground, it is not always reliable-- for example, a light-colored object on a dark background. The algorithm can be modified to keep track of the two possible assignments of sign and to guarantee that all the pairings in an interpretation have consistent assignments of sign. This approach, however, causes a noticeable degradation in the performance of the algorithm, since it reduces the pruning power of the constraints. Fortunately, we can use another form of the constraints to reduce the effect of this ambiguity.

As long as two edges do not cross or are not collinear, at least one edge must be completely within one of the half planes bounded by the other. This means that the components along one of the edge normals of all possible separation vectors will always have the same sign. Given a tentative pairing of two measured edge fragments and two model edges, we can use this property to pick the sign of one of the normals. The angle constraint between normals can then be used to consistently select the signs for other edges in that interpretation. Of course, the sign assignment is predicated on the initial pairing being correct, which it may not be, so we have lost some pruning power in any case.

We have also tested the algorithm in situations where the sign of the filtered image could be used to determine the edge normal reliably. The algorithm performs substantially better under these circumstances.

With or without the complete normal, the algorithm succeeds in locating the desired object in images where the edge data from any single object is very sparse (see Figures 1 and 2). To test the reliability of the algorithm on real data, we ran the following set of tests. A carton containing a total of eight parts selected from three different types of parts (two types are shown in Figures 1 and 2) was placed under a camera. The contents were arbitrarily perturbed to randomly orient and overlap the parts, and the recognition process was then applied. This process was repeated 100 times, and in each case an instance of a selected object was correctly identified and located in the image. The number of nodes of the interpretation tree actually explored in solving this problem was found to vary by up to an order of magnitude, depending on the difficulty of the image, but in all cases a correct interpretation was found.

The time to perform the recognition on this class of problems using the method described in this paper varies significantly, depending on the complexity of the problem, whether the complete normal is available, and whether the heuristic techniques of section 4.2 are employed. In all cases, the preprocessing to obtain edges is common. In our current implementation it takes approximately 30 seconds of elapsed time on a Symbolics 3600 Lisp Machine to process an image to obtain 80 edge fragments; this time, however, can be reduced by an order of magnitude by using existing hardware. We do not include this time in the times quoted below (all times are for our Lisp implementation).

When the sign of the normal is unknown and without using the Hough preprocessing, difficult cases such as in Figures 1 and 2 require a minute or more of matching time. In situations where the overlapping is slight, the matching time is closer to 30 seconds. This is almost twice as long as the performance of the algorithm on the same images when the sign of the normal is available. In this case, the typical matching time for lightly overlapped parts is around 10 to 15 seconds, with the worst-case times ranging from 30 seconds to minutes.

The effect of the Hough preprocessing is to make the recognition time nearly independent of the complexity of the scene. In our testing, we used the full set of x,y,η parameters for clustering the model/data edge-pairings. The Hough preprocessing itself takes on the order of seven or eight seconds for 80 data edges and 30 model edges. The recognition time after that is only from two to four seconds. The total recognition time is usually around 10 seconds. This is slightly longer than the time required by simple cases without the Hough preprocessing, but an order of magnitude better than the time required for the worst cases.

4.6. Range Data from Structured Light

We have applied our algorithm to locating a simple three-dimensional object in cluttered scenes, using relatively dense range data. The range data were obtained from a laser-striping system developed by Philippe Brou at our laboratory. The sensor is a vidicon camera above the scene. On either side of the camera there is a low-power laser at a known distance and angle from the camera. Each laser beam is converted into plane of light by a cylindrical lens; the resulting planes are scanned across the field of view by computer controlled mirrors. A Lisp machine identifies the xy position of points on the scene illuminated by the laser. The xy positions of these points are then used to compute the height above the plane of points intersecting the plane of light. Two lasers stripes, originating on opposite sides of the camera, are used to allow sensing in areas that would be shadowed for a single laser.

The data obtained from the ranging system are dense along each stripe, but the distance between stripes is under program control. The data used in our experiments (see Figures 3 and 4) were taken at a resolution of 0.03 cm in the vertical direction and 0.12 cm in the horizontal direction. The resolution in depth of our data is approximately 0.025 cm.[1]

Once the depth map is obtained, we preprocess the data to obtain planar patches. A least-squares planar fit is done at every data point. Regions are then formed whose normals are within a user-specified angle from some "seed" normal. Many techniques have been developed for obtaining planar regions for range data, e.g., [9]; any of these would also be applicable here.

As in the edge-fragment case described earlier, we can exploit our knowledge of the extent of the planar patches to more tightly constrain the matching process. We do this by selecting, within each planar region, four representative points that span the xy range of the region (see Figures 3 and

---

[1]The sensor has a depth resolution of about 1 part in 500 over a range of 12 cm.

4). The matching algorithm is applied to these representative points and their corresponding normals. As in the case of edge fragments, we require that all four points be assigned to the same model face.

In some cases, the algorithm will produce several very different interpretations that account for the same number of data points. In those cases some type of verification is required. Two simple types of verification tests available for range data are: (a) test that the computed position and orientation of the model does not have it penetrating the known support surface, and (b) make sure that there are no known patches whose xy projection lies on the localized object but whose z value is *less* than that indicated by the model. These tests are relatively easy to implement and are effective in many cases (Figure 10).

Our testing with the range data has been limited to a few objects, such as the wedge in Figure 10, in rather complex environments. The combined preprocessing and recognition time for these examples is approximately two minutes but, typically, only about 30 seconds of that is recognition time. It is the case, however, that the matching time grows fairly rapidly with the complexity of the model. This growth is due to the relative weakness of the three-dimensional constraints compared to the two-dimensional constraints. We expect that the Hough preprocessing will help reduce this problem, but we have yet to test the three-dimensional preprocessing.

4.7. Range Data from an Ultrasonic Sensor

Michael Drumheller [8] has developed a modified version of the algorithm described above and applied it to range data obtained from an unmodified Polaroid ultrasonic range sensor. The intended application is navigation of mobile robots. The system matches the range data obtained by a circular scan from the robot's position towards the walls of the room. The robot has a map of the walls of the room, but much of the data obtained arises from objects on the walls, such as bookshelves, or between the robot and the

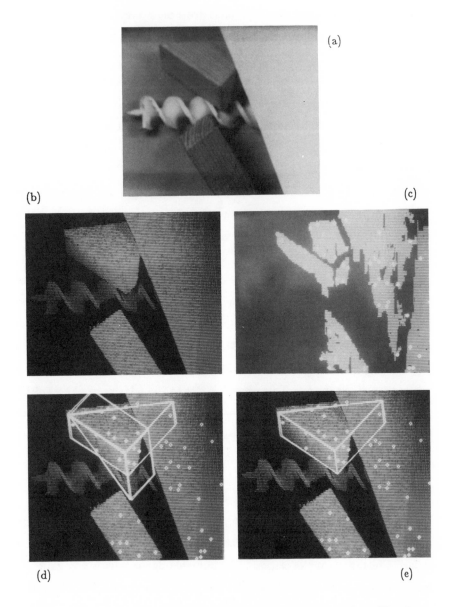

Figure 10: Three dimensional range data with partial verification.
(a) Original scene; (b) range data where brightness encodes height; (c)
planar patches with representative points; (d) legal interpretations
superimposed on range data (filled in circles are data points accounted
for); (e) interpretations that pass verification.

walls, such as columns. The algorithm first fits line segments to the range data and attempts to match these line segments to wall segments. After matching, the robot can solve for its position in the room.

5. Coupled Constraints

As we noted earlier, the decoupled constraints typically prune most of the non-symmetric interpretations of the data, but they are not guaranteed to reject all impossible interpretations. Consider Figure 11, for example. Consider matching point P_i to face f_u, point P_j to face f_v and point P_k to face f_w. These assignments are pairwise consistent, and the sections of the faces that are feasible locations for the sensed points are indicated by the sections labeled ij, etc. The assignment is not globally consistent, however, as indicated by the fact that the segments for face f_u and f_w do not overlap. Thus, since the points are pairwise consistent with the candidate faces, they are accepted as part of a feasible interpretation, even though clearly they are not. Using the decoupled constraints, it is only after the model test is applied to interpretations surviving pruning that all the available geometric constraints are exploited. For the case of a single object, this merely implies some inefficiency. For the case of multiple, overlapping objects, we may actually miss a correct interpretation. For example, a locally consistent (but globally inconsistent) interpretation of length M will cause us to ignore a globally consistent interpretation of length $m < M$. We would not discover our error until the model test is applied after pruning of the interpretation tree.

One solution is to interdigitate the model test with the tree generation stage. That is, whenever we reach a leaf of the interpretation tree, we apply the model test to ensure that the interpretation is globally consistent. If it is, then we update our global counter *MAX*, and continue. If it is not, then we continue our search with the current value of *MAX*. The problem with this method is that it may be computationally expensive. As we stated, the

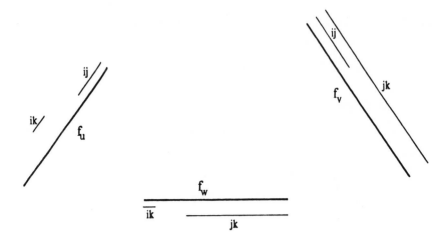

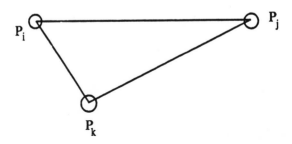

Figure 11: The constraints are decoupled. Consider matching point P_i to face f_u, point P_j to face f_v and point P_k to face f_w. These assignments are pairwise consistent, and the sections of the faces that are feasible locations for the sensed points are indicated by the sections labeled ij, etc. The assignment is not globally consistent, however, as indicated by the fact that the segments for face f_u and f_w do not overlap.

purpose of finding effective local constraints is to enable us to avoid applying an expensive model transformation, except when necessary.

An alternative solution is to find constraints that maintain global consistency without requiring an explicit model transformation. One such set of constraints is developed below for the two-dimensional case, and then extended to three dimensions.

5.1. The Coupled Constraints in Two Dimensions

Suppose we consider two edges of an object, oriented arbitrarily in sensor coordinates, as shown in Figure 12. With each edge we will associate a base

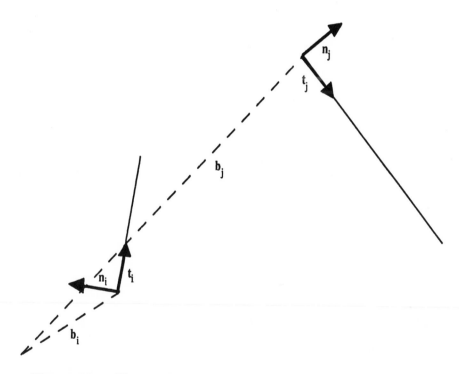

Figure 12: The constraints are recoupled. With each face, we associate a base vector $\mathbf{b}_i$, a tangent vector $\mathbf{t}_i$ and a normal vector $\mathbf{n}_i$. Then any point on a face can be represented by $\mathbf{b}_i + \alpha \mathbf{t}_i$ for some α between 0 and the length of the edge.

point, defined by the vector $\mathbf{b}_i$, a unit tangent vector $\mathbf{t}_i$, which points along the edge from the base point, and a unit normal vector $\mathbf{n}_i$, which points outward from the edge. Thus, the position of a point P_1 along edge f_i in this coordinate system is given by

$$p_1 = b_i + \alpha_1 t_i, \quad \alpha_1 \in [0, l_i]$$

where l_i is the length of the edge. Similarly, a point P_2 on face f_j can be represented by

$$p_2 = b_j + \alpha_2 t_j, \quad \alpha_2 \in [0, l_j].$$

The vector between two small measured patches is given by

$$p_1 - p_2 = d_{12} = b_i + \alpha_1 t_i - b_j - \alpha_2 t_j. \tag{1}$$

As in the earlier case, we know that we can measure d_{12}. Because of measurement error, however, the measured points P_1 and P_2 may not lie exactly on the object edges and as a consequence, what we can measure is

$$d_{12}^* = b_i + \alpha_1 t_i + u_1 - b_j - \alpha_2 t_j - u_2$$

where u_1 and u_2 are measurement errors whose size can be bounded. We can also measure the surface normal at the point P_1, say n_i^*, which in the case of perfect data would equal n_i. In general, we will only know that n_i^* is within some specified angle of n_i.

We can compute

$$d_{12}^* \cdot n_i^* = m_{12}$$

based on our measurements. We know m_{12} is an estimate of $d_{12} \cdot n_i$. We can compute bounds on the range of errors about the measured value so that we know that the true value of $d_{12} \cdot n_i$ lies in the range

$$\mathbf{d}_{12}\cdot\mathbf{n}_i \in [m_{12}-\epsilon,\, m_{12}+\epsilon]$$

where ϵ can be computed straightforwardly [14].

From Equation (1) we have

$$\mathbf{d}_{12}\cdot\mathbf{n}_i=(\mathbf{b}_i-\mathbf{b}_j)\cdot\mathbf{n}_i-\alpha_2(\mathbf{t}_j\cdot\mathbf{n}_i)\,. \tag{2}$$

The first term on the right is a constant and is a function of the object only, independent of its orientation. Thus, Equation (2) provides us with a constraint on the value of α_2. In particular, if $\mathbf{t}_j\cdot\mathbf{n}_i=0$, then this assignment of patches to faces is consistent only if

$$(\mathbf{b}_i-\mathbf{b}_j)\cdot\mathbf{n}_i \in [m_{12}-\epsilon,\, m_{12}+\epsilon]\,.$$

If this is true, then α_2 can take on any value in its current range. If it is false, then the assignment of these patches P_1,P_2 to these faces f_i,f_j is inconsistent and can be discarded.

In the more common case, when $\mathbf{t}_j\cdot\mathbf{n}_i \neq 0$, we have

$$\alpha_2\mathbf{t}_j\cdot\mathbf{n}_i \in [(\mathbf{b}_i-\mathbf{b}_j)\cdot\mathbf{n}_i-m_{12}-\epsilon,\, (\mathbf{b}_i-\mathbf{b}_j)\cdot\mathbf{n}_i-m_{12}+\epsilon]\,.$$

Thus, we have restricted the range of possible values for α_2 and hence the set of positions for patch P_2 that are consistent with this interpretation.

Similarly, by using the estimates for $\mathbf{d}_{12}\cdot\mathbf{n}_j$ obtained from the measurements, we can restrict the range of values for α_1 and, thereby, the position of P_1.

We can also consider the coordinate-frame-independent term

$$\mathbf{d}_{12}\cdot\mathbf{t}_i=(\mathbf{b}_i-\mathbf{b}_j)\cdot\mathbf{t}_i+\alpha_1-\alpha_2(\mathbf{t}_j\cdot\mathbf{t}_i)\,. \tag{3}$$

As before, we can place bounds on the measured value for $\mathbf{d}_{12}\cdot\mathbf{t}_i$ when error in the sensory data is incorporated. Then, given a legitimate range for α_1 we can restrict the range of α_2 and vice versa. A similar argument holds for $d_{12} \cdot t_2$.

These constraints allow us to compute intrinsic ranges for the possible assignments of patches to faces. The key to them is that we can propagate these ranges as we construct an interpretation. For example, suppose that we assign patch P_1 to face f_i. Initially, the range for α_1 is

$$\alpha_1 \in [0, l_i] \quad .$$

We now assign patch P_2 to face f_j, with

$$\alpha_2 \in [0, l_j]$$

initially. By applying the constraints derived above, we can reduce the legitimate ranges for these first two patches to some smaller set of ranges. We now consider adding patch P_3 to face f_k. When we construct the range of legal values for α_3, we find that the constraints are generally much tighter, since the legal ranges for α_1 and α_2 have already been reduced. Moreover, both α_1 and α_2 must be consistent with α_3, so the legal range for this patch is given by the intersection of the ranges provided by the constraints. Finally, the refined range of consistent values for α_3 may in turn reduce the legal ranges for α_1 and α_2, and these new ranges may then refine each other by another application of the constraints, and so on. In other words, the legal ranges for the assignment of patches to faces may be relaxed via the constraint equations, and in this manner, a globally consistent assignment is maintained. Of course, if any of the ranges for α_i becomes empty, the interpretation can be discarded as inconsistent without further exploration.

We thus have the basis for a second recognition and localization technique. As before, we generate and prune a tree of interpretations, by assigning sensed patches to faces of an object. Here there are two types of constraints. The first is that the angle between two sensed normals, modulo error in the sensor, must be consistent with the angle between the corresponding face normals, as in the previous case. The second involves the relaxation of mutual constraints on the range of positions on a face consistent with points of contact on those faces, as described above.

5.2. Extensions to Three Dimensions

The constraints derived in the previous section for the two dimensional case can be extended to three dimensions as well. In this case, we represent points on a face by

$$b_i + \alpha u_i + \beta v_i$$

where b_i is a vector to a designated *base* vertex of the face, and u_i and v_i are orthonormal vectors lying in the plane of the face. Furthermore, α and β are constrained to lie within some polygonal region, defined by the shape of the face. In the simplest case,

$$\alpha \in [0,A] \quad \beta \in [0,B] \quad .$$

Given two points of contact, P_1 and P_1, assigned to faces f_i and f_j, the vector between them is given by

$$d_{12} = b_i - b_j + \alpha_1 u_i + \beta_1 v_i - \alpha_2 u_j - \beta_2 v_j \quad .$$

As before, we can measure the component of this contact vector in the direction of the surface normals recorded at each patch. This leads, for example, to an estimate for

$$d_{12} \cdot n_i = (b_i - b_j) \cdot n_i - \alpha_2 (u_j \cdot n_i) - \beta_2 (v_j \cdot n_i) \quad . \tag{4}$$

As in the previous case, the term on the left hand side of Equation 4 is measurable and, given bounds on the errors in the sensor, actually defines a range

$$\mathbf{d}_{12} \cdot \mathbf{n}_i \in [l, h] \quad .$$

The first term on the right hand side of Equation (4) is a constant, defined independent of the coordinate system by the relationship between the two faces. Thus, Equation (4) essentially reduces to a linear constraint of the form

$$\alpha_2(\mathbf{u}_j \cdot \mathbf{n}_i) + \beta_2(\mathbf{v}_j \cdot \mathbf{n}_i) \in [(\mathbf{b}_i - \mathbf{b}_j) \cdot \mathbf{n}_i - h, \ (\mathbf{b}_i - \mathbf{b}_j) \cdot \mathbf{n}_i - l] \quad .$$

A similar expression holds for the other normal $\mathbf{n}_j$.

These constraints actually describe a region in a two-dimensional space spanned by α and β, as illustrated in Figure 13. Given a current polygonal region of consistency for α and β, we can intersect the region with this new range, to obtain a tighter region of consistency, as shown in the figure. Similar to the two dimensional case, as additional sensed patches are considered, the constraints they generate may be propagated among one another. If any polygonal region corresponding to a sensed patch vanishes, the interpretation is inconsistent and the procedure can stop exploring that portion of the interpretation tree.

Clearly, both the two-dimensional and the three-dimensional constraints developed here can be extended to deal with *null faces* in the same manner as the first algorithm.

5.3. Testing

We have tested the range propagation method on simulated data as well as on actual data. We will first describe these tests and then discuss our conclusions from these experiments.

We have done extensive testing of the algorithm with simulated two-

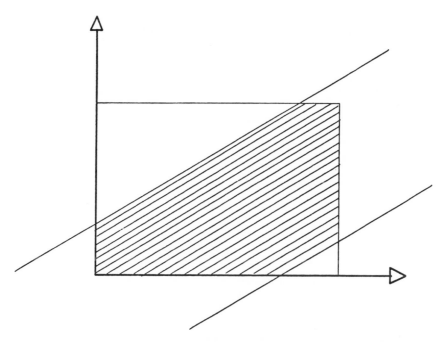

Figure 13: The intersection of legitimate face ranges for three dimensional data.

dimensional data of the type illustrated in Figure 8. As with the results reported in Section 4.2, a number of polygons, representing the outlines of parts, were overlapped at random, with random orientation, and with random offset relative to the origin of the sensor's coordinate system. The position and orientation of a predetermined number of data points were determined by tracing along equally-spaced rays from the origin of the coordinate system, and computing the outermost intersection of those rays with the polygon boundaries. The position and normal information were then corrupted by random errors designed to simulate the effect of imperfect sensors. The number of interpretations obtained under these sensing conditions, using the propagation of ranges of feasible contact, were very similar to those obtained in Section 4.2.

One of the interesting questions to consider for the range propagation technique is to what extent the implicit recoupling of the local constraints reduces the amount of explicit exploration of the interpretation tree. To test this, we ran the following set of simulations. For each of several ranges of sensing error, 100 simulations of the normal recognition and localization process were run. The number of nodes in the interpretation tree that were explicitly explored were recorded for each run. Then, using the same sensory data, a second run of 100 simulations was performed, now using the range propagation technique of Section 5. Again, we recorded the number of nodes of the interpretation tree that were explicitly explored. In Table 3, we record the median number of nodes explored, over the set of 100 runs, for both the normal and the range propagation techniques. Finally, we consider the ratio of the number of nodes explored using the range propagation technique, to the number of nodes explored for the same data using the normal technique. We record the minimum, mean, median and maximum values for this ratio, computed over the entire 100 runs, as well as the standard deviation of this distribution of ratios.

We note that, since the error ranges associated with each constraint differ between the normal constraints and the coupled constraints, it is possible for the coupled constraints to actually be less effective in removing portions of the interpretation tree. This is especially noticeable for large values of error in the surface normal. Overall, we see from the results shown in Table 3 that the average number of nodes explored in the interpretation tree is not significantly reduced from the normal method. Given the additional overhead associated with computing and intersecting the ranges of feasible positions along edges, it may not be worth while to use the range propagation method. We note that these results may differ when considering objects whose faces do not all form right angles with one another. To test this, we also ran a smaller set of simulations with a randomly constructed object, shown in Figure 15. The results, summarized in Table 3, suggest that this is not a critical factor.

6. Extensions

In this paper we have described a framework for a class of recognition algorithms. We considered two major variations depending on the class of constraints employed; some minor variations were employed in dealing with different types of sensors, notably grey-scale edges and laser range data. These variations, however, share many common assumptions as to the structure of the search for consistent matchings. We have assumed, for example, that we match some subset of the data elements against all the model elements at once; that we obtain all (longest) consistent interpretations; that the objects have comparable numbers of degrees of freedom and measurements. Beyond these algorithmic assumptions, we have preserved some assumptions about our domain. We have assumed, for example, that the data are made up of simple local measurements such as surface patches; that the model is made up of planar faces; that the dimensions of the objects are fixed and known *a priori*. All of these assumptions can be relaxed while retaining the characteristic flavor of the approach presented here. In this section we briefly explore these extensions. We have implemented all of these extensions with relatively minor modifications to the program code.

6.1. Partial Models - Features?

The size of the search space given an object with n faces (and the null face) and s sensed patches is $(n+1)^s$. We can reduce the size of the space by reducing the number of sensed patches and/or the number of model faces. If the input data are relatively dense, as for grey-scale edges, one interesting approach is to define small subsets of the model faces that are distinctive, and match them against all the available data patches. Because of the non-linear nature of the increase in search space size, it is usually more efficient to do two matches with half as many faces than a single match with all the faces. This may not be true, however, in cases where the pruning constraints are very efficient.

Having found one match for a partial model, the rest of the model can be used to predict where other faces may be found. The model test can be used to verify these predictions. One difficult problem encountered here involves deciding what counts as positive and negative evidence for evaluating a hypothesis, and how to combine this evidence to accept or reject a hypothesis. These decisions depend strongly on how reliable the sensor is and *a priori* expectations about the data.

The use of a partial model gains some of the combinatorial advantages of feature-based systems without requiring preprocessing of the data. Of course, this approach also inherits some of the disadvantages of feature-based systems. In particular, if the data are very sparse then there is a risk that none of the data corresponds to the selected model subsets.

Figure 14 shows some examples of matching edge data to partial models.

6.2. Choosing the First Interpretation

Most approaches to recognition seek a single interpretation of the data, either the first consistent one or the best under some measure. In contrast, the approach to recognition described in this paper has opted for finding all longest consistent interpretations of the data. This choice was made both for reasons of generality and to avoid having to define an absolute measure for "goodness of match". This approach may require considering substantially more pairings of data and model elements than an approach that settled for the first acceptable match.

We have experimented with a variant of the recognition algorithm that returns the first interpretation longer than a specified threshold that passes the model test. The difference in the number of search-tree nodes examined for this variant compared to the one that returns all interpretations is shown in Table 3. The table also indicates the number of times that the modified algorithm failed to find the correct interpretation while the original

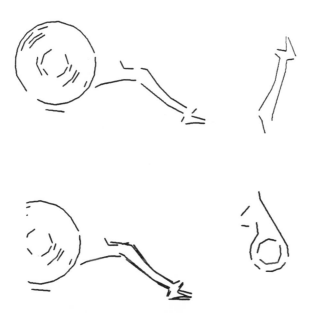

Figure 14: Examples of matches to partial models in grey-scale edges. (a) a set of grey-level edge fragments; (b) a partial model and part; (c) the interpretation of that partial model, overlapped with the original data; (d) a second partial model, for which no interpretation was found

algorithm did find it. A sample of the data used in these tests is shown in Figure 8.

We note that in the data reported for this case, we used a slightly different sensing technique. Each of the objects was rotated by some arbitrary amount, and then displaced by an arbitrary vector, whose magnitudes were less than some bound. Then a predefined number of equally spaced rays were traced from the origin of the coordinate system, and the outermost intersection of each ray with any of the objects was taken as a sensed point. The position and orientation at this point was then corrupted by noise, and these corrupted points formed the input to the recognition algorithm.

It can be seen from the data of Table 4 that, while the number of nodes actually explored in this first-acceptable mode is significantly reduced, it

No. of Nodes – Regular vs Range Propagation									
Object	Angle	Dist	Med	Med	Min	Mean	Med	Max	Dev
Lip	$\pi/10$	0.10	3630	3382	.334	.949	.908	2.479	.281
		0.05	2802	2528	.382	.915	.938	1.428	.157
		0.01	2351	2194	.199	.925	.924	2.286	.218
	$\pi/20$	0.10	2457	2323	.217	.954	.950	1.593	.196
		0.05	2044	1896	.488	.937	.948	1.318	.101
		0.01	2022	1729	.574	.954	.957	1.303	.094
Random	$\pi/10$	0.10	5276	3738	.048	.805	.738	3.757	.513
		0.05	3761	2682	.237	.744	.746	2.328	.231
		0.01	2642	1961	.164	.730	.716	2.070	.208
	$\pi/20$	0.10	1646	1360	.154	.863	.878	1.521	.227
		0.05	1251	1050	.155	.841	.868	1.405	.188
		0.01	1257	1081	.187	.872	.863	1.587	.180

Table 3: Comparison of statistics in the performance of the normal localization process, and the process in which the technique is terminated once an interpretation with a sufficient number of matched points is obtained. Each row lists parameters of a histogram, based on 100 trials with the object randomly oriented with 3 degrees of freedom. As well, the object was overlapped with one other object at random (similar results hold for more than one occluding object). In each case, 20 sensory data points were used. In the table, the *angle* column lists the angle of the error cone about the measured surface normal; the *dist* column lists the error range of the distance sensing. The next three columns list the minimum, median and maximum number of nodes actually explored in the interpretation tree for the normal case, taken over the 100 simulations. To test the effects of terminating the localization process, a cutoff of 10 matched points was used. The next three columns list the minimum, median and maximum number of nodes actually explored when the process was terminated in this manner. The next three columns list the minimum, median and maximum obtained by taking the ratio of the number of nodes explored in the truncated case over the number of nodes explored in the normal case. Finally the last column lists the percentage of cases in which terminating the process after 10 points were matched, caused an incorrect interpretation to be accepted. The process was run using the *lip* and *random* objects illustrated in Figures 9 and 15 respectively.

Obj	Angle	Dist	No. of Nodes – Regular vs First Acceptable									
			Min	Med	Max	Min	Med	Max	Min	Med	Max	Miss
Lip	$3\pi/20$	0.10	104	3516	15811	9	16	5358	0.004	0.01	0.34	0.44
		0.05	165	3581	12462	9	274	5113	0.007	0.08	0.46	0.34
		0.01	317	2912	15386	9	414	7795	0.009	0.14	0.56	0.21
	$2\pi/20$	0.10	287	2930	10092	9	172	5110	0.007	0.06	0.51	0.38
		0.05	196	2682	9541	9	137	4582	0.009	0.05	0.60	0.21
		0.01	317	2462	9129	10	466	5754	0.010	0.19	0.72	0.11
	$\pi/20$	0.10	186	2178	8614	9	112	3149	0.009	0.05	0.42	0.23
		0.05	249	1919	10302	10	337	4786	0.011	0.18	0.56	0.12
		0.01	243	1759	6817	9	358	3291	0.011	0.20	0.51	0.04
Rand	$\pi/10$	0.10	91	5285	23943	9	1326	9013	0.005	0.25	0.47	0.66
		0.05	91	3634	17176	10	1790	6965	0.011	0.47	0.57	0.22
		0.01	134	2713	9043	11	1635	6896	0.027	0.56	0.76	0.04
	$\pi/20$	0.10	182	1946	6408	9	787	2664	0.034	0.42	0.60	0.01
		0.05	481	1249	4104	10	586	2182	0.021	0.45	0.56	0.00
		0.01	402	1353	5301	11	416	2218	0.027	0.32	0.42	0.00

Table 4: Comparison of statistics in the performance of the normal localization process, and the process in which the technique is terminated once an interpretation with a sufficient number of matched points is obtained. Each row lists parameters of a histogram, based on 100 trials with the object randomly oriented with 3 degrees of freedom. As well, the object was overlapped with one other object at random (similar results hold for more than one occluding object). In each case, 20 sensory data points were used. In the table, the *angle* column lists the angle of the error cone about the measured surface normal; the *dist* column lists the error range of the distance sensing. The next three columns list the minimum, median, and maximum number of nodes actually explored in the interpretation tree for the normal case, taken over the 100 simulations. To test the effects of terminating the localization process, a cutoff of 10 matched points was used. The next three columns list the minimum, median, and maximum number of nodes actually explored when the process was terminated in this manner. The next three columns list the minimum, median, and maximum obtained by taking the ratio of the number of nodes explored in the truncated case over the number of nodes explored in the normal case. Finally, the last column lists the percentage of cases in which terminating the process after 10 points were matched, caused an incorrect interpretation to be accepted. The process was run using the *lip* and *random* objects illustrated in Figures 3 and 12, respectively.

appears to be at the expense of finding the correct answer. This is with a relatively high cutoff on the number of matches needed for an acceptable interpretation, in this case, 10 out of 20. Clearly, as we reduce this cutoff, the number of nodes explored will decrease, but the number of incorrect interpretations is liable to rise further.

There are several factors that can lead to the high percentage of incorrect interpretations. One possibility is that the symmetric nature of the object used leads to incorrect interpretations. To test this, we ran a similar set of simulations of the sort illustrated in Figure 15, using a non-symmetric object.

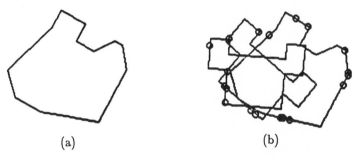

(a)   (b)

Figure 15:   Examples of data used when choosing the first acceptable interpretation. (a) a random, non-symmetric part; (b) an example of data taken from a number of overlapped parts, similar to Figure 8.

The results, indicated in Table 4, show that even in this case, a significant number of incorrect interpretations can still occur. This suggests that partial symmetries are not the only cause of incorrect interpretations when terminating on the first acceptable interpretation. The basic problem is that the length of an interpretation is not a reliable indicator of quality of match. Many of the matches, for example, may be from one part of the object and leave another part unconstrained.

6.3. Constrained Degrees of Freedom

We have assumed that if the objects are constrained to lie on a plane, then the data on each face are two-dimensional, and if the objects are completely unconstrained in position and orientation then the data are three-dimensional. In many applications, ·however, we can obtain three-dimensional data on objects constrained to be stably supported by a known plane, for example, a worktable. If we know the repertoire of the object's stable states, then we can exploit this knowledge as additional constraint to the matching process. Given a single data patch, the only candidate model faces for matching to it are those with similar values of the dot product between the face normal and the support plane normal. This constraint has the effect of drastically reducing the possible matches. This constraint is applicable even if we know that the object is not flush on the plane, but there is a known bound on its tilt relative to the plane.

6.4. More Distinctive Features

If distinctive features, such as the location of holes or corners, are readily available from the data, then the algorithm described here can still be applied to exploit the geometric constraints between the positions and orientations of these features. The resulting algorithm is similar in effect to the Local-Feature-Focus method [3].

6.5. Curved Objects

We have indicated that our algorithm is applicable to recognizing curved objects, approximated as a set of planar faces. The tree-pruning part of the algorithm can be extended readily to recognize objects composed of planar and curved faces from sparse point-like measurements. The decoupled constraints described in Section 3 merely compare measured values to the ranges of distances and dot products of normals between faces. These ranges can be computed for curved faces; the main difference is that one gets a range of dot products between normals, instead of a single dot product as with planar faces.

The difficulty comes when we try to solve for the position and orientation of the object given a candidate interpretation. The problem is that normal measurements are coupled to position on the face. Solving for orientation and position requires, in general, solving a numerical optimization problem. This process is slow and error prone. In fact, planar approximations to the object model are probably the best way to get a good initial guess for the numerical solution. Having found this solution, it is questionable whether the possible improvement is worth the cost of the numerical iteration.

We note that the examples of grey level edges, shown in Figures 1 and 2, illustrate the robustness of the technique in the presence of curved objects. The original object contains a number of slowly and rapidly curving edges. The model was constructed by sensing the object in isolation, and fitting straight line segments to the recorded grey level edges (zero crossings of the Laplacian of a Gaussian-smoothed image). As a consequence, the model is really an approximation to the actual object, as can be seen in Figure 1. When sensing the object in different positions, there is no guarantee that the same linear approximation to the edges will be obtained, and our experience is that the straight line approximations obtained under different sensing conditions are almost always somewhat different. Nonetheless, the recognition technique is robust enough to treat the differences as additional noise, and still recognize the object, as shown in Figure 1.

6.6. Free Parameters

We have assumed, throughout this paper, that models are metrically accurate, so that measured dimensions corresponded to model dimensions. This might not be true for two different reasons: we might be ignorant of some parameters in the sensing operation, such as viewing distance, or we might be dealing with variable objects, such as a family of motors. The general recognition approach we have described can be extended to deal with some of this variability. The basic idea is that for a match of a measurement to a model entity to be valid, we must make some assumptions about the values of all unknown parameters, such as object scale. All the matches in a (partial) interpretation must imply consistent values for the parameters, otherwise the interpretation can be pruned.

The application of this method to global parameters such as scale is straightforward. It might not be so evident that this also applies to model parameters such as the distance or angle between two faces; these will be expressed as functions of the model parameters. Applying the method above requires us to be able to invert these functions to solve for the range of legal parameters on the basis of tentative matches between data elements and model elements. The more complicated the relationship between model elements, the more difficult this is to do. In particular, angular variation requires solving trigonometric equations.

We have extended the recognition algorithm straightforwardly to allow for a linear scale factor, as illustrated in Figure 16. As might be expected, we find that the number of nodes of the interpretation tree actually searched is increased significantly from the comparable case of a known scale factor. This increase in the search space can be as large as an order of magnitude, depending on the amount of error inherent in the sensory data. The mean number of interpretations, given the same number of data points, is slightly higher in the case of an unknown scale factor than the case of a known one. Also, as shown in Figure 16, including an additional parameter in the recognition process may lead to multiple interpretations, in which different values of the parameter lead to different feasible interpretations.

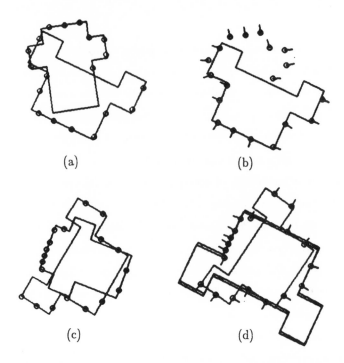

(a)   (b)

(c)   (d)

Figure 16: Examples of parameterized two-dimensional models. A scaled version of one of the models is intermixed with another model. The recognition algorithm correctly identifies the object, and determines its scale factor as well as its position and orientation. (a) a set of sampled data; (b) the interpretation of that data; (c) a second set of sampled data; (d) indicates that several interpretations of the data may be feasible

7. Recognition as Search

In our view of recognition as a search for a match between data elements and model elements, the crucial questions are the size of the search space and the constraints present between assignments. Much of the variation between existing recognition schemes can be accounted for by the choice of descriptive tokens to match. Some methods rely on computing a few very distinctive descriptors (*features*) that sharply constrain the identity and/or location of the object. Others use less distinctive descriptors and rely more on the relationships between them to effect recognition and localization.

The use of a few distinctive features sharply constrains the size of the search space. The resulting interpretation tree is very narrow and the search process can be very efficient. As an extreme example, to recognize a soft drink can from visual data, we could process the image to obtain the UPC bar code, which would uniquely identify the type of can. Moreover, knowing the position of the UPC label on the can and in the image would allow us to determine the position and attitude of the can in the scene. Of course, not all features will be as distinctive as a UPC label. Simpler examples might include corners, holes, notches and other local features. The idea is that very few such distinctive features should be needed to identify the object, and the search space can be effectively collapsed. Examples of techniques in this vein include the use of a few extended features [24], [1], and the use of one feature as a focus, with the search restricted to a few nearby features [28], [15], [27], [3], [4].

The approach of using a few distinctive descriptors is common to many commercial systems (see, for example, [2], [12], [19], [25]). These systems characterize both the measurements and models by a vector of global properties of binary images, such as area, perimeter, elongation and Euler number. Because of their global support, these descriptions do not extend well to overlapping or occluded parts.

Another type of recognition method relies on building elaborate high-level descriptions of the measured data before matching to the model. These approaches also rely on reducing the size of the search space by matching on a few distinctive descriptors. Examples of this approach include, for example, [22], [23], [21], [6], [5].

Approaches that rely on a few distinctive features have some weaknesses. First, the cost of the search has been greatly reduced, but at the expense of global preprocessing of the sensory data. Sensors do not provide distinctive descriptors directly; the descriptors must be computed from the mass of local data provided by the sensor. In some sensing modalities, such as tactile sensing, searching for data to build distinctive descriptors can be very time consuming. Second, heavy reliance on a few features can make recognition susceptible to measurement noise. If the imaging device is out of focus, for example, so that the image of the UPC bar code is blurred significantly, recognition may be altogether impossible. In this case, degradation in the presence of error is not graceful. Third, useful features are by definition sparse or they cease to be distinctive; this sparsity may be a problem when dealing with occlusion. In our UPC label example, if some other object occludes the UPC bar code from the sensor, we will not be able to recognize the can. This may occur even though virtually all of the rest of the can is available to the sensor.

An alternative approach to recognition relies more on the geometric relationships between simpler descriptors, rather than on a few distinctive features. The idea is that these descriptors are densely distributed and not particularly distinctive taken individually; for example, surface normals fit into this category. In these circumstances, the search space is large and constraints to prune it are critical. While the size of the search space explored by these methods will be larger than in the feature-based methods, the expectation is that the individual tests are very efficient. Representative examples of such schemes include [16], [17], [18], [10], [11], [14], [26], [7].

The key difference between matching on these low-level descriptors and on distinctive features lies in the availability of descriptors. The simpler sensor measurements are likely to be dense over the object. As a consequence, recognition schemes based on such simple measurements should be applicable to sparse sensors, and should be less sensitive to problems of occlusion and sensor error, since an input description can always be obtained and matched to the model. In this paper, we explore a recognition scheme that uses very simple sensor primitives that can be computed over the entire object. We rely on the power of geometric constraints to keep the combinatorics of the search process reasonably controlled.

8. Summary

We have presented a recognition technique based on a search for consistent matches between local geometric measurements and model faces. The technique offers a number of advantages: it is very simple yet efficient; it can operate on sparse data; it is applicable to a wide range of sensors and choice of features; it degrades gracefully with error. In addition to the advantages of the particular technique, the framework within which it has been developed has proven useful both to analyze expected performance of this method and to model a number of other methods. In summary, we believe the approach described here represents a useful tool in a wide variety of recognition situations.

Acknowledgments

This report describes research done at the Artificial Intelligence Laboratory of the Massachusetts Institute of Technology. Support for the Laboratory's Artificial Intelligence research is provided in part by a grant from the System Development Foundation, and in part by the Advanced Research Projects Agency under Office of Naval Research contracts N00014-80-C-0505 and N00014-82-K-0334.

The idea of using a null face to handle multiple objects was first suggested to one of us (TLP) by V. Milenkovic of CMU; we are very thankful for his remark. The image processing was done on a hardware/software environment developed by Keith Nishihara and Noble Larson. We thank Philippe Brou for kindly providing the laser ranging system with which we obtained the data reported in Figures 3 and 4. We also thank Dan Huttenlocher for his comments on an earlier draft.

A revised version of this paper has been submitted for publication in IEEE Transactions on Pattern Analysis and Machine Intelligence.

References

[1]   Ballard, D.H.
      Generalizing the Hough transform to arbitrary shapes.
      Pattern Recognition 13(2):111-122, 1981.

[2]   Bausch and Lomb.
      Bausch and Lomb Omnicon Pattern Analysis System
      Analytic Systems Division, Rochester, NY, 1976.

[3]   Bolles, R. C. and Cain. R. A.
      Recognizing and locating partially visible objects: The local-feature-
          focus method.
      International Journal of Robotics Research 1(3):57-82, 1982.

[4]   Bolles, R. C., Horaud, P., and Hannah, M. J.
        3-DPO: A three-dimensional part orientation system.
        In *First International Symposium of Robotics Research*. Note:
            Also in Robotics Research: The First International Symposium,
            edited by M. Brady and R. Paul, MIT Press, 1984, pp. 413-424,
            Bretton Woods, NH, 1983.

[5]   Brady, M.
        Smoothed local symmetries and frame propagation.
        In *Proc. IEEE Pattern Recog. and Im. Proc.*. 1982.

[6]   Brooks. R.
        Symbolic reasoning among 3-dimensional models and 2-dimensional
            images.
        Artificial Intelligence 17:285-349, 1981.

[7]   Brou, P.
        Finding the orientation of objects in vector maps.
        Int. J. Rob. Res. 3(4):89-175, 1984.

[8]   Drumheller, M.
        Robot localization using range data.
        S.B. thesis, MIT, Dept. of ME, 1984.

[9]   Faugeras, O. D., Hebert, M., and Pauchon. E.
        Segmentation of range data into planar and quadratic patches.
        In *CVPR*, pages 8-13. 1983.

[10]   Faugeras O.D., Hebert, M.
        A 3-D recognition and positioning algorithm using geometrical
            matching between primitive surfaces.
        In *Proc. Eighth Int. Joint Conf. On Artificial Intelligence*, pages
            996-1002. Los Altos: William Kaufmann, August, 1983.

[11]   Gaston, P.C. and Lozano-Perez, T.
        Tactile recognition and localization using object models.
        Technical Report AIM-705, MIT Artificial Intelligence Laboratory,
            1983.
        Also in IEEE Trans. Pattern Anal. Machine Intell., PAMI-6, No. 3,
            pp 257-265.

[12]   Gleason, G. and Agin, G. J.
        A modular vision system for sensor-controlled manipulation and
            inspection.
        In *Proc. Ninth Int. Symp. Industrial Robots*, pages 57-70. Society
            of Manufacturing Engineers, Dearborn, MI, March, 1979.

[13]   Grimson, W. E. L.
       The combinatories of local constraints in model-based recognition
          and localization from sparse data.
       Technical Report MIT AI Lab Memo 763, MIT, March, 1984.

[14]   Grimson, W.E.L. and Lozano-Perez, T.
       Model-based recognition and localization from sparse range or tactile
          data.
       International Journal of Robotics Research 3(3):3-35, 1984.

[15]   Holland, S. W.
       A programmable computer vision system based on spatial
          relationships.
       Technical Report GMR-2078, General Motors, Detroit, February,
          1976.

[16]   Horn, B. K. P.
       Extended Gaussian images.
       Technical Report AIM-740, MIT, AI Lab., 1983.

[17]   Horn, B. K. P. and Ikeuchi, K.
       Picking parts out of a bin.
       Technical Report AIM-746, MIT, AI Lab., 1983.

[18]   Ikeuchi, K.
       Determining attitude of object from needle map using extended
          Gaussian image.
       Technical Report AIM-714, MIT, AI Lab., 1983.

[19]   Machine Intelligence Corp.
       Model VX-100 machine vision system
       Mountain View, CA, 1980.

[20]   Marr, D. and Hildreth, E. C.
       Theory of edge detection.
       Proc. R. Soc. London B 207:187-217, 1980.

[21]   Marr, D. and Nishihara, K.
       Representation and recognition of the spatial organization of three-
          dimensional shapes.
       Proceedings of the Royal Society of London B200:269-294, 1978.

[22]   Nevatia, R.
       Structured descriptions of complex objects.
       In *Proc. 3rd Int. Joint Conf. AI*, pages 641-647. Stanford, CA,
            Aug., 1973.
       Also Ph.D. thesis, Stanford Univ.- AI Lab, 1974 (AIM 250).

[23]   Nevatia, R. and Binford, T. O.
       Description and recognition of curved objects.
       Artificial Intelligence 8:77-98, 1977.

[24]   Perkins, W. A.
       A model-based vision system for industrial parts.
       IEEE Trans. Comput. C-27:126-143, 1978.

[25]   Reinhold, A. G. and VanderBrug, G. J.
       Robot vision for industry; the auto-vision system.
       Robotics Age Fall:22-28, 1980.

[26]   Stockman, G. and Esteva, J. C.
       *Use of geometrical constraints and clustering to determine 3-D
            object pose.*
       Technical Report TR84-002, Michigan Stat Univ., Dept. of CS, 1984.

[27]   Sugihara, K.
       Range-data analysis guided by a junction dictionary.
       Artificial Intelligence 12:41-69, May, 1979.

[28]   Tsuji, S. and Nakamura, A.
       Recognition of an object in a stack of industrial parts.
       In *Proc. 4th Int. Joint Conf. Artificial Intell.*, pages 811-818.
            Tbilisi, USSR, Sept., 1975.

PART IV: SYSTEMS AND APPLICATIONS

Producing Space Shuttle Tiles
with a 3-D Non-Contact Measurement System

T. E. Beeler[1]

Boeing Company
Seattle, Washington

Abstract

Replacing damaged or missing tiles, or custom-fitting new tiles on space shuttle vehicles has been a labor-intensive operation. It is time consuming and can increase the interval between successive shuttle missions. Since the cost of having space shuttle vehicles in a non-productive status is enormous, the shuttle manufacturer has aggressively sought ways to speed up production of these tiles.

This article describes the Laser Cavity Digitizing System (LCDS), developed by Technical Arts Corporation, which dramatically reduces the time required to produce tiles. The LCDS is a precision non-contact 3-D measurement system used to measure a tile cavity (a space left by a missing tile). The system produces a mathematical model of a tile which is sent to an N/C postprocessor. A milling machine then produces a new tile.

1. Background

The structure of space shuttle vehicles is enclosed in a thermal protection system capable of withstanding re-entry temperatures in excess of 2300 degrees F. This thermal protection system consists mostly of tiles made of a nearly pure silica material weighing between 8 and 22 pounds per cubic foot. There are over 23,000 tiles on each vehicle, few of which have exactly the same dimensions. Over 95% of the tiles are generally rectangular in shape with planar sides and curved upper and lower surfaces which match the spacecraft body contours. Tiles range in thickness from 0.50 to 4.75 inches

---

[1]formerly of the Technical Arts Corp., Redmond, Washington

depending on the maximum temperatures they are expected to encounter during re-entry. Their lengths vary from 3 to 8 inches. Roughly 70% of the tiles are pre-manufactured in arrays of 20 or more which are shipped to the final assembly site where they are bonded to the spacecraft. Since the process of bonding the arrays to the spacecraft cannot locate the tiles precisely, long rows of spaces are left for the last tiles which must be custom made to correct for the accumulated error in installing the pre-manufactured tiles. These tiles are called 'close-out' tiles. Each shuttle has some 4000 close-out tiles which must be custom manufactured in a slow manual process. In addition to close-out tiles, some tiles on existing shuttle vehicles must be periodically replaced using the same manual process. These tiles are replaced due to thermal or mechanical damage or engineering modifications. Figures 1 and 2 show a typical row of close-out tiles and a tile cross section, respectively.

The manual method of measuring and making patterns for close-out and replacement tiles is incongruous with the technological progress symbolized by the space programs, and lengthens the time required to refit a shuttle vehicle for successive missions.

Since the problem was identified, the shuttle manufacturer has sought to use non-contact measurement techniques for making tiles. The Laser Cavity Digitizing System (LCDS), manufactured by Technical Arts Corp. is such a system. It can reduce the time required to produce a tile from 13 hours to 90 minutes. Subsequent sections describe the LCDS, how it is designed, and the theory behind its operation.

2. General Description of LCDS

The LCDS produces mathematical models of close-out and replacement tiles. It transcribes a six-sided cavity into digital information in a 3-dimensional rectangular coordinate system. This information is used by a postprocessor to produce numerical control (N/C) programs to manufacture

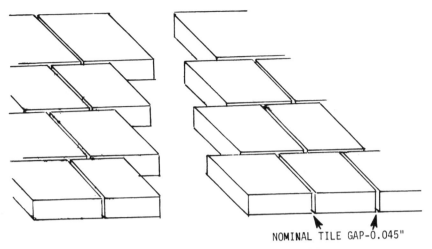

NOMINAL TILE GAP-0.045"

Figure 1: A Typical Row of Close-Out Tile Cavities

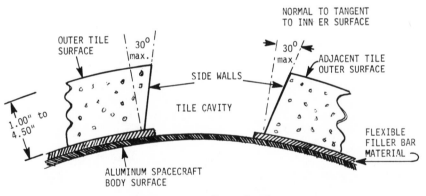

Figure 2: Typical Tile Layout Cross-Section

tiles. The system includes a control console, a microcomputer subsystem, and a sensor unit (scan head).

The control console, Figure 3, consists of a Visual-500 terminal interfaced to the microcomputer and a CRT which monitors the video output from the scan head. It also includes a movable metal desk which has a dust-free compartment for the microcomputer subsystem.

The microcomputer subsystem consists of an Intel 8086 CPU with an 8087 floating point processor, 1048 Kbytes of RAM, and a 20 megabyte Winchester disk. A digitizer card provides the interface between the microcomputer and the cameras in the scan head. The microcomputer also

Figure 3:   LCDS Control Console

includes a card which controls laser positioning, and a module for communications with a Data General Eclipse postprocessor. The system operates under the iRMX-86[2] operating system with most of the software written in Pascal and PL/M-86. The scan head, Figure 4, contains five pairs of cameras and lasers mounted rigidly in the scanning unit. Each laser projects a beam of light which is spread into a plane by an oscillating mirror mounted to a galvanometer scanner. The planar beam is projected to the cavity and is positioned by a mirror attached to a second galvanometer scanner. Each camera senses the light projected on the tile cavity by its associated laser and this information is transmitted directly to the digitizer where it is encoded in digital form. The scan head is connected to the control console by a 50 foot cable allowing it to be moved freely under the spacecraft. Figure 5 shows the interrelationships of the principal components of the LCDS.

3. Overview of LCDS Operation

Each camera/laser pair acts as a separate sensing unit under control of the microprocessor, and is used to collect data from a specific region of the cavity. Four of the camera/laser pairs are used to collect data from the sides of the cavity and portions of the outer surface of the adjacent tiles. A fifth pair collects data from the bottom (inner surface) of the cavity.

Each planar laser beam may be positioned at from 20 to 40 locations on the cavity at spacings from 0.10 to 0.25 inches apart. When the laser is placed at each position, its corresponding camera will transmit a video image to the microprocessor subsystem. For a sidewall camera, the image will look like Figure 6.

For each laser position, the camera will digitize as many as 240 data points from the laser image in a period of 32 milliseconds: one point for each raster line of the camera. These points are stored on the Winchester disk

---

[2] iRMX-86 and PL/M-86 are trademarks of Intel Corporation.

Figure 4:   LCDS Scan Head (Side Covers Removed)

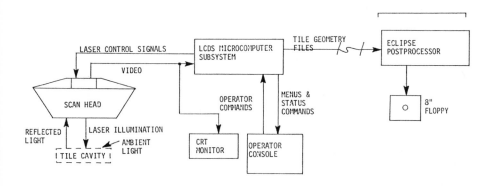

Figure 5: Top-Level Context Diagram

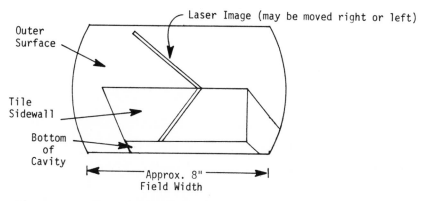

Figure 6: Typical Data Image Seen by Sidewall Camera

immediately after collection. This is repeated for each laser position, resulting in about 4,000 data points for each camera/laser pair.

When all the data are collected, the scans are repeated and a second set of data is obtained. Data from the first and second scans will be identical except for system repeatability errors and possible mechanical movement of the spacecraft or scan head. Any corresponding data points from the first and second scan differing by more than a predetermined amount (between 0.003 and 0.008 inches) are rejected. The data are also subjected to other validation checks to ensure that erroneous data caused by ambient light and other error sources are rejected.

When data validation is completed, the microcomputer performs a task known as 'cavity analysis'. First, all of the valid data are converted from the individual coordinate systems of each camera into a single rectangular coordinate system. When this is done, a parsing algorithm separates the data from each of the tile cavity surfaces into distinct groups:
1. One for each of four cavity sides
2. One for the cavity bottom (inner surface)
3. One for the outer surface of the adjacent tiles

When the data parsing is complete, a least-squares fit is performed on each sidewall data set to obtain the equation of a plane best fitting the sidewall surface. Similarly, a least-squares fit is performed on the inner surface data to calculate the best fit of a quadratic surface. This surface is represented by the equation

$$Z = C_0 + C_1 X + C_2 Y + C_3 X^2 + C_4 XY + C_5 Y^2 \ .$$

The outer surface of the missing tile is assumed to be a continuous extension of the outer surface of the adjacent tiles. This surface is therefore determined by fitting the data from the outer surface of the adjacent tiles to a quadratic equation using the least squares method.

When all six surfaces of the cavity are determined, their intersections are calculated by solving sets of simultaneous equations. These points are the corners of the missing tile. They are saved in files for later transmission to the postprocessor. Files containing points closely spaced along the tile perimeter are also generated. Finally, the cavity analysis task computes a fine grid of points for both the top and bottom surfaces of the tile. These points will determine the path of a milling machine tool as it cuts a tile from a block of silica material.

The final activity performed by the LCDS is to transmit the files containing corner points, edge points, and surface grid points to the postprocessor system. When the postprocessor has received the data, it can then produce a floppy disk containing the commands to cause an N/C milling machine to produce a replica of the missing tile. The total time consumed in this process is as follows:

Set-up activities	25 min.
Scanning and validation	15 min.
Cavity analysis	5 min.
Data transmission(4800 Baud)	5 min.
Postprocessor activity	15 min.
Milling of tile material	25 min.

This section has provided an overview of how the LCDS works. A more detailed understanding requires some knowledge of the Technical Arts 100X White Scanner. The 100X is described in the next section. Section 5 will return to a discussion of the LCDS system design.

4. How the 100X White Scanner Works

4.1. Basic Concepts

The 100X White Scanner is a structured-light 3-D measurement device. The light source projects a plane of light which illuminates only a profile of an object lying in its path. The remainder of the object remains unilluminated. Figure 7 illustrates a situation like this. When the object is moved slightly up or down along the Z-axis, the viewer will perceive a movement to the right or left in his field of view. If the viewer had 'calibrated' his field of view with an object of known dimensions and shape, and if he knew the basic geometry of the situation (D,λ, and γ in Figure 7), he could use simple trigonometry to deduce the X and Z coordinates of any portion of the object illuminated by the plane of light. The Y coordinate of any point can be obtained by equally simple considerations. Referring to

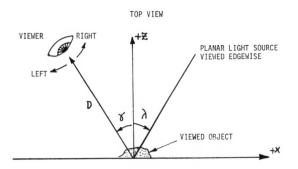

Figure 7:   Viewing an Object Illuminated by a Planar Beam

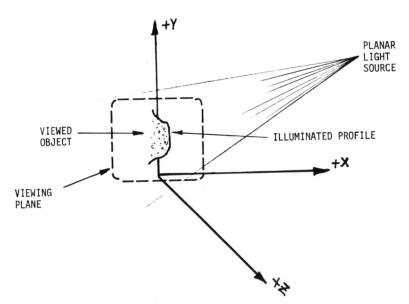

Figure 8:   Perspective as Seen by Viewer

Figure 8, a perspective view, one can see that the Y coordinate of a point may be obtained by noting its angular displacement upward or downward in the field of view and transforming this angular displacement into a Y value. This transformation will have been previously determined during a calibration process.

This is the essence of 3-D measurements with a 100X Scanner. At any time we are actually measuring the coordinates of a two-dimensional subset of 3-space. In order to measure all of the 3-space coordinates of an object, a new element must be introduced. Moving the object incrementally by known amounts along the X-axis and obtaining a profile view of the object for many X positions will provide a true 3-dimensional data set describing the object. Alternately, multiple profile views of an object can be obtained by leaving the object stationary and directing the planar light source to many positions on the object.

The same principle is used by other optical measurement systems using an even more constrained light source which illuminates only a single point on an object. This type of system was not appropriate for the LCDS because of the need to collect large amounts of data in a short time. The following sections describe the specific operating principles of a 100X system.

4.2. Principal 100X Components

The 100X is comprised of three major components; a sensor, a digitizer and a microcomputer.

4.2.1. Sensor

The sensor (Figure 9) consists of a helium-neon laser light source, a General Electric video camera, a galvanometer scanner (spreader galvo) which spreads the light into a plane, and galvanometer scanner (director galvo) which moves the plane of light laterally in the camera's field of view.

A system having galvanometer scanners will also have a Laser Control System (LCS) card installed in the microcomputer. Under control of the microcomputer, the LCS will cause the spreader galvo to oscillate, producing a plane of light of the desired amplitude and vertical location (Y direction) in the camera's field of view. The LCS will also control the director galvo to position the laser beam laterally (in the X direction). The accuracy to which the director galvo can be positioned is crucial to the accuracy of X and Z

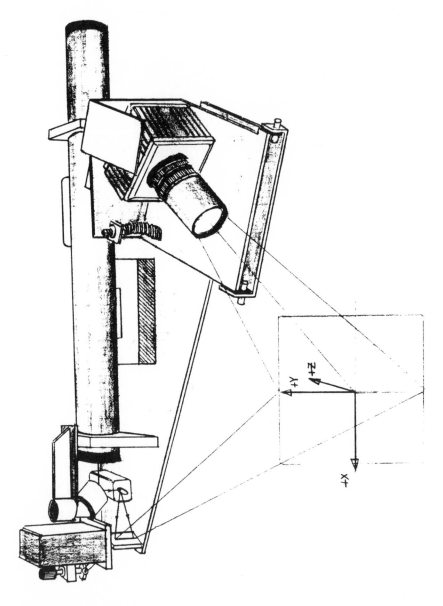

Figure 9:   100X Sensor

measurements, and it is currently accurate to 18 arc seconds, or 90 microradians.

The video camera is a GE model 4TN-2505 which uses a CID (charge injection device) imager array having a pixel grid of 248 rows by 388 columns. Critical to system accuracy is its·ability to locate the center of a light pulse on each raster line (row of pixels). Since there are only 388 pixels in a row, one might assume that the resolution would be only 1/388 of the camera's field of view (FOV). But if the light pulse is spread over 4 or more pixels, the digitizer can locate its center to an accuracy exceeding 1/20,000 FOV.

4.2.2. Digitizer

The digitizer is a single circuit card installed in the microcomputer. It processes video signal input from the camera and produces information describing the location of each light pulse sensed on each row of pixels (Figure 10). The digitizer can be described as a separate component of the

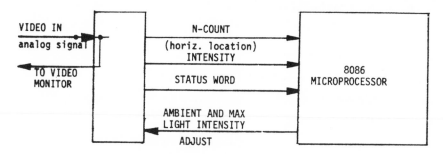

Figure 10: Digitizer Interfaces

system because it performs a function unique to the 100X. It is a continuously operating device, collecting data without being externally driven by the microcomputer. One data point is generated for each of the 240 raster lines, resulting in a minimum collection rate of 14,400 points per second. A data point is defined to be the location of the center of the light pulse on a raster line, expressed as a pair of numbers. The first of these two numbers is the coarse location of the pulse expressed as an integer called "coarse-N". The second number is a correction factor which is applied to "coarse N" to yield a floating point "fine N" with 24 bits of resolution.

Since the digitizer might confuse ambient light with the light produced by the laser, some gain adjustment is made to compensate for ambient light. The digitizer gain is also adjusted depending on the maximum light intensity expected. This adjustment is made by the microprocessor which outputs two numbers, "peak" and "threshold", to the digitizer for each raster line. These numbers would have been generated by previously scanning the background image.

In addition to the N-count value, the digitizer produces an integer number containing the intensity of the light pulse and another word containing status bits which allow the validity of the data to be determined. The status bits may contain any of the following information:
- video blank line
- no light pulse
- good data
- more than one light pulse
- data too bright

4.2.3. Microcomputer

The microcomputer subsystem, essentially the same as that described in Section 2, is the third major component of the 100X. Of its megabyte of RAM, 300 Kb is dedicated to the "100X Library" which remains resident during system operation. The 100X Library, written in PL/M-86, performs all of the software functions essential to the scanner:
- set-up and initialization
- calibration
- scanning
- adjusting for background light levels
- transmitting data to application program or host computer
- receiving data from application program or host computer

The 100X Library may be used in any of three modes:
1. a stand-alone mode stimulated by commands from the operator console

2. stimulated by an application program resident on the 100X system
3. stimulated by commands received from an interface to a remote host computer

The 100X microcomputer may also be operated with the 100X Library inactive. It has 750 K of memory available for use by the iRMX-86 operating system and a large application program, a language processor (Pascal, PL/M, etc.), or a utility such as a text editor. The 100X can thus be used to develop its own applications and may be operated with a printer to produce source listings. Figure 11 is a block diagram which describes the logical organization of the 100X system.

4.3. Linearization and Calibration

Two elements crucial to the accuracy of the 100X system are camera linearization and calibration.

Linearization is the process by which aberrations in the camera lens and its CID imager array are corrected. The linearization process produces a data file which, in conjunction with the 100X Library, maps every pixel in the camera's focal plane to a perfectly flat coordinate system. If the lens were optically perfect and the camera imager array had no imperfections, the linearization file would be an identity map (no correction necessary). The present linearization process corrects a camera/lens pair to an accuracy of 1/7500 FOV (field of view). However, since the linearization process itself uses the 100X system, its accuracy is limited by errors in the remainder of the system. With future enhancements to the linearization process, it is not unreasonable to expect linearization accuracy to approach 1/20,000 FOV.

The calibration process determines a coordinate transformation which maps any pair of coordinates in the camera's focal plane to a Cartesian 3-space coordinate system located in or near the viewed object. The calibration process involves placing an object of precisely known geometry in the camera's field of view. This object is called a calibration gauge, or

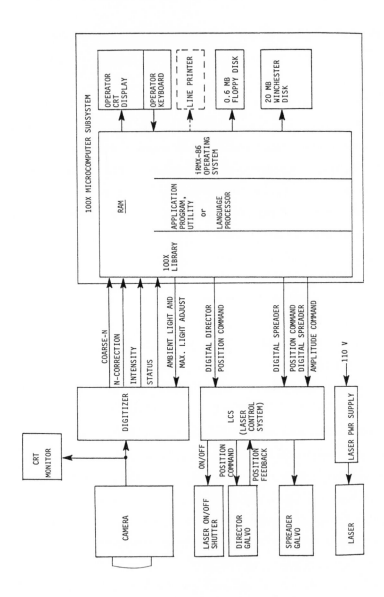

Figure 11: Logical Organization of the 100X system

standard. The calibration gauge used in the LCDS has a cross-sectional shape as shown in Figure 12(a). The 100X coordinate system on the calibration gauge is established as in Figure 12(b). Calibration is valid when the laser strikes the gauge parallel to the Y-axis, which is vertical in the camera's field of view. The X-axis must also be horizontal in the camera's field of view. Deviations in camera and laser alignment are called "pitch" and "roll"; they are calculated by the 100X calibration program and, if they exceed prescribed limits, vernier adjustments are made to the camera mount or director galvo mount to reduce them to acceptable values.

When pitch and roll are acceptable (less than 1 degree is usually adequate), the 100X will scan the gauge and collect data from the gauge as shown in Figure 12(c). A least-squares fit is then performed on each of the straight line segments of the gauge image. The angular relationships of these straight line segments and geometry of the gauge, previously input by the operator or application program, provide sufficient information to allow the 100X Library to calculate the transformation from camera coordinates to 3-space. This nonlinear transformation may be stored in RAM or disk for later use.

When the calibration gauge is moved from the camera's field of view, the coordinate system will remain; measurements of any object will be expressed in that coordinate system.

This completes the description of linearization and calibration. Figure 13 shows how these functions integrate with other 100X functions to produce 3-D measurements. Section 5 will describe how the LCDS is designed using the 100X system as its basis.

5. LCDS System Design

The LCDS is basically a 100X system with five sensors. The mechanical, electrical, and software design are discussed in this section.

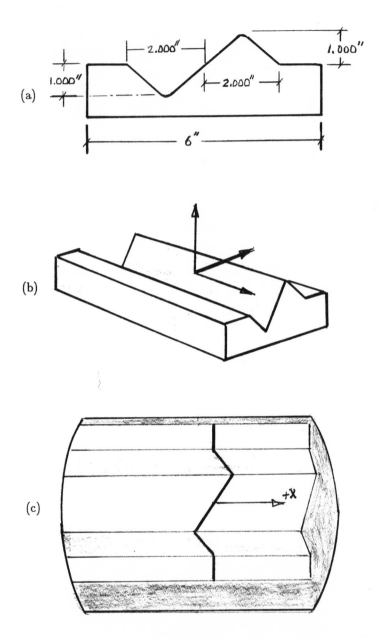

Figure 12: (a)   Gauge Cross-Section;   (b)   Calibration Gauge
Coordinate System; (c) Camera's View of Laser Illuminating Gauge

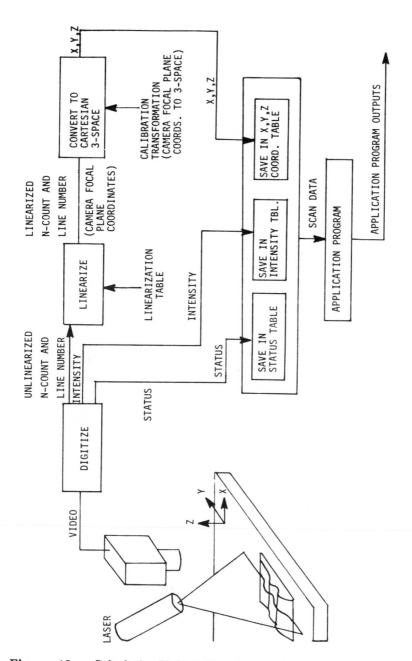

Figure 13: Calculating X-,Y-,Z-Coordinates: Data Flow Diagram

5.1. LCDS Mechanical Design

Perhaps the principal distinguishing mechanical feature of the LCDS is the design of the scan head. The mechanical design is dictated by the geometry of the tile cavities and the need for near perfect structural rigidity in a lightweight structure. The primary structure of the scan head is a truncated pyramid made of tubular aluminum. Figure 14 is a simplified cross section of the scan head. Four of the sensors are oriented to take data from the cavity sidewalls and a fifth sensor is oriented to take data from the inner surface.

Each sidewall sensor is a separate sub-assembly as shown in Figure 9, fastened securely to the tubular aluminum structure. The scan head components must remain rigid to ensure accuracy; a 0.001 inch movement in parts of the scan head can cause measurement errors of 0.004 inches. The tubular structure must be able to support the weight of the scan head, 130 pounds, and that of the calibration fixture, which weighs 43 pounds, without flexing more than 0.00025 inches.

Each sensor has several critical structures which must be rigid. They are the mounting plate, the camera mount, the laser mounts, and the galvo mounts. The camera mount must allow adjustment of the camera on its pitch and roll axes while bearing the weight of the camera and lens. The director galvo mount must enable alignment of the plane of the laser beam. All of these structures allow operation of the LCDS in various attitudes without introducing unacceptable errors.

Another mechanical challenge was the design of the calibration fixture. This fixture is a rigid box-like structure which has five separate gauges fastened securely inside it as shown in Figure 15. The system is calibrated with this fixture held in place on the scan head by four locating pins. Each gauge is machined to an accuracy of ± 0.0005 inches. The sidewall gauges are mounted at a 45 degree angle to the bottom of the gauge fixture. Since the coordinate systems for each of the sensors are located on the faces of

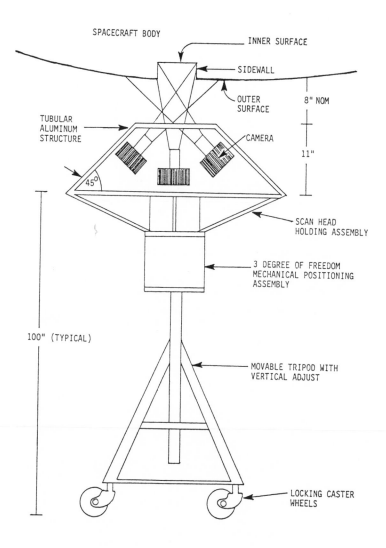

Figure 14: Simplified View of Scan Head and Positioning Fixtures

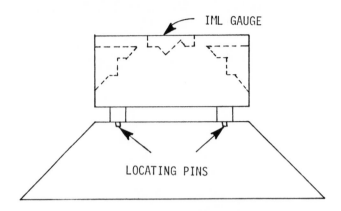

Figure 15:   Calibration Fixture on Scan Head

their gauges, they are all related to each other by simple linear transformations. This allows all measurements to be expressed in a single coordinate system. The common coordinate system chosen for the LCDS was that of the inner surface sensor (Figure 15).

5.2. LCDS Electrical Design

The salient features of the LCDS electrical design are its Laser Control System (LCS), the multiplexer, postprocessor interface and galvo temperature controllers.

The LCS controls the director galvo position and the position and amplitude of the spreader galvo oscillators. It is a single card in the microcomputer chassis. Inputs to the LCS from the microprocessor are digital values expressing director galvo position, spreader galvo amplitude and spreader galvo offset. Outputs from the LCS consist of a constant analog signal to drive the director galvo and an oscillating signal to drive the spreader galvo (recall that the spreader galvo converts the light from the laser into a plane by means of a small, rapidly oscillating mirror). The LCS also receives a position feedback signal from the director galvo. Figure 16 illustrates the LCS interfaces.

The multiplexer, located in the scan head, allows the microcomputer to select one of five sensor subsystems. It uses a digital signal from the microcomputer to switch between cameras, galvos, and lasers. A laser is

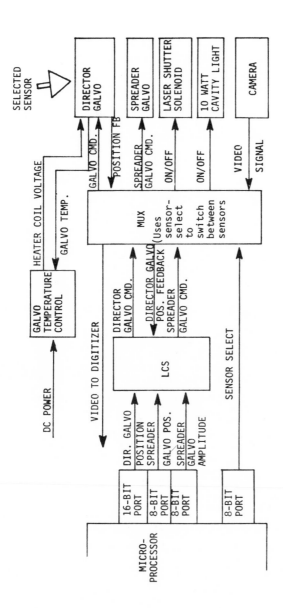

Figure 16: Laser Control System and Multiplexer Interfaces

selected simply by opening a solenoid-shutter which blocks the laser beam when it is in the closed position. Multiplexer interfaces are also shown in Figure 16.

The postprocessor interface consists of an Intel SBX-351 serial interface module mounted directly to the CPU board. Output from the SBX module goes to another module which allows switching between an RS-232 or current loop interface. When the postprocessor is connected directly to the LCDS, files are transmitted at 9600 baud; when an RS-232 connection is made via a modem-phone link, it is operated at 1200 baud. The baud rate in either situation is software-selectable.

The galvo heater controllers were not in the original LCDS design. These devices were designed to correct for a tendency of the director galvos to drift with changing ambient temperatures. In fact, a change in ambient temperature of one degree Fahrenheit was observed to cause measurement errors of 0.006 inches. Since the LCDS is intended to operate in a building similar to an aircraft hangar where temperatures exceed 105°F during the summer months, this situation was clearly unacceptable. The solution to this problem was to wrap each galvo with a heater coil and insulating blanket, and heat the galvos to a constant temperature of 130°F. This is accomplished with the heater controller, a small circuit located near each director galvo.

5.3. LCDS Software Design

The LCDS software consists of a single application program, written in Pascal, which interfaces to the resident 100X Library. Except for a small I/O task used for transmitting files, the LCDS program operates as a single task under control of the iRMX-86 operating system. A multi-task operation was not feasible because nearly 100% of the CPU capability is consumed while scanning and collecting data.

The LCDS software is menu-driven, allowing the operator full control over the various processes, and to repeat any process if errors are detected. The main program menu consists of the following six selections:

1. Calibration
2. Jobsheet Entry
3. Cavity Framing
4. Cavity Scanning
5. Cavity Analysis
6. Transmit Data (to postprocessor)

These selections are the six basic steps used to scan a tile cavity. Each menu selection activates a separate program overlay which is stored on the Winchester disk. Data are passed between these overlays entirely through files written to the disk; passing volatile stack variables between program segments was avoided. Thus, if system power is interrupted, only a limited amount of data will be lost. A description of each of the main menu selections follows.

Calibration

The calibration process generates a transformation from camera coordinates to 3-space coordinates for each position of the laser. This is achieved by scanning the calibration gauge fixture. These coordinate transformations, also discussed in Section 4.3, are stored on disk and retrieved for processing each set of scan data. The calibration process is performed at least once every eight hours to correct for system errors due to thermal expansion and electronic drift.

Jobsheet Entry

This module allows the operator to enter information describing the general physical characteristics of the tile to be scanned and to identify tiles adjacent to the cavity. An operator can also enter parameters controlling the distance between scans, data validation parameters, and the baud rate of the data link. Data are entered by simple edit screens and the system validates input values.

Cavity Framing

The cavity framing process is performed to ensure that each of the five sensor subsystems collects data from the proper region of a cavity. This process consists of two steps. First, the scan head is positioned directly in front of the tile cavity about 6 inches from the spacecraft. When the scan head is positioned properly, the inner surface of the cavity will be roughly centered on the CRT monitor screen and cavity sidewalls will be parallel to the sides of the screen. Second, the console operator uses an edit screen to select each camera and identify laser positions where scanning will begin and end. The operator will also establish the laser intensity and vertical position by entering commands changing the spreader galvo parameters. A series of scans is then made to determine parameters which adjust for background light. When this is done, the framing process is complete and all relevant parameters are written to a disk file for later reference by the cavity scanning module.

Cavity Scanning

Cavity scanning causes each sensor to scan the cavity between laser positions established during cavity framing. Data are stored in 2-D camera coordinates on the Winchester disk. The cavity is scanned a second time to obtain a nearly identical set of data which is also stored on the disk. The two sets of data are converted to 3-space coordinates using the transformations obtained during calibration, and the data are then validated. The CRT displays a running tally of valid data points collected for each camera and the percentage of data points which were repeatable within a specified tolerance as illustrated below.

Occasionally workers inside the spacecraft near a tile cavity will cause very small flexures in the spacecraft body, and data between the two sets of scans will not agree as closely. The percentage figures in the above display would have reflected this discrepancy, alerting the LCDS operator to repeat

Camera	No. of Valid Points	Percent Differing by Less Than .003	Percent Differing by Less Than .008
1	1447	98.1	99.9
2	2015	98.0	99.9
3	2241	97.0	99.8
4	3002	97.9	99.8
5	2695	99.0	99.9

the scans.

When cavity scanning is complete, a large file containing validated 3-space coordinates exists for each camera. These files are ready for processing by the cavity analysis module.

Cavity Analysis

The cavity analysis module processes data produced by the cavity scanning module to create a set of files defining a tile which fits the cavity precisely. First, the data are separated into six distinct sets as follows:

- Four sets containing points from each planar sidewall
- One set containing points only on the outer surface of adjacent tiles
- One set containing points on the inner surface of the cavity

The cavity analysis module fits each set of sidewall data to a plane using the least squares method. The outer and inner surface data are fit to a quadratic surface using a least squares fit. This process, also described in Section 3, produces a set of files ready for transmission to the postprocessor.

During the cavity analysis, information is displayed on the CRT which informs the operator of the program's progress and allows a skilled operator to trace errors. When the task is complete, a summary of cavity dimensions is written to the CRT.

Data Transmit

The data transmit module may be used either to transmit a set of files containing tile cavity data directly to the postprocessor or to save the data files for later transmission. The LCDS can save as many as 15 sets of tile data on its Winchester disk. When data are transmitted, a CRC (cyclic redundancy check) word is sent with each data block, allowing transmission of files over long distance telephone lines.

6. Concluding Comments

The LCDS is one of the first successful applications of a high-speed 3-D vision system. In contrast to other high accuracy laser vision devices, the LCDS enables a tile cavity to be completely digitized in a matter of a few minutes.

LCDS accuracy in an operational environment has fallen somewhat short of original expectations. At the writing of this article, the system can locate cavity corners to an accuracy of about ± 0.020 inches, limiting the usefulness of the system in a production environment. However, many correctable error sources were identified, including camera mount movement, galvo mount movement, laser positioning errors, calibration errors and director galvo drift.

A follow-on project has been planned to correct these deficiencies and produce a system which will locate tile corners to an accuracy of ± 0.0085 inches. When this goal is achieved, the LCDS will be usable as a primary tool for the manufacture of shuttle tiles.

In the course of developing the LCDS, a wealth of knowledge was gained in identifying and solving numerous problems encountered. Without listing all of the problems, it is a fact that virtually all of the LCDS subsystems posed problems which at one time or another limited system accuracy. Probably the most significant unexpected problem was the effect of ambient temperature variations on accuracy. The galvanometer scanners were extremely sensitive to temperature variations and the pointing accuracy of the lasers was found to be affected by small temperature gradients in the laser tube housing. Even a critical oscillator in the digitizer was found to be temperature sensitive.

While the accuracies in the follow-on project will be sufficient to qualify the LCDS for tile manufacturing, further significant advances in systems of this type can be made in the foreseeable future. Improvements will be made in speed, accuracy, and size. A digitizer employing VLSI and array-processing technologies could vastly increase speed. Cameras specifically made for vision system applications and improvements in galvanometer scanners will improve measurement accuracy. The use of non-metallic materials less sensitive to temperature changes will also provide improvements. And, as engineering experience is accumulated, it will be possible to build smaller and lighter systems.

Such improvements will open the door for more industrial and scientific applications. Indeed, if the vision system industry matures at the same rate experienced by computer graphics and CAD/CAM, we will be seeing a mature industry in 12 years. Applications such as the LCDS will seem prosaic. An LCDS will be little more than an oversized camera, capable of "photographing" a cavity in seconds and locating cavity corners to better than ± 0.001 inches. Accuracies in other applications will routinely be better than 0.0003 inches, and costs will be but a fraction of what they are today.

Acknowledgment

I am greatly indebted to Mr. George Bechtold for the efforts he has expended in helping me to assemble this article.

Three-Dimensional Vision Systems Using the Structured-Light Method for Inspecting Solder Joints and Assembly Robots

Yasuo Nakagawa
Takanori Ninomiya

Production Engineering Research Laboratory
Hitachi, Ltd.
292 Yoshida-cho
Totsuka-ku, Yokohama, 244
Japan

Abstract

The structured light method is effective for detecting a sectional view or range data for industrial applications. This article presents two three-dimensional vision systems developed at the Production Engineering Research Lab of Hitachi, Ltd., Japan. One system is for inspecting solder joints and the other was for automatic part assembly by robots. The inspection system can detect and correctly judge the shape of solder joints. The assembly robot vision system consists of a fixed 3-D detector and a small X-type slit detector attached on the robot hand. The former detects the overall range data of robot work-fields, and finds and detects the position and posture in 3-D space of a particular object. The latter detects the precise position and posture of an object.

1. Introduction

Solder joints of printed circuit boards have defects such as excess of solder, shortage of solder in some areas, and blowholes, all of which cannot be detected by electric circuit testers because electrically they have contact. Visual inspection for these defects is therefore necessary to eliminate these defects. Conventionally this inspection is done by humans. However, it is very difficult both to eliminate human error and increase productivity. Thus, automated inspection systems are required to achieve a high inspection rate and high reliability in inspection.

For a robot to handle an object which does not have a pre-established position and posture, such as a connector with a cable in assembling electrical products, detection of the three-dimensional position of the object is required. This is difficult to do with gray-scale images obtained by conventional TV cameras because of the enormous computational cost. Therefore, in the vision systems we developed for an automatic inspection task [1] and an automatic assembly task we used the structured light method to directly obtain three-dimensional range information.

In the past, the structured light or light-section method has been used for measuring optical shapes such as the contours of a section of mechanical parts for turbine blades. A light-section microscope with a slit approximately 3 μm wide can detect a concave-convex shape to within an accuracy of 7 μm. Recent applications of this technique of detecting 3D shapes include the H shape for the steel rolling process [2], the beveling shape in arc welding [6], and the width of high-speed train rails [5]. In robotics research, the structured light technique is used for detecting silhouettes of parts on a conveyor belt [8] and for detecting the position and orientation of parts on a table [7].

2. Light-Section Method for Shape Detection

The light-section method can be configured in different ways depending on the relationship between a slit projector and an image detector. Figure 1 illustrates several possible configurations. In Figure 1(a), a slit projector and an image detector face each other so that their optical axes form a 90° angle, and a slit projected from the oblique direction is detected from the opposite direction. In Figure 1(b), a slit is vertically projected onto the object surface. In Figure 1(a) the detector detects the direct reflection of the slit image, and in Figure 1(b) it detects the diffused reflection. In either case the projected slit on the object is in the objective plane of the image detector.

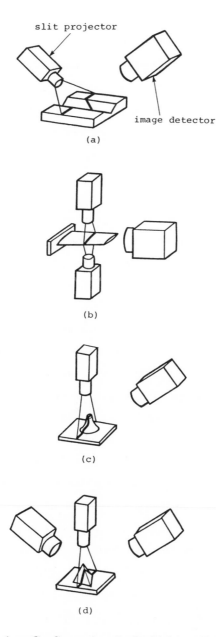

Figure 1: Various Configurations in the Light-section Methods

Thus, these configurations have an advantage in obtaining a good focussed image of the slits, and are used widely as the basic configurations in applying the light-section method. These configurations are effective for examining objects whose sectional shapes do not change quickly in the plane including the optical axes of the projector and the detector. Otherwise, however, an unnatural sectional shape may be obtained due to the oblique sectioning in Figure 1(a) or the object may obstruct detection of the projected slit in Figure 1(b).

These shortcomings must be overcome for the two applications, solder inspection and part assembly, to be discussed in this paper. Thus we adopted the configuration of Figure 1(c) for the solder inspection task; where the slit is projected from the vertical direction and is detected from the oblique direction. Using this method the solder shape can be steadily detected without being influenced by neighboring joints. Also, defocusing is not a serious problem when the image is focussed at the middle of the solder surface. One difficulty in using the configuration of Figure 1(c) is occlusion of projected slits by the object itself (dead angle), which must be minimized. To resolve this problem in the part assembly task by robots we used the two detector and one projector configuration, as shown in Figure 1(d). In this case, in order to get range information over the maximum field of view, we combine the images obtained by two detectors.

The following two sections will describe the two vision systems, solder inspection and part assembly, which use the light-section method to obtain the three-dimensional range data.

3. Automatic Inspection of Solder Joints on Printed Circuit Boards

An automatic visual inspection technique for solder joints on printed circuit boards has been developed. To detect the solder joint shape, structured light is used as the optical system and a simple waveform processing is applied to the extracted shape for judging the shape of a solder joint.

3.1. Solder Joint Defects

Although types of solder joint defects can be classified in detail into more than ten categories, there are five fundamental types, four of which are

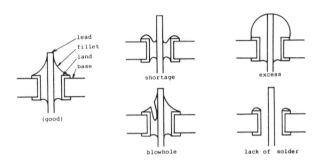

Figure 2:   Examples of Solder Joint Defects

illustrated in Figure 2. The sectional shape of solder joints are shown where the soldered surface is on top and the loaded electrical parts are underneath. The fifth type of defect is a solder bridge, which can be detected by an electric tester and is thus excluded from detection by automatic visual inspection. The lack-of-solder type, shown in Figure 2, can also be detected by the tester if there is no contact, but in general the lead of the part often touches the land of the board resulting in electric contact. Thus visual inspection is necessary for this type.

3.2. Basic Approach

In solder joint inspection, the gloss and blur of the solder surface can be the major cause in the difficulty of detecting the shape of the solder joint correctly.

We have adopted the configuration of Figure 1(c) where the slit is projected from the vertical direction and is detected from the oblique direction. Using this method the solder shape can be detected stably, not influenced by neighboring joints, and focusing remains good over the whole range when the camera is focussed at the middle of the solder surface. The 3-D shape inspection of solder joints is done sequentially. First, a light-section waveform is extracted from a gray-scale image obtained by an image detector. Next, the waveform is analyzed, and the data are transformed into more meaningful information, such as, "this waveform has some irregularity". Then, by moving the XY table stepwise, a sequence of cross-sections is detected and processed in the same manner. Finally, the meaningful information is summarized and judged.

In the case of dual in-line IC leads, lead contour is an oblong, and gaps exist between the lead and through-hole of a printed circuit board on both

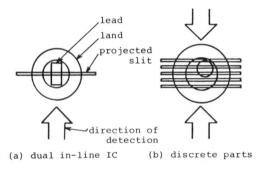

(a) dual in-line IC   (b) discrete parts

Figure 3:   Difference in Detection Methods for Dual In-Line IC and Discrete Parts. (a) Dual In-line IC; (b) Discrete Parts

sides of the lead (see Figure 3(a)). The real defects exist only in these gaps. But for discrete parts like condensers and registers, judgment of defect must be made by examining a number of waveforms by moving the table stepwise, because the location of the gap is not known beforehand. Also, for discrete parts, if the projected slit is detected by an image detector from only one side, a defect behind the lead cannot be found due to the dead angle. For this reason, we must use two detectors to detect the projected slit from both sides, as shown in Figure 3(b).

3.3. Judgment Algorithm

The algorithm to detect defects and to judge their types should be as simple as possible and also not too sensitive to the noise of the slit image. With an eye on these points, we developed the following algorithms.

3.3.1. Extraction of Light-Section Waveform

As shown in Figure 4(a), the projected slit is bright in a light-section image. The video signal of each vertical sweep, shown as a dot-dash line in

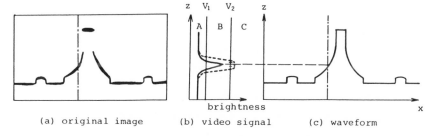

(a) original image   (b) video signal   (c) waveform

Figure 4:   Waveform Extraction. (a) Original Image; (b) Video Signal; (c) Waveform

Figure 4(a), is examined. Referring to Figure 4(b), if the video signal has a peak in range "B" (the case shown by the solid curve in Figure 4(b)), its peak z-coordinate is recorded as the z-value at that x-position of the waveform: see the dash line leading to the corresponding position in Figure 4(c). When a signal has a peak in range "C" (that is, above the threshold V_2 : the case shown by the dotted curve), the average value of two points where the video signal intersects with the threshold level V_2 is recorded as the z-value. When the video signal stays in range "A" (that is, below threshold V_1), the z-value is undetermined. This operation is done for each x-position of the image, and the sequence of recorded z-values produces a light-section waveform, as shown in Figure 4(c). All of this operation is done in real time by hardware. Instead of using a simple peak detection in the video signal of each vertical sweep, processing the signal in three different levels of brightness A, B, and C as explained above allows for stable detection of solder joint shapes.

3.3.2. Preprocessing of the Waveform

Preprocessing of the waveform consists of histogramming and noise cleaning. A histogram of z-values is used to find out the base level Z_b which corresponds to a printed circuit board in the detected view field. As shown

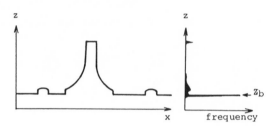

Figure 5:   Detection of Base Level Z_b

in Figure 5, the value Z_b is determined from the histogram as the value where the first peak is found when the histogram is scanned from low to high in z values. The base level Z_b is used as the reference of height in the next evaluation process. The height change due to warping of a printed circuit board is compensated for by adjusting Z_b.

Noise cleaning consists of median filtering and interpolation of broken parts of the waveform. In the former process, each pixel's z value $z(x)$ is replaced by the median of itself and its neighbors' values; i.e., by the median of $\{z(x\text{-}1), z(x), z(x\text{+}1)\}$. This removes tiny irregularities included in the waveform. Next, broken parts of the waveform is interpolated by the stepwise interpolation as shown in Figure 6.

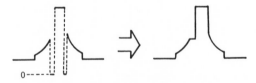

Figure 6:   Stepwise Interpolation of Broken Waveform

3.3.3. Evaluation of the Waveform

The final process is the evaluation of a waveform. This is done by checking several parameters. In calculating these parameters, a particular evaluation zone is set up for each parameter, as shown by the rectangles in

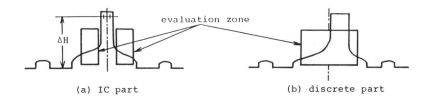

(a) IC part   (b) ·discrete part

Figure 7: Establishment of Evaluation Zone. (a) IC Part; (b) Discrete Part

Figure 7, and the counting is done only in these zones. For dual in-line IC parts, x-position of these zones is determined at the position of a lead, as shown in Figure 7(a). For discrete parts, x-position of the zone is fixed in the view field, because the posture of their leads is not constant. The z-positions of these zones are determined by Z_b.

We used four parameters for evaluation: areas S_l, S_h, S_b, and broken length W. (See Figure 8 for definitions.) These parameters are useful to detect the four fundamental defect types mentioned previously, respectively.

IC solder joints are evaluated by the parameter values obtained from one waveform, and if one of these parameters exceeds an established allowance, the solder joint is judged as a defect. Solder joints of discrete parts are evaluated by summing the values of each parameter over a number of waveforms obtained while moving the table. The broken part is defined by the logical sum of the broken parts of waveforms detected by two image detectors.

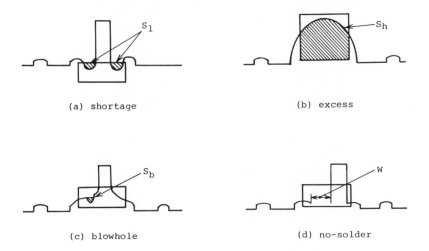

(a) shortage   (b) excess

(c) blowhole   (d) no-solder

Figure 8: Parameters for Judgment. (a) shortage; (b) excess; (c) blowhole; (d) no-solder

3.4. Results and Conclusions

Figure 9 shows the final system organization of a prototype inspection

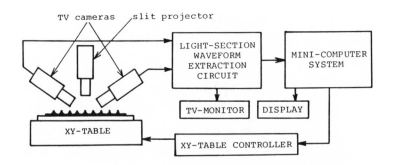

Figure 9: System Organization of a Prototype Inspection System

system, and Table 1 shows the specifications of the system. Figure 10 is the photograph of the prototype system. Image detectors (TV cameras) take images of a projected slit from left and right at a 35° angle from the horizon.

Figure 11 shows examples of detected shapes of solder joints: the pitch of y-directional movement is 0.1 mm, sensitivity of height is 18 μm, and the

slit projector

light source	:	halogen lamp,300W
slit	:	0.2mmW, 10mmL
projection lens	:	macro-lens, f=90mm, mag. 1X

image detector

TV camera	:	2/3 inch vidicon
focusing lens	:	f=95mm, mag. 0.5X
pixel size	:	$10\mu m$/pixel (x) $18\mu m$/pixel (y)

light-section waveform extraction circuit

data point	:	240 pixels

Table 1: Specifications of Solder Joint Inspection System

Figure 10: Prototype Inspection System

resolution in the x-direction is 10 μm with the total sample points of 240.

Figure 12 shows a TV-monitor image of a projected slit on a soldered

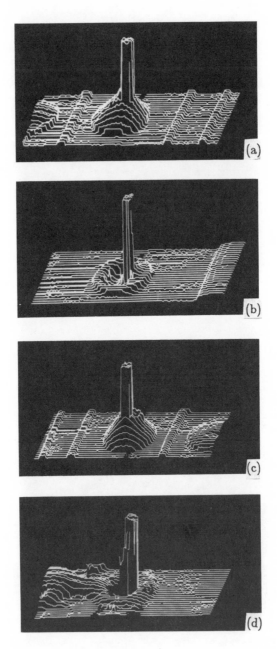

Figure 11: Examples of Detected Solder Joints. (a) good; (b) shortage; (c) blowhole; (d) no-solder

Figure 12: Monitored Slit Image

surface. Figure 13 shows an example of judgment, where the solder joint is judged as a defect of shortage of solder.

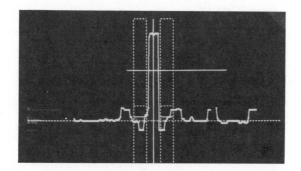

Figure 13: An Example of Judgment

Figure 14 shows the distributions of parameter values S_h, and $S_l + S_b$ for good joints and joints with various types of defect. An inspection speed of 10 points per second has been achieved for dual in-line ICs, but it takes about 1 point per second for discrete parts.

4. Robot Vision System

Range data is one of the most suitable types of sensory information for a robot, because a robot works in 3-D space. As a feasibility study, we developed a 3D assembly robot vision system. This system consists of two

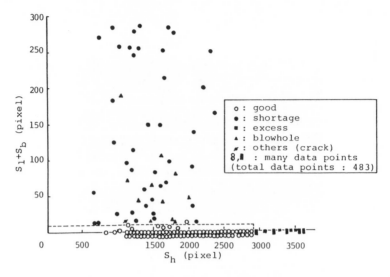

Figure 14: Distribution of Parameter Values for Various Types of Defects. (IC Parts)

kinds of sensors as shown in Figure 15: a fixed range finder and a compact sensor at the robot's hand. The former detects the overall range data of the robot's workspace to find and detect the position and posture of a particular object in 3-D space, and the latter detects the precise position and posture of the object. Both sensors use the structured light method and are described in Sections 4.1 and 4.2.

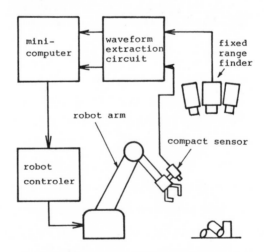

Figure 15: Organization of Robot Vision System

4.1. Fixed Range Finder

A fixed range finder is used to overlook the robot's workspace as shown in Figure 16. The slit projector and the two solid-state TV cameras are placed on a moving table driven by a stepping motor. A light slit is projected from above the objects. Two light-section waveforms from the two cameras are extracted simultaneously and are combined into one waveform to eliminate the shadow. Light-section waveforms are detected sequentially while the table is moving to obtain the range image of the whole scene.

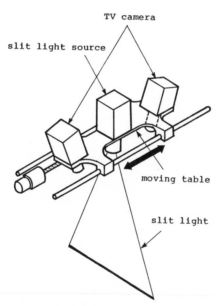

Figure 16: Fixed Range Finder

The slit light source is formed by a line of infrared LEDs. The range image is 256 x 256 pixels, with the resolution of 0.3 mm/pixel, and is acquired in 4.3 seconds. Figure 17 shows an example of a range image obtained for a connector with a cable on a circuit board.

From such range data, the planar surfaces of parts are extracted. We describe the algorithm for it in Figure 18. The first step is to detect jump edges, where range values jump abruptly from one value to another.

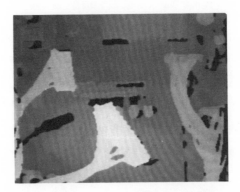

Figure 17:   Range Image of Connectors on a Circuit Board

Individual parts or parts placed on other parts form regions whose heights are well above the circuit board. Therefore, it is possible to extract each part from the range data by detecting the jump edge and segmenting the closed regions. The Roberts gradient [3] is used for detecting the jump edge, and the standard search and propagation process [4] is used for segmentation.

These segmented regions are then classified into individual parts like connectors, ICs, registers, board plane, etc. For this process, the average height of each segment is calculated from the raw range data included in the labeled region segment. In finding a connector, for example, we used the properties where the average height of a connector is higher than that of a part mounted directly on the board, and that the area of the connector's segment has a certain value.

The next step is an extraction of the major planar object surface. This is done by extracting the maximum region surrounded by roof edges and jump edges. We used the following operator to detect the roof edge.

$$ANG(i,j) = |tan^{-1}[\frac{\{V(i+n,j)-V(i,j)\}}{n}] - tan^{-1}[\frac{\{V(i,j)-V(i-n,j)\}}{n}]|$$

$$+|tan^{-1}[\frac{\{V(i,j+n)-V(i,j)\}}{n}] - tan^{-1}[\frac{V\{V(i,j)-V(i,j-n)\}}{n}]| \qquad (1)$$

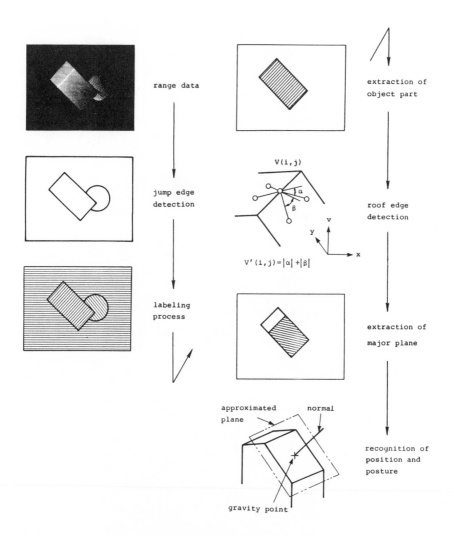

Figure 18:   Sequence of Range Data Processing for Extracting Planar Surfaces

where, $V(i,j)$ is a value of point (i,j) and $n \geq 2$.

After detection of roof edges in the selected segment, the segmentation process is applied again to the segment to extract the major plane of the object surface.

Finally, the gravity point (x, y, z-coordinates) and the surface normal of

the major plane are computed. The shape of the plane observed from its
normal direction can be obtained by geometrical transformation, and two-
dimensional shape-recognition techniques can be applied to recognize the
part and to detect its rotation angle.

As an example of the result, Figure 19 shows the jump edges of the whole

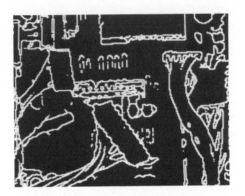

Figure 19:   Jump Edges of the Range Data

Figure 20:   Extracted Jump and Roof Edges of a Connector

image of Figure 17, and Figure 20 shows the jump and roof edges of the
extracted connector. Figure 21 is the identified object where the plus (+)
mark indicates the gravity point of the major plane of the object surface of
the connector and the straight line shows the normal direction of the surface
in trigonometry. The outline in the upper left of the photograph is the

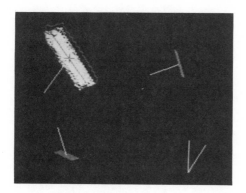

Figure 21:   Identified Connector

contour line of the connector after being transformed to a view from the normal direction. In this way, we can determine the 3-D position and posture of parts, together with parts identification. A robot can now approach a particular part appropriately.

4.2. Compact Sensor on Robot Hand

Figure 22 shows the principle of a compact sensor on a robot hand, which consists of two slit projectors and a TV camera. The two slit projectors are arranged for projecting the X-type slit light by illuminating two diagonal slits. The X-type slit image is able to determine the x, y, z-coordinates and the slant angles of objects. Each diagonal slit light is illuminated one at a time, and two light-section waveforms are extracted separately. The extracted slits are transformed to polygonal lines.

Distance calibration maps, as shown in Figure 22, are calculated beforehand and are stored in the system. The 3D coordinate of the turning points of slits are found by referring to the map. The position and posture of the object can be calculated from these values.

Figure 23 illustrates the actual arrangement of the sensor. The X-type projectors and a TV camera are placed in a row above the robot fingers.

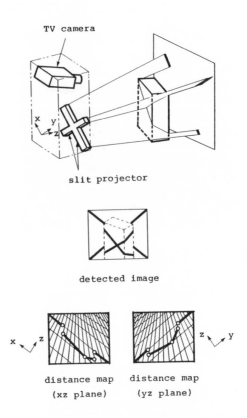

TV camera

slit projector

detected image

distance map   distance map
(xz plane)   (yz plane)

Figure 22:   Principle of Compact Sensor

This enables detection of an object while grasping it.   Figure 24 shows a photograph of the compact sensor on the robot hand.   The slit light source is a line of infrared LEDs, and only the sensing head of the solid-state TV camera is placed on the robot hand.   The sensor only weighs 550 grams. Detection time of the object position and posture is 0.1 second and detection accuracy is 0.1 mm.

5. Concluding Remarks

We have demonstrated applications of the structured-light method for inspection of solder joints and a robot vision system.   As it is possible to control the illumination environment in many industrial applications, an

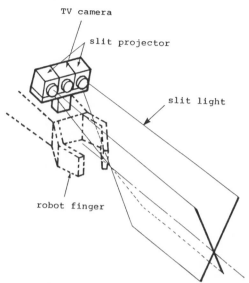

Figure 23: Arrangement of Slit Projectors, a TV Camera, and Robot Fingers

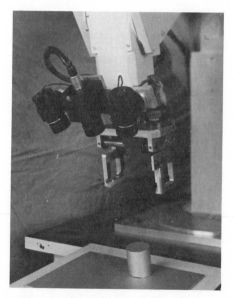

Figure 24: Photographs of Compact Sensor on a Robot Hand

active range finder such as the light-section method is very effective. A combination of range information and brightness image will be more effective for some cases. The range finders suitable to industrial applications must be compact, inexpensive, fast, and accurate, and their operation must be insensitive to the color, gloss, and darkness of objects.

Acknowledgments

I wish to thank Hirotoshi Kurabe and Isamu Kimura (Hitachi Totsuka Works) for helpful discussions and the samples required to develop and test the inspection system. Also thanks are due to Hiroshi Makihira (a co-worker) for design of the waveform extraction circuit.

References

[1] Nakagawa, Y.
     Automatic visual inspection of solder joints on printed circuit boards.
     In *Proceedings of SPIE*, pages 121-127. Arlington, 1982.

[2] Okada, S.
     Welding machine using shape dectector.
     Mitsubishi-Denki-Giho (in Japanese) 47-2:157, 1973.

[3] Roberts, L. G.
     Machine perception of three-dimensional solids.
     Optical and Electrooptical Information Processing.
     MIT Press, 1965, pages 159-197.

[4] Rosenfeld, A. and Kak, A. C.
     Digital picture processing.
     Academic Press, 1976.

[5] Takeshita, M.
     Optical rail warp measurment system.
     Japan Patent, TKS S51-143358.

[6] Tsukada, M.
     Sectional shape detection apparatus.
     Japan Patent, TKS S49-58864.

[7] VanderBrug, G. J., Albus, J. S., and Barkmeyer, E.
     A vision system for real time control of robots.
     In *9th International Symposium on Industrial Robots*, pages
        213-230. Washington, DC, 1979.

[8] Ward, M. R., Rossol, L., and Holland, S. W.
     CONSIGHT: A practical vision-based robot guidance system.
     Technical Report GMR-2912, General Motors Research, 1979.

A Semantic-Free Approach to 3-D Robot Color Vision

R. A. Jarvis[1]

Department of Electrical Engineering
Monash University
Victoria, Australia

Abstract

This paper argues the case for approaching 3-D robot vision from the semantic-free end of the analysis spectrum with the immediate goal of making explicit the volumes and shapes occupying space, along with their surface properties, and in a form supportive of subsequent high level vision processing. Expedience and robustness are emphasized, with real time sensory-based robotic task environments in mind as the application domain.[2]

1. Introduction

Imagine a conglomeration of objects in the work space of a robotic manipulator which is to be guided by a vision system to carry out sorting, stacking or assembling tasks in that restricted environment. The exact volumetric and surface description of each separable object in the scene, along with placement and orientation, represents a rich database upon which to base object recognition, trajectory planning (including obstacle avoidance), gripper positioning, transport and subsequent placement at a specified location and orientation within the goal structure and constraints

---

[1]Previously in the Department of Computer Science, Australian National University

[2]Many of the figures in this paper were originally in color, but had to be printed as gray-scale pictures which may be misleading. In particular, Figures 5(a), 5(b), 6(a), 9(d), 11(b), and 11(d) may need special attention from the reader to correctly interpret the contents.

of a semantically oriented frame of reference. Rather than using the semantic information available as a means of disambiguating scene interpretations, a powerful alternative is to apply whatever sensory transduction mechanisms are necessary to gain as complete an unambiguous description of the scene as possible prior to binding this with the available semantic knowledge. This promotes robustness and generality while reducing the complexity of higher level processing stages of analysis.

Clearly, this 'semantic-free' approach to preliminary vision processing is not appropriate in many situations, nor will it necessarily be the usual approach when massive computational resources become available at little cost. Currently, however, in terms of practical robotic guidance applications, it seems well worth investigating as a near term solution.

It would be ideal if the exposed surfaces of objects could be sprayed with some magic powder, which, when gathered up, could, particle by particle, be interrogated as to where it had been in space and what the properties of surface it touched were. This would represent the true 3-D information sought to support the higher level processing stages mentioned earlier. That this magic powder had no capability of capturing global shape information, structural independence of the objects, or their identity would be a disadvantage but not a fatal flaw.

In fact, getting the sort of information provided by the magic powder described above is by no means easy, and a great deal of research has gone into estimating it from lower dimensional data. That vision has dominated as the means by which this type of description is attempted is no surprise given its importance in supporting human manipulation of, and movement through, our three-dimensional world of solid objects. By providing analysis at a distance, that is, prior to contact, vision is a very powerful remote sensing mechanism which provides dense data at high speed.

2. Overall Robotic Vision System Hierarchy

It is convenient to regard a vision driven robotic system as having a function analogous to human hand-eye coordination. This analogy conveys the idea of the need for tight coupling between seeing and physical action. It also suggests that the most appropriate means of testing a scene interpretation is to attempt a robotic action based on that interpretation and decide whether or not it was successful. The descriptive components of the hypothesis can be directly related to the type of robotic action required to fulfill a given subgoal within some hierarchical goal structure. Using robotic manipulation as an active part of the scene analysis, as in picking up an object to test a support principle or exploring a scene with a robot hand-held roving camera to capture data obscured in the initial view, is very much part of the hand-eye coordination model.

One system structure for vision driven robotics consistent with the hand-eye coordination analogy is given in Figure 1. Image (and perhaps also range) data acquisition is followed by a segmentation phase which, even if not semantically driven, should separate portions of the scene in a way which has a high probability of being consistent with the isolation of semantically distinct objects in that scene. Such a seeming contradiction can be resolved if the natural structure of the data supports a partitioning process such that no two visually distinct components are combined. Having more, rather than less, segments than there are physical objects is often the safer approach since it is usually simpler to combine components at higher levels of analysis than it is to break too large segments into appropriate smaller components. The properties of visually adjacent segments and their mutual boundaries can be used to suggest safe merges; an example will be offered later. Stage 3 consists of taking relevant measurements, for each segment, of the size, shape, color, position, and adjacency conditions with respect to other components. 'Relevance' is determined by the extent to which the measurements permit discrimination between the permissible

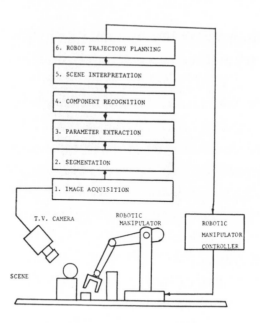

Figure 1:   Robotic Vision System Hierarchy

objects in the scene at the component recognition phase (4). Then follows a
scene interpretation which should include those aspects such as component
identity, position, orientation, juxtaposition, support principles and so on
which are, in turn, relevant to the task of trajectory planning for robotic
action. Trajectory planning is concerned not only with the task of driving to
a point for grasping and moving an object, but also with the problem of
avoiding collision with other objects while fulfilling some optimality
requirement such as shortest path, least expended energy, or shortest time.
The scene is subsequently altered by the robotic action and can either be
subjected to complete reanalysis or merely to some process of checking the
correctness of the action and modifying the scene interpretation. Thus,
within the cycle of analysis, each stage directly supports the next higher and
is required to provide only what is necessary to do this effectively and
unambiguously. A more sophisticated system would no doubt include
feedback loops among the stages to either correct flaws of lower level

analysis or to provide more detail in support of some higher level decision-making process. Such additions would, while presumably improving the quality of the result, increase the computational complexity considerably. In the service of simplicity, such looping structures should be avoided when possible. More complete analysis at any stage is preferred to developing weak hypotheses requiring verification, unless that verification process is easy to apply. These guidelines are offered in the context of complexity and time constraints which are assumed to be severe given the current state of affordable technological support for practical robotic tasks, not as inviolate principles of system design.

3. 2-D Image Segmentation

Semantic-free image segmentation can be regarded as an unsupervised pattern recognition task or cluster analysis. The aim is to tag each pixel of an image array with a label indicating, but not naming, the component to which it belongs such that the same label does not appear in two distinct segments. It is often reasonable to carry out a simple and efficient 'first partition' of the image into many more segments than would be necessary to isolate the semantically distinct components. Then, structured merge techniques, using the richer information that can be extracted from region and border properties of adjacent segments, can be applied to reduce the total number of segments (but not to cause false amalgamations). It is also possible to combine merges and splits in an elegant manner as given in [3].

Two examples of 2-D segmentation are presented. Figure 2 shows some results of an image segmentation process based on clustering. The shared near neighbor clustering procedure [6] is used for a first partition, followed by a structured merge procedure [7] using region and border properties of the initial segments to suggest a merge. This process is applied repeatedly to the structure until a satisfactory result is achieved. Devising stopping rules such that the least number of segments without a false merge is produced is

(a)      (c)

(b)   (d)

Figure 2:   Image   Segmentation   Using   Shared   Near   Neighbor
Clustering.   (a) Original Image; (b) An Initial Segmentation; (c) A
Segmentation Merge   Refinement; (d) A Further Merge Refinement.

a problem inherent to all clustering procedures. The computational expense
of extracting the neighbors, forming the first partition, deriving the
database for merge decisions and updating after a merge is quite high; some
minutes of VAX-11/750 processor time are required for a typical 64×64
quantized color image. No semantic information was used directly, although
it can be argued that some semantically based ideas are indirectly embedded
in the merge decision rules applied.

Figure 3 shows an example of the application of a much faster, though
cruder (less reliable), segmentation procedure [8] based on carrying out
binary connectivity [24] on slices of the image defined in relation to 'keys'
and tolerance limits. A 'key' is simply the red, green, and blue components
of the next unlabelled pixel in raster sequence. A slice is made up of those as
yet unlabelled elements of the image with each color component within the
tolerance limit of the corresponding key component. The cycle of key
selection, slicing, connectivity labelling, and removal from further analysis is
applied to exhaustion. Typically, ten to twenty cycles are required
depending upon the tolerance values used. A typical segmentation on a 64
$\times$ 64 color image on a Data General Nova (on which the example of Figure 3

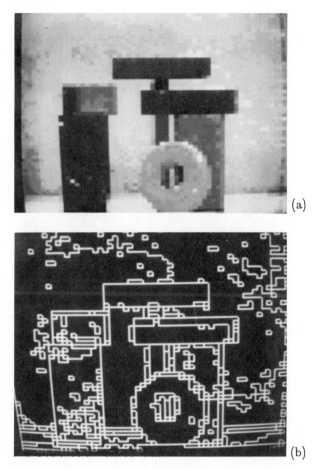

(a)

(b)

Figure 3: A Fast Image Segmentation Method Using Threshold Slicing and Connectivity. (a) A Typical Scene; (b) An Initial Segmentation

was computed) took several seconds. On a VAX-11/750 at least one order of magnitude of speed up can be achieved. Specialized hardware could produce satisfactory results perhaps another order of magnitude faster. Figure 3(b) is the first partition achieved using the above procedure applied to Figure 3(a), while Figure 3(c) is the result of additionally applying a trivial merge rule based on a single pair of pixels across a border being similar within a specified tolerance for each color component. The 2-D segmentation of a more complex image is shown in Figure 4. From the segmentation of Figure 4(b), the enclosing rectangles and centers of area of segments within a

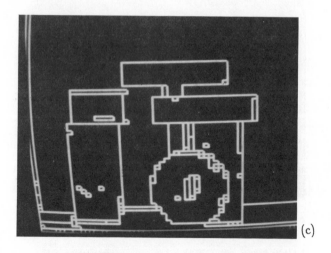

(c)

Figure 3: (c) A Merge Refinement.

specified size range (to exclude small fragments and the large background area) are given in Figure 4(c). In some cases, carrying out a 'hue transform' [9] on the original image before segmentation can lead to a more reliable result, as it tends to remove intensity variations due to lighting position and shadow effects. A hue transform can be achieved by simply dividing each of the red, green, and blue components of each pixel by the sum of the three components and scaling up the result to a typical intensity value for each component. The resulting image, as one might expect, looks very 'flat' and a bit noisy. An example of a hue transform is shown in Figure 5.

Since this paper is primarily aimed at 3-D robotic vision, why are the above two 2-D segmentation methods presented here? There are two reasons. The first is that the inadequacies of 2-D image analysis for robot vision are a good springboard for considerations of the 3-D aspects of scenes. The second is that, as will be seen later, 2-D segmentation can play a useful role as a refinement at a stage of analysis after segmentation of the distinct 'solids' in a scene, where each might consist of several objects in touching contact.

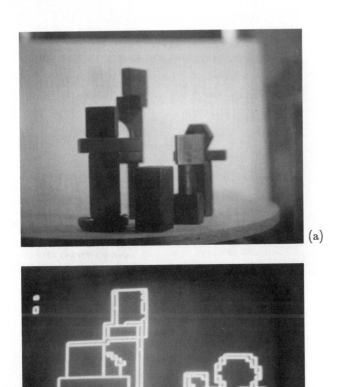

Figure 4: A More Complex Example of Fast Segmentation Method. (a) Original Image; (b) A Merge Refined Segmentation

4. 2-1/2 D Scene Analysis

If each pixel of a regular array of intensity values (or color intensity components) constituting a 2-D image is associated with a range value with respect to the viewing position, a 2-1/2 D scene representation results. The extra 1/2 D corresponds to the specification of the depth position in 3-D space of only those parts of the scene viewable from a single direction. Surfaces of the objects in the scene which are either facing away from the viewing position or obscured by other surfaces (of objects) are not specified in the third (range) dimension. In general, about half of the exposed surfaces

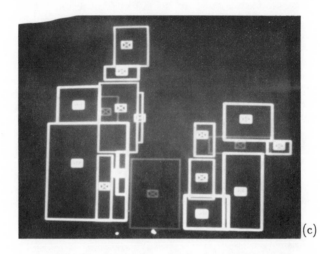

Figure 4: (c) An Enclosing Rectangle/Center of Area Scene 'Summary'.

have a third dimension specified.

If, from a single viewing position, several objects of the same intensity and color are in visual juxtaposition, semantic-free 2-D segmentation methods such as those described in the previous section will be unlikely to properly distinguish between them, since neither semantic guidance nor appropriate sensory data is available. For example, the second 2-D segmentation method of the previous section was unable to separate the several red objects in the scene of Figure 6(a). The addition of range data as a fourth component of the specification of each pixel (red-green-blue-range) resulted in the pleasing result of Figure 6(b) using the same slice and connectivity label cycle as before except for the extra component in the key, tolerance and pixel component values. Fragmentation on the edges of objects as seen from the viewing position is a result of inexact registration of range with color values and the 'jump boundary' problem inherent in ranging points on a step function range change interface. Details are given in [10].

The literature on various attempts to a reliably acquire this extra 1/2 D

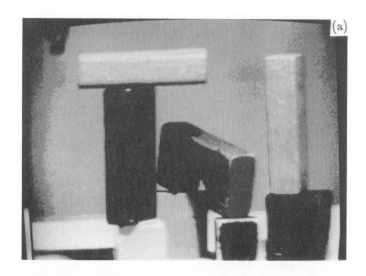

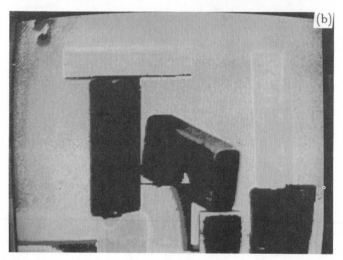

Figure 5: 'Hue Transform' Example. (a) Original Image; (b) 'Hue Transform' (a).

of a scene description is quite plentiful [11]. The best known indirect methods are due to researchers at MIT in the mid-sixties and early seventies [23], [2], [27], who worked on projects which later were collectively referred to as the 'blocks world' experiments. These were essentially model-based (semantically driven) methods which relied on the restricted domain of valid

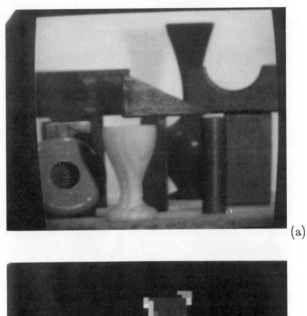

(a)

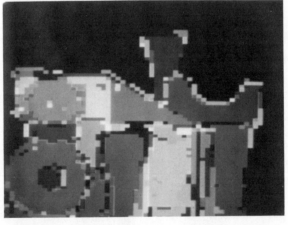

(b)

Figure 6: (a) Scene with Similar Colored Objects Visually Juxtaposed; (b) Scene Segmentation Using Color and Range Data

objects and the way parts of them might appear to reduce the search for matches against known attributes of the prototypical models. They were then, and are perhaps still now, the most impressive examples of semantically driven interpretation of scenes. The strong reliance on correct extraction or provision of clean and correct straight line drawings (hidden lines removed) for the successful application of the methods is considered to be one serious weakness of these approaches to 3-D scene analysis.

More recently, alternative and somewhat more direct (still image-based) ranging methods have been explored. These include passive stereo, [28] striped lighting [21], texture gradient [1], Moire fringe analysis [5], and optical flow [22].

Passive stereo and striped lighting methods are essentially triangulation based, and as such suffer from both the 'hidden parts' problem and the loss of accuracy with range. In the first of these, the 'hidden parts' problem arises from not being able to see all the same parts of a scene from the two viewing positions. In the second, it manifests itself as parts of stripes of light from the source not being viewable from the laterally displaced camera position.

An effective edge extraction method with strong parallels in biological vision was recently devised by Marr and Hildreth [18], and has stimulated renewed interest in edge-based passive stereo ranging. One particular attraction of this approach is that passive stereo methodology is applicable (can be used for naturally-lit scenes). Another is that the family of edges extracted using different sized convolution $(\nabla^2 G)$ masks can be used for 'pyramid' style computation of disparity; gross line structure matches can be used to guide the matching of finer line structures, thus lowering the computational effort considerably while reducing mismatches leading to incorrect ranging values. However, without specialized hardware, the convolution processes required for edge extraction are computationally expensive. Figure 7 shows a left and right image pair $\nabla^2 G$ zero crossing result for a simple colored scene. Figure 8 shows a $\nabla^2 G$ zero crossing result for a natural scene. In all given examples, the evaluations for each color component were treated separately.

Considering the stereo pair of line extractions of Figure 7, it is clear that disparity based ranging can only be applied at edges and surface landmarks and not to uniform intensity surfaces. Thus, dense ranging to all visible (spatially quantized) points of the scene cannot be extracted in this way;

Figure 7: $\nabla^2 G$ Zero Crossing Edge Stereo Pair

using '2-1/2 D' to describe the result is being overly generous.

The sort of problems which arose when dealing with image based ranging, some of which have been described above in relation to passive stereo, led researchers to consider direct triangulation and time of flight monocular methods.

A simple infrared diode source triangulation range finder from an inexpensive Canon camera is shown mounted on the carriage of a digital

Figure 8: (a) Natural Image; (b) $\nabla^2 G$ Zero Crossing Edge Image for (a)

plotter in Figure 9(a). The scene of Figure 9(b) was scanned from the viewpoint indicated in Figure 9(c) to produce the pseudo color rangepic result of Figure 9(d). The no signal return case of the background space is shown as red. Figure 9(e) shows the same ranging device mounted on the wrist of a robotic manipulator.

The two main monocular range-finding candidates are ultrasonic ranging

Figure 9: Simple Triangulation Range Finder. (a) range finder mounted on plotter carriage for scanning operation; (b) plan view of scene

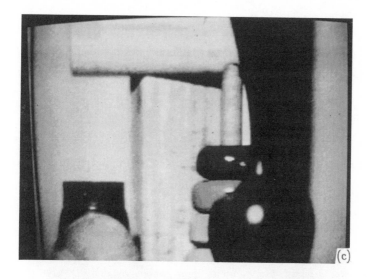

Figure 9: (c) front view of scene; (d) pseudo-color range picture of scene of (c)

and laser ranging. Whether such methods are legitimately part of computer 'vision' is another matter, but this will not concern us here.

The velocity of sound in air being relatively low makes ultrasonics an attractive ranging energy source candidate; only inexpensive equipment is necessary to measure transit times accurately. However, there are two serious problems with using ultrasonic techniques for dense ranging. The

Figure 9:   (e) range finder mounted on wrist of robotic manipulator.

first is that it is difficult to produce a sufficiently narrow energy beam for high resolution 2-D spatial quantization ranging (e.g., 128 × 128). The use of arrays of ultrasonic transducers is one approach to solving this problem. The second difficulty arises from a fundamental property of wave motion energy sources. Any surface with undulations of a scale comparable to or finer than the wavelength of the source will cause specular reflectance of the beam. Thus, a surface which is smooth in relation to the ultrasound wavelength will reflect the beam in a direction which equalizes the incident and reflected angles with respect to the surface normal. A smooth surface whose normal is significantly inclined to the incident beam might not return the beam to the capture aperture of the receiving transducer so that no range reading is available. Worse, a secondary reflection back to the receiver would result in a false range reading. An ultrasonic range scanner built in the Computer Vision and Robotics Laboratory at the Australian National University (A.N.U.) using a Polaroid device is shown in Figure 10. With laser range finders [19], [17], [12] there are no problems with the spread

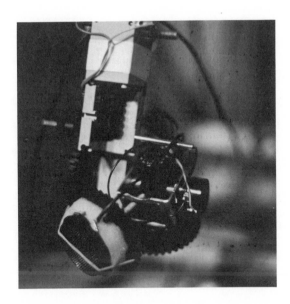

Figure 10:   Ultrasonic Range Scanner

of the beam and the specular reflectance effect is restricted to mirror smooth surfaces. However, the electronic instrumentation needs to be very sophisticated; given that the energy source velocity is 30 centimeters per nanosecond, sub-centimeter ranging accuracy requires time resolutions in the vicinity of ±30 picoseconds. The SRI [19] laser rangefinder is based on phase shift detection and is capable of recovering albedo as well as range data. However, some two hours were taken to collect a dense range picture (128 × 128) with 7-8 bits of range accuracy. The CALTECH [17] instrument was based on direct measurement of the time-of-flight of a pulse of infrared laser light. Under ideal conditions it was estimated that it would be possible to collect a rangepic of 128 × 128 resolution with 2 centimeter accuracy in a 1-3 meter range in three minutes. The A.N.U. [12] instrument built by the author in the CALTECH style is capable of capturing a crude rangepic of 64 × 64 resolution in four seconds, and one of ±1/4 centimeter relative accuracy in forty seconds. However, this last instrument, despite the use of a sensitive constant fraction discriminator to determine the instant of the return of the pulse independently of the returned energy (dependent on surface properties and range), gives varying ranges for objects of different colors at the same distance. This effect could probably be compensated by color calibrating the instrument and using the visual color information which is collected in registration with the range data, but this is not an elegant solution; a better quality constant fraction discriminator would provide a more satisfying result. Figure 11 provides a study of a ranging experiment using the A.N.U. laser rangefinder. This rangefinder was also the source of the range data used for the example in Figure 6.

Another interesting dense ranging method worthy of further study because of its speed, simplicity and wide applicability is based on the square law of light intensity with distance for point light sources [13]. A simple experiment used a 35mm projector as an effective point light source positioned at two locations between a single video camera and the scene (but just outside the line of sight). It was demonstrated that a trivial calculation involving the pixel by pixel intensity ratio for the close lit and far lit conditions could result in fairly reliable dense ranging which was applicable to multicolored, uniform, or textured surfaces equally, provided the video camera could either measure intensity accurately or be calibrated to allow correction of its output to the proper intensity reading at each quantized point in the image. Being a monocular method, no hidden parts problems are encountered. Also, range and intensity information are registered by construction. Problems could arise where secondary reflections confound the brightness measurements. Only a few seconds of VAX-11/750 time were necessary to evaluate the appropriate formula for each point of a 128 × 128 resolution range picture. Accuracy analysis has not yet been carried out. Better quality cameras would no doubt improve the accuracy considerably. Problems would also arise for this method if non-lambertian scattering surfaces were being ranged. Figure 12 shows an example of rangepic results

Figure 11: A.N.U. Laser Time-of-Flight Range Scanner Examples. (a) Teapot and Milk Jug Scene; (b) Pseudo-color Range Picture of (a) using the range-from-brightness method.

5. More Than 2-1/2 D Scene Analysis

All the ranging techniques dealt with in the previous section can, at best, provide 2-1/2 D scene data for supporting robot vision. How can the dimension be extended beyond 2-1/2 D towards the true 3-D representation? One conceptually quite simple approach is to place the objects of a scene on

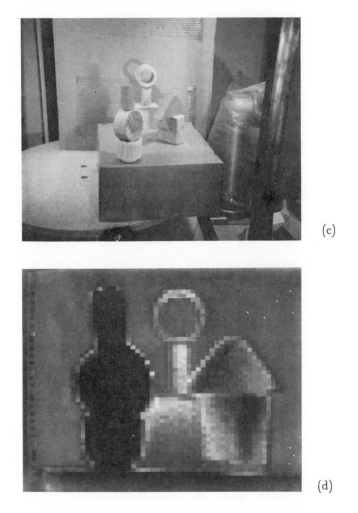

Figure 11: (c) Wooden Blocks Scene; (b) Pseudo-color Range Picture of (c).

a transparent supporting surface and apply direct dense ranging from all six directions, each normal to a face of a virtual cube enclosing the scene. Such an approach is reported in a preliminary way in [20]. However, robotic access to the objects in the scene may be hindered by the instrument placement configuration. This may not be a serious objection since the experimental set up could include a means of retracting some of the sensors after data acquisition.

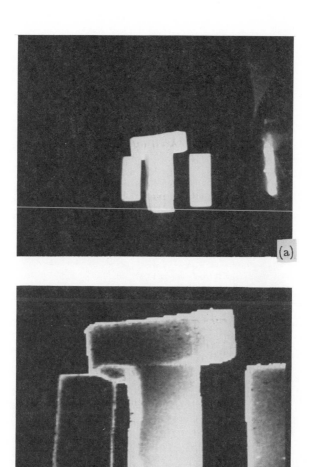

Figure 12: Range from Brightness. (a) Scene of Foam Plastic Blocks; (b) Intensity Modulated Range Picture of (a).

Another approach is the main topic for most of the remainder of this paper. It has the advantages of simplicity, robustness, wide application and completeness (in the sense of permitting shape extraction, recognition, obstacle avoidance, gripper placement and change detection to be carried out very easily). Much of the preliminary computation could be implemented on special purpose chips to permit high speed processing. Also, the data acquisition, lighting and camera placement requirements are simple and inexpensive. In a phrase, the method has 'real industrial application potential'.

6. Space Cube Modelling From Multiple Projections

Three pairs of N × N images are acquired as the data input to the modelling process. Each pair consists of a full color image (scene flood lit) and a back lit silhouette image. One pair is a front view of the scene, the second a plan view and the third a side elevation view. An experimental set up utilizing an underlit rotating table and inclined mirror arrangement is

Figure 13:   Space Cube Construction Experimental Set-up

shown in Figure 13. Underlighting the table not only permits the plan silhouette view to be acquired, but also serves to visually isolate the objects from their support plane so that they are not seen as connected from below. Further details are provided in [14]. Fixing three cameras in the three orthogonal viewing positions would, with suitable back and flood lighting, be preferable as it would obviate the need to rotate the supporting table for the side elevation view.

The three silhouette images are conceptually projected, from appropriate directions, through an $N \times N \times N$ cube, marking the ray intersections as occupied cells. This is a computationally trivial process but requires a large memory for appropriate values of N (32 to 128). A simple extension of Rosenfeld's [24] binary connectivity algorithm which looks at the $3 \times 3 \times 3$ neighborhood of each cell while evaluating connectivity in a strict raster sequence (say left to right, top to bottom, front to back), can generate connected component labels for each occupied cell of the $N \times N \times N$ space cube. Each isolated 'solid', perhaps made up of several objects in touching contact, can be extracted from the cube and used to evaluate volumes, shapes, center of volumes, etc. In fact, a whole range of binary 2-D operations [26], [4], [16] can easily be extended into the 3-D domain where and when required for pattern recognition, volume intersection, union or exclusive OR types of calculations, among others.

2-D connectivity analysis is an N^2 complexity operation, four of the eight neighboring cells being considered to determine the connectivity status of each cell in turn in the specified raster sequence. In the 3-D connectivity case, thirteen of the twenty-seven neighbors in the $3 \times 3 \times 3$ neighborhood need to be examined for each of the N^3 cells. Thus, the complexity ratio of 3-D to 2-D connectivity is $(13/4)N$, which is quite significant for N in the range 32 to 128.

Fortunately, and this was discovered after completing reference [14], it is perfectly valid (leads to identical results) if binary connectivity analysis is carried out on each of the orthogonal silhouette views first and the connectivity labels themselves projected from appropriate directions through the $N \times N \times N$ cube. Thus, instead of $13 N^3$ evaluations, only $12 N^2$ are required, a saving ratio of $(13/12)N$. For $N = 64$, which was the case for the experiments to be reported later in the paper, this represents a speed up of approximately 69, with a typical time of four seconds on a

VAX-11/750.

For each 2-D connectivity result, the labels were compacted to fill the gaps which could result when regions, seemingly distinct at an earlier part of the scan, merge later; the lower label is retained and the cells having the higher label relabelled. If it is assumed that no more than 255 different labels can arise from any one 2-D connectivity evaluation, three bytes of memory associated with each cell of the $N \times N \times N$ space cube can be used to store the three labels at ray intersection points in the cube. Empty cells are simply labelled '0' for all three bytes. Another label compaction can then produce a continuous set of numerical labels for the distinct 'solids' in the cube.

The observation concerning the computational complexity advantage of carrying out 2-D connectivity analysis before projection, instead of 3-D connectivity after projection, is more significant than it may at first appear. There are already available on the market specialized systems for binary connectivity; these are capable of high speed connectivity labelling calculations. Even faster devices will no doubt soon appear. Three such devices, one associated with each of the orthogonally placed cameras, can be used to channel 2-D connectivity labels into a processor which does the label projection and higher level analysis. This could, with some care, lead to a 3-D vision system with direct industrial application potential at quite low a cost.

If three views are insufficient to capture the necessary details of the shapes of objects in the scene, the space cube projection scheme can be easily extended to accommodate more.

7. Painting the Solids

The full color images are used in the following manner. Using an $N \times N \times N$ cube where space occupancy is defined, each colored pixel of each viewing position is projected into the cube from the appropriate direction and is positioned in a color cube at the cell of first contact with occupied space. The result is to effectively paint, from the three orthogonal directions, the solids represented in the space cube. Some surfaces facing directions from which the color images were not taken (bottom, back, other side) together with surfaces obscured by other 'solids' are left blank (or black). Each painted solid can be isolated from the scene using the connectivity label data derived earlier. These painted 'solids' can be viewed from any direction by applying rotation transforms on all the cell positions making up the solid and moving the painted and unpainted patches into appropriate positions of a 2-D viewing window. Blank (or black) patches indicate surfaces that were either facing away from the original viewing positions or were obscured by parts of other 'solids'.

8. Refinement and Manipulation

For each isolated, painted 'solid', the type of 2-D image segmentation procedures described in Section 3 can be applied for each of the three original views. The segmentation process is simplified by the fact that each view of each 'solid' can be dealt with in isolation to separate contiguous objects. The segmentation labels from the three views for each isolated solid can then be combined in an attempt to define the 3-D extent of each of the objects in touching contact which make up that isolated solid. Features suitable for pattern recognition purposes can be extracted for each object.

Even if the refinement described above is not carried out, it is possible to guide robotic pick and place action using only space occupancy and 3-D connectivity labelling data, avoiding the problems of unknown support relationships by always dismantling the scene from the top, object by object, and using repeated 3-D analysis cycles to guide subsequent manipulation. If the robotic manipulator hand is fitted with transparent fingers, each object, once picked up, can be held up to a camera for identification in isolation, as many views as necessary being provided by having the robotic manipulator adjust the position of the object relative to the camera viewing direction. Thus post-manipulation recognition rather than the usual pre-manipulation recognition can be applied with ease.

Pick up points can be easily selected by finding the center of volume of some specified slice off the top of a 'solid' and checking to see whether the robotic manipulator fingers could fit around the object at that position while not impinging upon any other part of the scene. Simple 3-D binary cube logical operations are all that are required for this last evaluation. Just how much of the 'solid' lifts up when the manipulator grips the selected point at an appropriate, prechecked, orientation may not matter much unless some complex interlocking of objects in the 'solid' is involved. Thus, the robotic manipulation can itself be used to resolve support relationships in an unambiguous manner. This is a rather nice attribute of the system and is particularly consistent with the thrust of the hand-eye coordination analogy.

9. Examples

A number of examples of the approach to 3-D robot vision described above are presented in Figures 14, 15, and 16.

Figure 14(a) shows a scene containing a small microscope and two simple block constructions. Figure 14(b) is a view of the 'solids' of (a) as modelled and painted in the solid space cube after a small isometric rotation (tilted forward and rotated counterclockwise).

Figure 15(a), (c), (e) are front, plan, and side elevation views after quantization of a toy figure scene taken under flood lighting conditions and masked by the silhouette views in (b), (d), (f), respectively (which cuts out all background variation). Figures 15(g), (h), and (i) are the distinctly painted figures separated by the projected connectivity labelling procedure described earlier. Figures 15(j), (k), and (l) are the corresponding isometric rotated views of enclosing rectangular prisms constructed to frame the three separate solids, and the pick up points corresponding to the centers of

Figure 14: Isometric Rotation View of Colored Space Cube Objects. (a) Photograph of scene; (b) Isometric rotated view of objects of (a) in space cube model.

volume of the occupied cells in the top portion of each solid (including all cells within 10 quantization intervals from the top).

Figure 16(a) is the front view of a block construction scene. Figures 16(b), (c), and (d) are the separated painted solids of the scene, each 'solid' being made up of more than one object. Separation of the objects in any one solid can be tackled by using multiple view 2-D segmentation on each 'solid'

Figure 15: Toy Scene Solid Space Cube Modelling Example. (a) Front view of floodlit scene; (b) Front view silhouette (backlit) in turn.

10. Collision-Free Trajectory Planning

The space cube solid model of the scene is also useful for planning collision-free, minimal path length trajectories. Consider the simple case of a compact (i.e., not multi-jointed) vehicle which is to find its way via the shortest collision-free path from a given free space point (start) to another

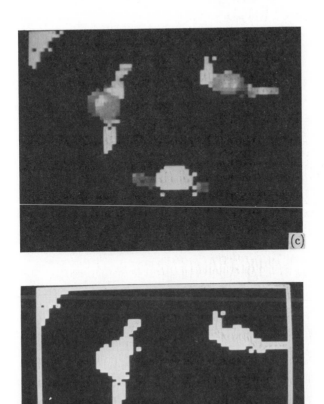

Figure 15: (c) Plan view of floodlit scene; (d) Plan view silhouette (underlit)

(goal) by 3-D translation only. With a nominated reference point identified on the vehicle, an elegant solution can be found by applying the following three stage process:

1. Grow the obstacles to account for the physical extent of the vehicle by placing its reference point at each surface voxel of the obstacles and filling in the extra volume occupied by a 'reversed' version of the vehicle (with reference point as point of rotation).

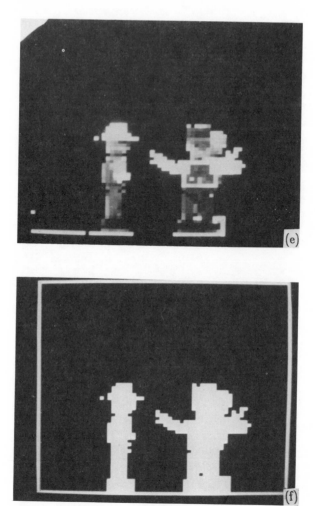

Figure 15: (e) Side view of floodlit scene; (f) Side view silhouette (backlit)

2. From the nominated goal point propagate distance values [25] throughout all free space around the grown obstacles and contained by some defined artificial enclosing space (an outside obstacle).

3. From the starting point, walk down hill along the steepest descent path in distance space to the goal. The shortest path will result. The path will exist if the goal is reachable from the starting point.

Figure 15: (g) Isolated left most figure (painted); (h) Isolated central figure (painted)

An example of this procedure is shown in Figure 17. Further details are available in [15].

If multiple goals are specified and distance propagation is carried out from each (the algorithm used automatically accommodates this), the shortest path to the nearest goal will be found. The method is robust in the sense that complex 3-D mazes will be accommodated by the procedure without modification.

Figure 15: (i) Isolated right most figure (painted); (j) Isometric rotated view of   enclosing rectangular prism and pick up point for left most figure

For more complex collision avoidance problems, for example, where a multi-jointed robotic manipulator is required to pick up an object surrounded by other objects (obstacles), a possible, but unexplored approach is as follows.

1. Hypothesize a trajectory for the end effector of the robotic manipulator using some general principles derived from the arm

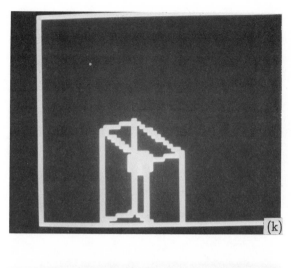

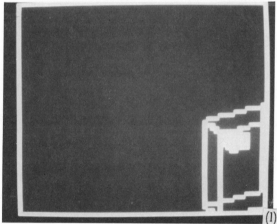

Figure 15: (k) Isometric rotated view of enclosing rectangular prism and pick up point for central figure; (l) Isometric rotated view of enclosing rectangular prism and pick up point for right most figure.

configuration and expected scene domain.

2. Calculate the swept volume for all parts of the manipulator for carrying out the suggested trajectory.

3. Map the calculated swept volume into a space cube of the same dimension and position as that used for the 3-D scene modelling.

Figure 16: Block Scene Solid Space Cube Modelling Example. (a) Front view of block scene; (b) Left most painted 'solid'

4. Mark the cells of a third space cube which correspond to points where the scene model and swept volume cubes indicate intersections of occupied space.

5. Use the volumes and shapes of the intersection components to suggest a refinement of the trajectory.

6. Cycle through steps 2 to 5 until a satisfactory solution is found.

Figure 16:   (c) Central painted 'solid'; (d) Right most painted 'solid'.

11. Eye-in-Hand Vision

The possibility of controlling the view of a scene with great accuracy by carrying a camera in the hand of a robotic manipulator is an exciting one; such an approach permits the combination of many stages of vision support for robot guidance in the one basic mode of operation.

Initial multiple viewpoint scene analysis can be carried out with the camera viewing the scene at a distance, with missing pieces of visual data

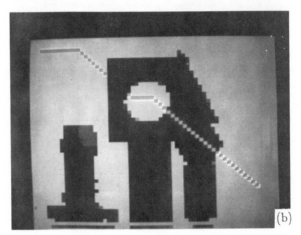

Figure 17: Example of Obstacle Avoidance Using Distance Transforms. (a) block scene; (b) minimal path length collision-free trajectory for translation of a $4 \times 4 \times 4$ vehicle (cube model is $64 \times 64 \times 64$).

being sought by actively exploring in and around the objects in the scene. Depth data can be derived from passive stereo, temporal stereo (analyzing a time sequence of frames as the camera moves into a scene), multiple projections, or carrying an active direct range finder in the hand along with the camera. The approach to a target object can be refined as the hand moves towards it to assume a gripping stance (visual servoing). Finally, the gripping process can be carefully monitored while in progress and fine adjustments made if necessary. This last point could be particularly important if complex, dextrous hands are used to grip delicate and complex-shaped objects.

Figure 18 shows a solid state camera mounted on the wrist of a Unimate

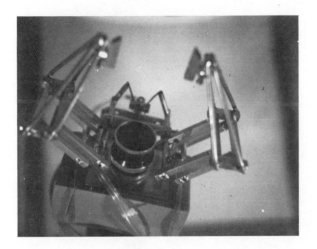

Figure 18: Solid State Color Video Camera Mounted 'In the Hand' of Robotic Manipulator

250 robotic manipulator in the Computer Vision and Robotics Laboratory at the A.N.U. The camera looks out between the two fingers of the hand and is able to see the fingers as they close on an object. Figure 19(a) shows a sequence of three images taken as the hand held camera is moved into the scene. Figure 19(b) shows two sets of edges for a pair of images taken in time sequence; the separation of corresponding edges indicates closeness of that part of the scene to the camera. More intensive research into the eye-in-hand approach to robotic vision is planned for the near future.

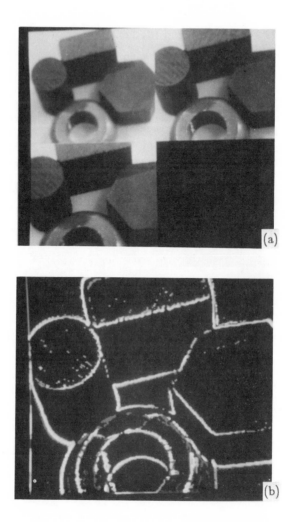

Figure 19: Eye-in-the-Hand Sequence Study. (a) Three views with camera moved into scene by robot; (b) Line extraction for a single view in a sequence

12. Conclusion

This paper has championed that approach to 3-D robot vision which favors semantic-free analysis based on rich sensory data sources being pushed to the limits of providing information useful for scene interpretation, trajectory planning and robotic manipulation before problem-specific details

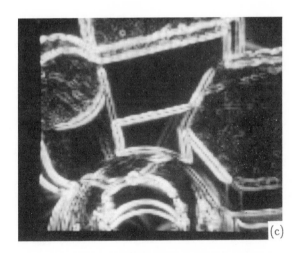

Figure 19: (c) Line extraction for two views in sequence (range data clearly available).

are used as constraints or guidelines. This approach has the advantage of simplicity and robustness while promising potential for near term application to a wide range of robotic guidance tasks in industry.

A number of issues and ideas concerning the extension of 2-D to 2-1/2 D, eventually approaching 3-D analysis, have been explored, particularly in relation to color scene analysis. The space cube modelling approach based on multiple projections has been emphasized as a means of realizing robust semantic-free analysis including scene segmentation, robotic pickup point evaluation and collision-free trajectory planning. This approach could be fairly easily adapted to the real time requirements of industrial vision robotics.

The brief coverage of eye-in-hand vision analysis indicates the way of the future in terms of integrated vision and manipulation. This promises considerable flexibility by combining initial analysis with approach and gripping strategies in an almost continuous process of time sequence frame analysis.

There is little doubt that vision will continue to play a leading role in providing sensory data for guiding robotic manipulators. The quest for a complete 3-D scene analysis system will become much more active both because of improving technological support in sensory transduction apparatus and computational resources, and because the potential applications of such a system are so diverse.

References

[1]   Bajcsy, R. and Lieberman, L.
        Texture gradient as a depth cue.
        Compt. Graphics Image Processing 5:52-67, 1976.

[2]   Guzman, A.
        *Computer recognition of three-dimensional objects in a visual
           scene.*
        Technical Report MAC-TR-59, Ph.D. thesis, Project MAC, MIT,
           Mass. , 1968.

[3]   Horowitz, S. L. and Pavlidis, T.
        Picture segmentation by a directed split-and-merge procedure.
        In *Proc. 2nd Joint Conf. Pattern Recognition*, pages 424-433.
           August, 1974.

[4]   Hu Ming, K.
        Visual pattern recognition by moment invariants.
        IRE Trans on Information Theory , 1962.

[5]   Idesawa, M., Yatagai, T., and Soma, T.
        A method for automatic measurement of three-dimensional shape by
           new type of Moire fringe topography.
        In *Proc. 3rd Int. Joint Conf. Artificial Intell.*, pages 708-712.
           Coronada, CA, November 8-11, 1976.

[6]   Jarvis, R. A.
        Region based image segmentation using shared near neighbor
           clustering.
        In *Proc. International Conf. on Cyber and Soc.*, pages 641-647.
           Washington, D.C., September 19-21, 1977.

[7]   Jarvis, R. A.
        Structured merge strategies for image segmentation.
        Proc. Compsac 77:472-485, November, 1977.

[8]   Jarvis, R. A.
        Fast image segmentation for robotic vision.
        In *Proc. Radio Research Board Seminar on Image Processing*,
            pages 28.1 - 28.8. School of E.E. and C. S., Univ. of N.S.W., June,
            1982.

[9]   Jarvis, R. A.
        A computer vision and robotics laboratory.
        Computer :8-24, June, 1982.

[10]   Jarvis, R. A.
        Expedient range enhanced 3-D robot color vision.
        Robotica 1:25-31, 1983.

[11]   Jarvis, R. A.
        A perspective on range finding techniques for computer vision.
        IEEE Trans. Pattern Anal. Machine Intell. PAMI-5:122-139,
            March, 1983.

[12]   Jarvis, R. A.
        A laser time-of-flight range scanner for robotic vision.
        In *IEEE Trans. on Pattern Anal. Machine Intell. PAMI-5*, pages
            505-512. September, 1983.

[13]   Jarvis, R. A.
        Range from brightness for robotic vision.
        In *Proc. 4th International Conf. on Robot Vision and Sensory
            Controls*, pages 165-172. London, October, 1984.

[14]   Jarvis, R. A.
        Projection derived space cube models for robotic vision and collision-
            free trajectory planning.
        In *Proc. 2nd International Symposium of Robotics Research*, pages
            294-301. Kyoto, August 20-23, 1984.

[15]   Jarvis, R. A.
        Collision-free trajectory planning using distance transforms.
        In *Proc. National Conf. and Exhibition on Robotics*, pages 86-90.
            Melbourne, August 20-24, 1984.

[16]   Jarvis, R. A. and Patrick, E. A.
        Interactive Binary Picture Manipulation.
        In *Proc. UMR - Mervin J. Kelly Communications Conference*,
            pages 20-1-1 to 20-1-8. Univ. of Missouri-Rolla, 1970.

[17] Lewis, R. A. and Johnston, A. R.
      A scanning laser rangefinder for a robotic vehicle.
      Proc. 5th Int. Joint Conf. AI :762-768, 1977.

[18] Marr, D. and Hildreth, E. C.
      Theory of edge detection.
      Proc. R. Soc. London B 207:187-217, 1980.

[19] Nitzan, D., Brain, A. E., and Duda, R. O.
      The measurement and use of registered reflectance and range data in
          scene analysis.
      Proc. IEEE 65:206-220, Feb., 1977.

[20] Page, C. J. and Hassan, H.
      Non-contact inspection of complex components using a rangefinder
          vision system.
      In *Proc. 1st International Conf. on Robot Vision and Sensory
          Controls*, pages 245-254. Stratford-on-Avon, April, 1981.

[21] Popplestone, R. J., Brown, C. M., Ambler, A. P., and Crawford, G. F.
      Forming models of plane-and-cylinder faceted bodies from light
          stripes.
      In *Proc. 4th Int. Joint Conf. AI*, pages 664-668. Sept., 1975.

[22] Prazdny, K.
      Motion and structure from optical flow.
      In *Proc. 6th Int. Joint Conf. Artificial Intell.*, pages 702-704.
          Tokyo, Japan, November 8-11, 1979.

[23] Roberts, L. G.
      Machine perception of three dimensional solids.
      Computer Methods in Image Analysis.
      IEEE Press, 1977, pages 285-323.

[24] Rosenfeld, A. and Pfaltz, J. L.
      Sequential operations in digital picture processing.
      J. Assn. for Comp. Mach. 13(4):471-494, October, 1966.

[25] Rosenfeld, A. and Pfaltz, J. L.
      Distance functions on digital pictures.
      Pattern Recognition 1:33-61, 1968.

[26] Saraga, P. and Woolons, D. J.
      The design of operators for pattern recognition.
      In *IEE NPL Conference on Pattern Recognition*. 1968.

[27]   Waltz, D.
       Understanding line drawings of scenes with shadows.
       The Psychology of Computer Vision.
       McGraw-Hill, New York, 1975, pages 19-91.

[28]   Yakimovsky, Y. and Cunningham, R.
       A system for extracting three-dimensional measurements from a
           stereo pair of TV cameras.
       Comput. Graphics Image Processing 7:195-210, 1978.